Frontiers in Neurobiology 1

Series Editors:

Anthony J. Turner (Leeds)

F. Anne Stephenson (London)

Neuropeptide Gene Expression

Edited by
Anthony J. Turner

Portland Press
London and Chapel Hill

Published by Portland Press Ltd, 59 Portland Place, London
W1N 3AJ, U.K.
In North America orders should be sent to Portland Press Inc.,
P.O. Box 2191, Chapel Hill, NC 27515–2191, U.S.A.

ISBN 1 85578 044 5   ISSN 1353-6508

British Library Cataloguing in Publication Data
A catalogue record for this book is available from the British Library

Typeset by Unicus Graphics Ltd, Horsham, Sussex and
Printed in Great Britain by Whitstable Litho Printers Ltd

Contents

Preface

The phenomenal advances in neuroscience over the last decade underpinned by molecular genetic technologies have been the impetus to develop this series highlighting Frontiers in Neurobiology. Nowhere is the complexity of the nervous system better expressed than in the multiplicity of neuropeptide signal molecules, of which the opioid peptides, the enkephalins, were the archetype some 20 years ago. Neuropeptides, therefore, provide a fertile field for the first volume in this series. Until the discovery of the enkephalins, the known neurotransmitters were but a handful — which should be more than sufficient for any simplistic model of neural excitation and inhibition. These basic assumptions are confounded by the realization that dozens of peptides have roles to play in neural signalling, either directly as transmitters, or in modulating the activity of other transmitters. Often these peptides are co-stored and released with other classical transmitters. Molecular cloning combined with immunohistochemical mapping of the peptide products has produced many new molecules of known structure and anatomical location but of unknown physiological function. Challenges for the next decade are in defining their functions and understanding what pathophysiological roles these compounds may fulfil. The discovery of the many neuropeptides has also provided revealing insights into more general mechanisms of eukaryotic gene expression. Thus, the chapters in the present volume while of most direct relevance to neurobiologists, also have lessons to provide for students of gene expression in general.

The multiplicity of mammalian neuropeptides means that it is impossible, in a short volume, to provide a comprehensive review of all the brain peptides. My rationale in selecting the individual topics and authors has been to choose some of those examples that have been instrumental in helping us understand the rules that govern the expression of neuropeptide genes and how this regulation can generate even greater diversity. The choice is a purely personal one but, I hope, is sufficiently diverse and interesting to attract the reader's attention. A general introduction on the role of RNA splicing with a number of specific examples is followed by more detailed instances of expression of individual neuropeptide genes, including calcitonin and CGRP, oxytocin and vasopressin, neuropeptide Y, vasoactive intestinal peptide (VIP) and neurotensin. Clinical and pathophysiological conditions affecting neuropeptide expression are also highlighted, particularly in relation to VIP. Post-translational, as well as post-transcriptional, events are equally important in defining the population of mature neuropeptides, and some of the enzyme systems involved in these processes are also considered. The most significant group, whose identity has been sought for over 25 years, are the 'prohormone convertases' or 'PCs', only revealed as members of the subtilisin family of serine proteinases in the early 1990s. Their discovery and exclusive expression in endocrine and neuroendocrine tissues are documented here. Metalloproteinases also feature in peptide processing and metabolism, and the best-characterized such enzyme at the molecular genetic level is angiotensin-converting enzyme, a fascinating two-domain enzyme whose structure and expression are considered in detail. Final generation of biologically active neuropeptide usually also involves a *C*-terminal amidation step catalysed by another fascinating enzyme, in this case an enzyme of dual specificity — a combined hydroxylase and lyase. Again, the expression of this enzyme is restricted to endocrine and neuroendocrine tissues, and the cloned and expressed enzyme may well have important applications in biotechnology. Finally, the diversity of peptide signal multiplicity is reflected in the heterogeneity of specific receptors and signalling systems. Most of the neuropeptide receptors fall into the family of G-protein-linked receptors and over 40 neuropeptide receptors have now been cloned. Their structural and genetic analysis and evolution is documented in the final chapter in this volume.

I would like to thank all the authors for their timely contributions and I hope that my editing has not obscured or altered their intent significantly! I would also like to acknowledge a debt to Dr Denes Agoston for many helpful discussions and suggestions in relation to this volume in its early stages. Unfortunately, he was unable to contribute to the final volume.

Tony Turner
University of Leeds, March 1994

Contributors

F. Alhenc-Gelas	Collège de France, INSERM U 367, 17 rue du Fer-à-Moulin, 75005 Paris, France
S. G. Amara	Vollum Institute for Advanced Biomedical Research, Oregon Health Science University, 3181 Sam Jackson Park Road, Portland, OR 97201, U.S.A.
N. Bhogal	The Department of Biochemistry and Molecular Biology, University of Leeds, Leeds LS2 9JT, U.K.
A. F. Bradbury	National Institute for Medical Research, The Ridgeway, Mill Hill, London NW7 1AA, U.K.
F. Checler	Institut de Pharmacologie Moléculaire et Cellulaire, C.N.R.S. UPR 411, Université de Nice Sophia Antipolis, 660 Route des Lucioles, Sophia Antipolis, 06560 Valbonne, France
M. Chrétien	J.A. DeSève Laboratory of Molecular Neuroendocrinology, Clinical Research Institute of Montreal, 110 Pine Avenue West, Montreal, Quebec, Canada H2W 1R7
E. Clauser	Collège de France, INSERM U 36, 3 rue d'Ulm, 75005 Paris, France
P. Corvol	Collège de France, INSERM U 36, 3 rue d'Ulm, 75005 Paris, France
A. Cox	The Department of Biochemistry and Molecular Biology, University of Leeds, Leeds LS2 9JT, U.K.
R. Day	J.A. DeSève Laboratory of Biochemical Neuroendocrinology, Clinical Research Institute of Montreal, 110 Pine Avenue West, Montreal, Quebec, Canada H2W 1R7
D. Donnelly	The Department of Biochemistry and Molecular Biology, University of Leeds, Leeds LS2 9JT, U.K.
R. B. Emeson	Department of Pharmacology, Vanderbilt University, Nashville, TN 37232-6600, U.S.A.
J. B. C. Findlay	The Department of Biochemistry and Molecular Biology, University of Leeds, Leeds LS2 9JT, U.K.
C. Hubert	Collège de France, INSERM U 36, 3 rue d'Ulm, 75005 Paris, France
C. R. Hurrell	The Department of Biochemistry and Molecular Biology, University of Leeds, Leeds LS2 9JT, U.K.

M. A. Kirschner Vollum Institute for Advanced Biomedical Research, Oregon Health Science University, 3181 Sam Jackson Park Road, Portland, OR 97201, U.S.A.

J. T. Lu Department of Pharmacology, Vanderbilt University, Nashville, TN 37232-6600, U.S.A.

C. D. Minth Wright State University, Department of Biochemistry, School of Medicine, College of Science and Mathematics, 234 Biological Sciences, Dayton, OH 45435-0001, U.S.A.

N. G. Seidah J.A. DeSève Laboratory of Biochemical Neuroendocrinology, Clinical Research Institute of Montreal, 110 Pine Avenue West, Montreal, Quebec, Canada H2W 1R7

D. G. Smyth Institute of Molecular Biology, Austrian Academy of Sciences, A-5020 Salzburg, Bilrothstrasse 11, Austria

F. Soubrier Collège de France, INSERM U 36, 3 rue d'Ulm, 75005 Paris, France

J. A. Waschek The Mental Retardation Research Centre and Department of Psychiatry, University of California at Los Angeles, CA 90024-1759, U.S.A.

L. Wei Collège de France, INSERM U 36, 3 rue d'Ulm, 75005 Paris, France

W. S. Young Laboratory of Cell Biology, National Institute of Mental Health, Bethesda, MD 20892, U.S.A.

Abbreviations

ACE	angiotensin I-converting enzyme
ACTH	adrenocorticotropic hormone
AI	angiotensin I
AII	angiotensin II
AMPN	anterior medial preoptic nucleus
ASF	alternative splicing factor
BK	bradykinin
BNST	bed nucleus of the stria terminalis
BPV	bovine papillomavirus
CAT	chloramphenicol acyltransferase
CCP	calcitonin C-terminal cleavage peptide
CDF	cholinergic differentiation factor
CGRP	calcitonin gene-related peptide
CHO	Chinese hamster ovary
CNA	central nucleus of the amygdala
CNF	ciliary neurotrophic factor
CNS	central nervous system
CRE	cyclic AMP responsive element
CRF	corticotropin-releasing factor
DHFR	dihydrofolate reductase
Dnp-F	1-fluoro-2,4-dinitrobenzene
DRB	5,6-dichloro-1-ribofuranosylbenzimidazole
DRG	dorsal root ganglia
FCDI	familial central diabetes insipidus
GPCR	G-protein-coupled receptor
GRE	glucocorticoid responsive element
GRF	growth hormone-releasing factor
HA	haemagglutinin
hGH	human growth hormone
hnRNP	heterogeneous ribonucleoprotein
hTPPII	human tripeptidyl peptidase II
iNN	NN-like immunoreactivity
iNT	NT-like immunoreactivity
LH	luteinizing hormone
LHA	lateral hypothalamic area
LH-RH	luteinizing hormone-releasing hormone
LIF	leukaemia inhibitory factor

LINE	long interspersed repeated DNA element
LPH	β-lipotropic hormone
MLP	major late promoter
MPN	medial preoptic nucleus
MSH	melanocyte-stimulating hormone
MTC	medullary thyroid carcinoma
N-proCT	N-terminal peptide derived from the calcitonin prohormone
NGF	nerve growth factor
NN	neuromedin N
NPY	neuropeptide Y
NRE	negative regulatory element
NSCLC	non-small cell lung cancer
NT	neurotensin
ORE	oestrogen responsive element
ORF	open reading frame
OT	oxytocin
PCR	polymerase chain reaction
PHI	peptide histidine isoleucine
PHM	peptide histidine methionine
PMSG	pregnant mare serum gonadotrophin
POMC	pro-opiomelanocortin
PP	pancreatic polypeptide
PPT	preprotachykinin
prepro-NT	prepro-neurotensin precursor
PRP	pre-mRNA processing
PVN	paraventricular nucleus
PYY	peptide YY
rPSS	rat prosomatostatin
RT-PCR	reverse transcriptase-polymerase chain reaction
SCN	suprachiasmatic nucleus
SF2	splicing factor 2
SHR	spontaneously hypertensive rat
snRNA	small nuclear RNA
snRNP	small nuclear ribonucleoprotein
SON	supraoptic nucleus
TPA	12-*O*-tetradecanoylphorbol-13-acetate
TRH	thyrotropin-releasing hormone
VIP	vasoactive intestinal peptide
VP	vasopressin
WKY	Wistar-Kyoto rat

The role of pre-messenger RNA splicing in neuropeptide gene expression

M. A. Kirschner and S. G. Amara

Vollum Institute for Advanced Biomedical Research, Oregon Health Science University, 3181 Sam Jackson Park Road, Portland, OR 97201, U.S.A.

Introduction

The discovery in the late 1970s that individual genes are encoded discontinuously within the genome provided a basis for complex gene regulation at the level of RNA processing (Berget *et al.*, 1977; Chow *et al.*, 1977; Klessig, 1977). In addition to requiring splicing machinery that accurately joined such interrupted coding sequences, it became apparent that the splicing process itself could ultimately direct the synthesis of more than one gene product from a single transcript. A wide array of gene products are encoded within complex transcription units that generate multiple, mature messenger RNAs (mRNAs) through the selective use of distinct start sites, splice sites and/or polyadenylation sites (Leff *et al.*, 1986; Breibart *et al.*, 1987). This mechanism of alternative exon usage is a central theme in neuropeptide gene expression. The calcitonin gene related peptide (CGRP) which is a spliced derivative of the calcitonin gene, and the multiple products of the preprotachykinin gene which are generated by alternative splicing, are noteworthy examples of how protein diversification, such as that observed in mammalian nervous systems, can be achieved by mRNA splicing.

In eukaryotes, RNA processing not only alters the potential number of products encoded by a single gene but can also selectively regulate gene expression in different tissues and at different developmental stages. In the nervous system many gene products in addition to neuropeptides are the products of pre-mRNA alternative splicing, including genes encoding receptors (Goldman *et al.*, 1987; Dal Toso *et al.*, 1989; Giros *et al.*, 1989; Monsma *et al.*, 1989), ion channels (Schwartz *et al.*, 1988), enzymes for neurotransmitter synthesis (Grima *et al.*, 1987), neural oncogenes (Martinez *et al.*, 1987) and structural proteins such as myelin basic protein (de Ferra *et al.*, 1985). Gene products with potentially important roles in neuronal development, N-CAMs (Murray *et al.*, 1984; Goridis *et al.*, 1985), the Alzheimer's amyloid precursors (Kitaguchi *et al.*, 1988; Ponte *et al.*, 1988; Tanzi *et al.*, 1988) and nerve growth factor (Edwards *et al.*, 1986) have also been shown to arise from alternatively processed transcripts. Clearly, the role played by RNA splicing in gene regulation is pivotal to understanding the more basic issues relevant to nervous system function and development.

In this review we shall focus on studies of the regulation of RNA processing events in genes encoding neuropeptides. To provide a background for considering these studies, we will first review the basic mechanisms of RNA processing and the components of the spliceosome. We will discuss a few systems in which mRNA splicing is well understood and then go on to consider in detail mRNA processing of neuropeptides.

Splicing mechanisms

The mRNA splicing process, as it is understood today, can be modelled as a two-step mechanism (Ruskin *et al.*, 1984; Padgett *et al.*, 1986). The first step entails cleavage at an intron 5′ splice site and formation of the lariat intermediate (Fig. 1). This lariat structure is the product of covalent bonding between the intron 5′ end and an adenosine residue, known as the branch point, 15 or more nucleotides upstream from the 3′ splice site. In the second step, the intron lariat is cleaved from the second exon and the two exons are ligated.

The molecular machine which both recognizes the splice site and excises the intron is known as a spliceosome. Spliceosomes are comprised of particles that contain both small nuclear RNAs (snRNAs) and protein components. These particles are known as small nuclear ribonucleoproteins or 'snRNPs'. Several of the RNA components of the spliceosome can base pair with sequences at or near splice sites. The discovery of autocatalytic or 'self-splicing' functions of RNA in simpler organisms (reviewed by Cech, 1990) has led to the proposal that spliceosome snRNAs are the catalytic agent in splicing reactions. Although spliceosome protein components have not been as well characterized, it has been suggested that their differential expression may be important for the tissue-specific splicing patterns found in higher eukaryotes, and in particular for the differential expression of neuropeptides. Spliceosome proteins can also play a role in the recognition of, and binding to, splice site sequences.

Sequences present at exon–intron junctions have been shown to be critical determinants of the interaction of nascent nuclear mRNA transcripts with the spliceosome. Comparison of these sites in a multitude of genes has identified three consensus sequences: one at the 5′ splice site, the second at the 3′ splice site and the third at the lariat branch point (Mount, 1982) (Fig. 1). Early events in the spliceosome assembly are known to be dependent on sequences at the 5′ end of the intron (Bindereif and Green, 1987; Cheng and

Fig. 1. Schematic illustration of pre-mRNA processing events

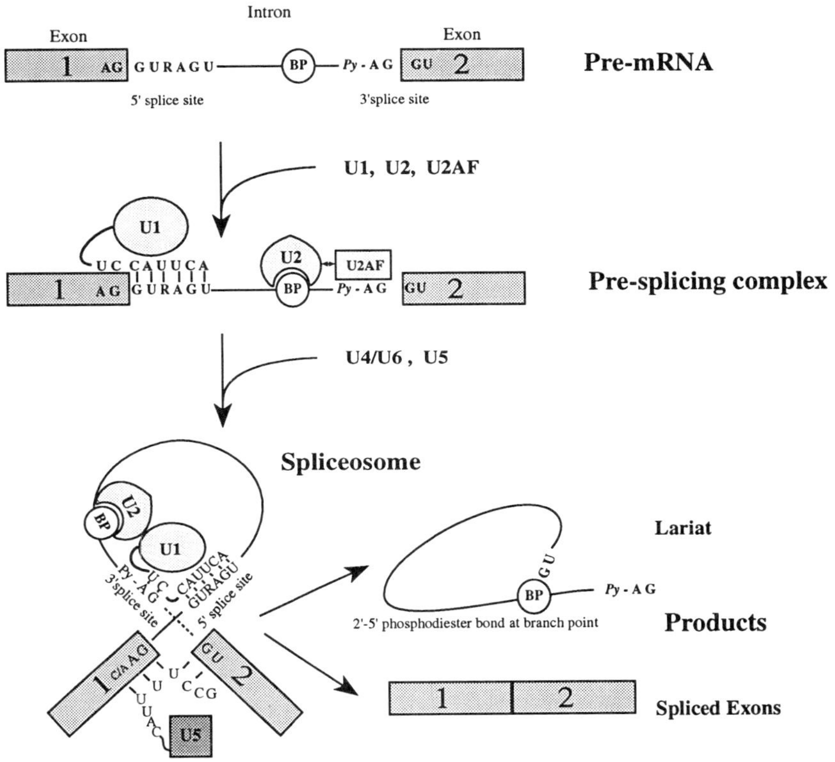

Splice site consensus sequences are shown (Mount, 1982). Recent evidence for interaction of U1 and U5 at both the 5′ and 3′ splice sites (Newman and Norman, 1991, 1992; Reich et al., *1992) is incorporated in a model of the spliceosome (adapted from Steitz, 1992). U2 binding at the branch point (BP) is dependent on an interaction with U2AF [which binds the polypyrimidine tract (py)] as well as ATP. U4/U6 is believed to serve as the catalytic component of the spliceosome.*

Abelson, 1987). It has been suggested that the initial assembly occurs through recognition of 5′ splice sites. Studies of deletion constructs both *in vivo* (Wieringa *et al.*, 1984; Fu *et al.*, 1988), and *in vitro* (Ruskin and Green, 1985) support a significant role for this region in pre-mRNA splicing. The 5′ splice site together with the 3′ splice site have been proposed to form an exon-defining spliceosome complex. It is also conjectured that the conformation of the spliceosome complex is then modified, delineating the intron and mediating its removal (Robberson *et al.*, 1990). Surprisingly, sequences around the branch point in higher eukaryotes show a greater degree of divergence than might be expected for regulatory sequences involved in such a critical process. Interestingly, in higher eukaryotes it has been shown that consensus divergence has a role in the control of splice site selection, as will be discussed later.

Pre-mRNA binding and catalysis

The spliceosome complex is made up of five unique snRNPs, designated U1, U2, U4, U5 and U6. Each snRNP in turn consists of U-specific snRNAs which associate with several specific polypeptides, some common to all of the snRNPs and some unique to particular snRNPs. For catalytic activity, all five of these snRNPs must be present in the spliceosome.

U1 snRNP binds to 5′ splice sites prior to the binding of any other snRNPs. This binding does not require ATP hydrolysis or additional factors (Legrain *et al.*, 1988; Seraphin and Rosbach, 1989) but does involve base pairing with the 5′ splice site consensus sequence (Zhuang and Weiner, 1986). The efficiency of U1 binding is dependent not only upon the integrity of the 5′ site but also on sequences around the branch point since the presence of point mutations in the branch point region greatly reduces U1 binding (Ruby and Abelson, 1988). Before a commitment to spliceosome formation is made, a branch point consensus element must be present (Legrain *et al.*, 1988; Serapin and Rosbach, 1989). In addition, studies have recently shown that the 3′ splice site can also interact directly with U1 snRNA (Reich *et al.*, 1992). Interestingly, mutations at the 3′ exon–intron boundary that block the first step of splicing (leading to the accumulation of linear precursor) can be over-ridden by compensatory base changes in the U1 snRNA.

5′ splice-site binding by U1 snRNP is followed by binding of the U2 snRNP to the branch point region (Mount *et al.*, 1983; Ruskin *et al.*, 1984; Bindereif and Green, 1987; Ruby and Abelson, 1988; Seraphin and Rosbash, 1989). The presence of U1 snRNP at the intron is a prerequisite for U2 binding (Barabino *et al.*, 1990). Like U1, the U2 snRNA base pairs with a consensus sequence at its binding site, the branch point (Parker *et al.*, 1987; Zhuang and Weiner, 1989). However, U2 snRNP binding is ATP dependent and requires an additional factor designated U2 auxiliary factor (U2AF). In mammals, U2 snRNP binds to the highest affinity site available within 18 to 35 or more nucleotides of the 3′ splice site. The U2AF binds to the 3′ splice site and its polypyrimidine tract and is necessary for accurate binding of the U2 snRNP to the branch point (Ruskin *et al.*, 1988; Nelson and Green, 1989).

Once U1 and U2 snRNP binding has occurred, snRNPs U4, U5 and U6 are incorporated completing the spliceosome assembly (Bindereif and Green, 1987). U4 and U6 snRNPs associate into a single complex by base pairing of their RNAs (Hashimoto and Steitz, 1984; Bringmann and Luhrmann, 1986; Siliciano *et al.*, 1987). This pairing is known to become destabilized late during spliceosome assembly just prior to the cleavage event at the 5′ splice site (Pikielny *et al.*, 1986; Cheng and Abelson, 1987; Lamond *et al.*, 1988). It has been suggested that U4 serves as a negative regulator of U6 activity, and its unwinding leads to the activation of U6 activity (Guthrie, 1991).

U6 snRNA plays an essential role in splicing *in vivo* and is postulated to serve as the catalytic subunit in the splicing reaction (Brow and Guthrie, 1988). Studies of self-splicing group I and II introns led to the hypothesis that the RNA moieties of the spliceosome and not the protein components are catalytically active and are essential for cleavage and ligation of exons (Guthrie, 1991). Evolutionarily, U6 snRNA is the most highly conserved spliceosomal RNA (Brow and Guthrie, 1988). The unique occurrence of an intron in U6 snRNA in *Schizosaccharomyces pombe* has been suggested to have arisen as a consequence of an aberrant splicing event in which an intron became incorporated into a proximal portion of the catalytic machinery of the spliceosome (Tani and Ohshima, 1989).

The role of polypeptides in splicing

If evolutionary considerations hold true and snRNAs function both in the recognition of intron consensus sequences as well as catalysis of the splicing reactions, what role(s) do the polypeptide components serve in the splicing process? Recent studies suggest that the more complex patterns of splicing observed in multicellular eukaryotes may actually hinge upon the differential expression of various spliceosome polypeptides. In the following sections we will first consider what is known about these polypeptides and then discuss specific examples of splicing in SV40-infected HeLa cells and in the sex determination of *Drosophila*.

Spliceosome core polypeptides

All snRNPs in mammalian spliceosomes contain seven core polypeptides designated B, B′, D, D′, E, F and G. Some of these core polypeptides are recognized by the human autoimmune antiserum, anti-Sm (Bringmann and Luhrmann, 1986). A consensus element in U snRNA, known as the Sm-binding site, is highly conserved among U snRNAs and is thought to be the binding site for the core polypeptides (Guthrie and Patterson, 1988). In addition there are protein components of the spliceosome unique to particular snRNPs such as U1-specific 70K, A and C, and U2-specific A′ and B″. The U1-70K protein has been found to contain an RNA recognition motif shared by several RNA-associated proteins (Query *et al.*, 1989). Although it is not yet well understood how these factors interact with snRNAs in the splicing reaction, certain protein factors have been shown to have a functional role in splicing. For example, U2AF is required (but not sufficient) for U2 binding of the 3′ splice site (see above) (Zamore and Green, 1989, 1991).

A neuron-specific snRNP polypeptide, referred to as ‘N’, has recently been identified in mammalian cells and has been proposed as a potential tissue-specific splicing factor based on its distribution (McAllister *et al.*, 1988; Schmausse *et al.*, 1992). While it shares many characteristics of other snRNP core polypeptides, including amino acid sequence homology and the Sm epitope, its functional significance has yet to be elucidated. Recent studies have demonstrated that N does not appear to be sufficient to direct the *in vivo* neuron-specific splicing of the α-calcitonin gene transcripts to CGRP (Delsert and Rosenfeld, 1992).

Pre-mRNA processing mutants (*Saccharomyces cerevisiae*)

Investigation of pre-mRNA processing (PRP) in mutants of *S. cerevisiae* has genetically identified a number of additional proteins which localize to the spliceosome. Some of these PRP genes have been isolated and characterized and shown to contain numerous conserved regions, including the DEA(D/H) box which is characteristic of a family of ATP-dependent RNA helicases (Company *et al.*, 1991; Wassarman and Steitz, 1991).

Analyses of the yeast PRP genes suggest that spliceosome proteins are required not only for the recognition of splice sites but also function as helicases and have proof-reading activity. One yeast gene, PRP5, which functions at an early stage in spliceosome assembly encodes a protein containing the DEAD motif (Dalbadie-McFarland and Abelson, 1990). PRP16, which was identified as a dominant suppressor of a branch point mutation, if depleted from yeast splicing extracts by immunoprecipitation, blocks processing and results in the accumulation of spliceosomes and splicing intermediates (Schwer and Guthrie, 1991). The addition of PRP16 and ATP alleviates the block. Further experimental evidence also suggests that PRP16 may serve a proof-reading function (Burgess *et al.*, 1990). The PRP22 gene, however, encodes a protein thought to facilitate disengagement of the mRNA from the spliceosome (Company *et al.*, 1991).

Mammalian splicing factors

Expression of SV40 small t antigen in HeLa cells requires the alternative use of an upstream 5′ splice site. In HeLa cell extracts the upstream 5′ splice choice generating the small t mRNA is regulated by a soluble protein factor designated ASF (alternative splicing factor) (Ge and Manley, 1990). ASF not only activates proximal 5′ splice sites in a concentration-dependent manner (Krainer *et al.*, 1990*a*, *b*) but also represses the large T splice choice. ASF is also essential for constitutive splicing activity.

The gene coding ASF (also known as SF2) has been independently isolated by two laboratories (Ge *et al.*, 1991; Krainer *et al.*, 1991). Analysis of the ASF/SF2 sequence revealed remarkable similarities with three of four known splicing regulators in the *Drosophila* sex-determination pathway (Ge *et al.*, 1991; Krainer *et al.*, 1991). The deduced amino acid sequence indicates that the C-terminus contains clusters of alternating arginine and serine residues similar to those found in the *Drosophila tra*, *tra*-2 and *su(w^a)* gene products (Boggs *et al.*, 1987; Chou *et al.*, 1987; Amrein *et al.*, 1988). The function of such residues is not well understood although R/S rich domains of *tra* and *su(w^a)* have been shown to target proteins to subnuclear compartments in Cos-1 cells where they are co-localized with other splicing proteins (Li and Bingham, 1991). Characteristic RNA binding motifs have also been identified in the N-terminus of ASF/SF2. Such motifs were initially identified in the yeast poly-A binding protein (Adam *et al.*, 1986; Sachs *et al.*, 1986) and are characteristic of a number of RNA binding proteins, including *tra*-2 and U1 snRNP 70K (Amrein *et al.*, 1988; Query *et al.*, 1989). Interestingly, three distinct ASF cDNAs having unique 3′ open reading frames have been isolated providing evidence of alternative 3′ splice site usage in the transcript of this regulator of 5′ splice site utilization (Ge *et al.*, 1991).

RNA splicing and *Drosophila* sex determination

The cascade of regulated splicing events that mediate sex determination in *Drosophila* illustrates regulatory principles that may also be important for the regulated splicing in other multicellular organisms (Baker, 1989; Maniatis, 1991). In *D. melanogaster* a number of genes, including *sex-lethal* (*sxl*), *transformer* (*tra*) and *doublesex* (*dsx*), are important for sex determination. The *sxl* and *tra* proteins are exclusively found in female flies although pre-mRNA is expressed in both sexes. In males, however, the transcripts are spliced differently and generate products which are inactive. The *dsx* transcript, however, is present in both sexes but is spliced to generate two different but related proteins in females and males.

In female flies the first three exons of *dsx* are spliced to a female-specific exon 4 and its polyadenylation site immediately downstream, while in males the first three exons are spliced to two male-specific exons and exon 4 is spliced out. The polypyrimidine tract of the female-specific 3′ splice site is anomalous because of interruption by purine residues and is inefficiently recognized by the splicing machinery (Burtis and Baker, 1989; Hedley and Maniatis, 1991). The *tra* and *tra*-2 proteins, which are expressed in female flies, act as splicing regulators in the sex determination cascade by interacting with the *dsx* gene transcript and activating the splice to exon 4 — the female-specific exon of *dsx* (Baker, 1989) (Fig. 2). *Tra* and *tra*-2 as well as other nuclear factors interact with repeat sequences in *dsx* exon 4 (Fig. 2) (Burtis and Baker, 1989; Nagoshi and Baker, 1990; Hedley and Maniatis, 1991; Hoshijima *et al.*, 1991; Tian and Maniatis, 1992).

Neuropeptide mRNA splicing

Mechanisms underlying alternative RNA processing in *Drosophila* somatic sex determination are likely to be broadly analogous to the mechanisms that influence tissue-specific pre-mRNA processing and are therefore relevant to the biology of neuropeptides. Two of the best studied examples of neuropeptide RNA processing are the calcitonin/CGRP and preprotachykinin systems. In both cases, regulatory mechanisms show commonality and are

Fig. 2. Model for the regulation of neuropeptide alternative splicing

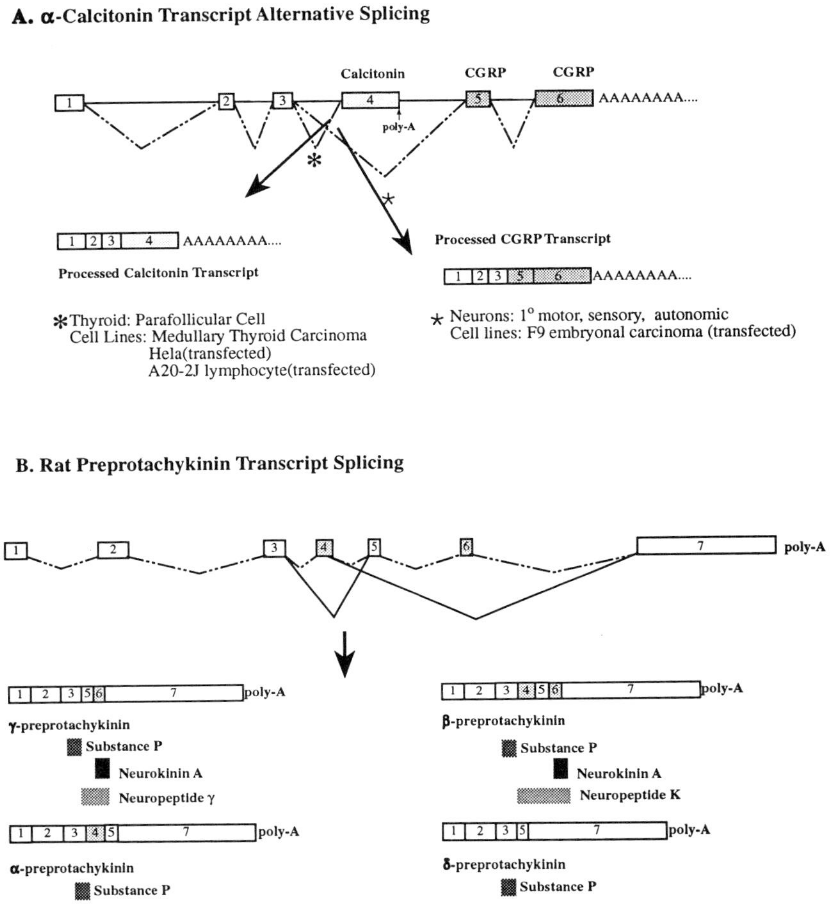

(A) Calcitonin/CGRP cell-specific splicing patterns. (B) The spliced products of the preprotachykinin (PPT) gene and protein products derived from the spliced transcripts; in rat, the γ-PPT transcript is the predominant species. Recently a unique product of RNA splicing, δ-PPT, has been detected in dorsal root ganglia.

likely to have more general significance for the neuronal post-transcriptional regulation of gene expression.

Calcitonin-CGRP

Northern blot analysis of poly-A RNA from rat medullary thyroid carcinoma (MTC) tumours that were ‘low producers’ of calcitonin led to the discovery of a second calcitonin gene product, a consequence of alternative RNA splicing (Amara *et al.*, 1982). The alternative splicing mechanism joins exon 3 to exon 4 to make calcitonin mRNA or exon 3 to exon 5 to make the CGRP mRNA (Amara *et al.*, 1982, 1984; Rosenfeld *et al.*, 1983) (Fig. 3). The resulting mRNAs have distinct 3′ exons each with a different polyadenylation site (Amara *et al.*, 1984). Recently, a third alternative RNA splicing pathway in the human MTC line has been described which yields a transcript coding for calcitonin that utilizes the CGRP poly-A signal (Minvielle *et al.*, 1991).

A key feature of calcitonin transcript splicing is that it is regulated according to tissue cell type *in vivo*; calcitonin is the predominant product of neural-crest-derived parafollicular C cells of the thyroid, whereas CGRP is made in a subset of neurons (Rosenfeld *et al.*, 1983; Amara *et al.*, 1985). Motor neurons in brain stem and spinal cord, sensory ganglia neurons, and neurons of the autonomic nervous system are the principle locales of CGRP expression. Another form of CGRP, β-CGRP, is encoded on a distinct gene but is also expressed in neurons (Amara *et al.*, 1985). The β-CGRP peptide differs by only one amino acid in rat and three amino acids in human. The β-CGRP gene has an overall structure similar to the calcitonin/α-CGRP gene, but lacks an alternatively spliced region capable of encoding a calcitonin-like hormone. In the region where the calcitonin exon would otherwise be, there exists a 'vestigial' exon with a coding sequence interrupted by multiple stop codons (Steenbergh *et al.*, 1986).

A mechanism common to many if not all neuronal populations appears to generate the neuron-specific splice of calcitonin to CGRP. Transgenic mice expressing the rat calcitonin/CGRP transcription unit under the direction of the metallothionein promoter synthesize mature CGRP mRNA in many neurons (including populations where it is not normally expressed such as hippocampal neurons), whereas the calcitonin mRNA, is the predominant product in non-neuronal tissues (Crenshaw, III *et al.*, 1987). Thus a generalized neuron-specific splicing mechanism that operates on a broader range of transcripts might be postulated. Other neuronal gene transcripts such as that of the preprotachykinin gene whose products include substance P (co-localized in sensory ganglia with CGRP) and substance K (the bovine peptide found in thyroid C cells with calcitonin) (Nawa *et al.*, 1984; Crenshaw, III *et al.*, 1987) may also be regulated by the specific factors common to many neuronal cell types (see the section on preprotachykinin).

Role of polyadenylation

The differential splicing of immunoglobulin mRNAs is influenced by the site of transcriptional termination and polyadenylation (Mather *et al.*, 1984; Gough, 1987; Galli *et al.*, 1988; Peterson and Perry, 1989). Since mature calcitonin and CGRP mRNAs have not only distinct 3′ termini but also unique polyadenylation sites, it seems plausible that these features may in some way determine sites for alternative splicing. Several experiments argue

Fig. 3. **RNA processing of neuropeptides: alternative splicing of the calcitonin and preprotachykinin gene transcripts**

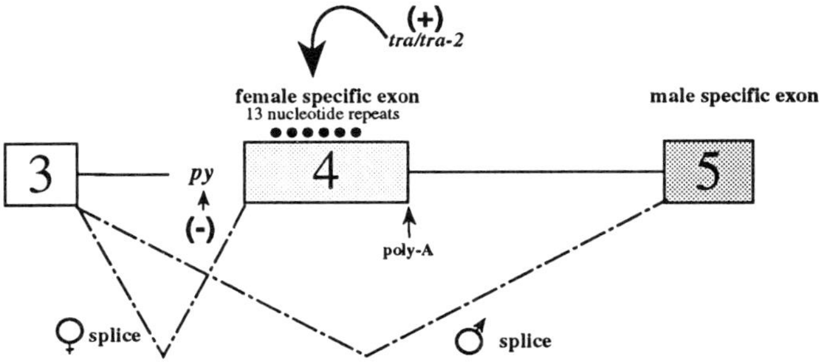

A. Model for *Drosophila dsx* pre-mRNA processing

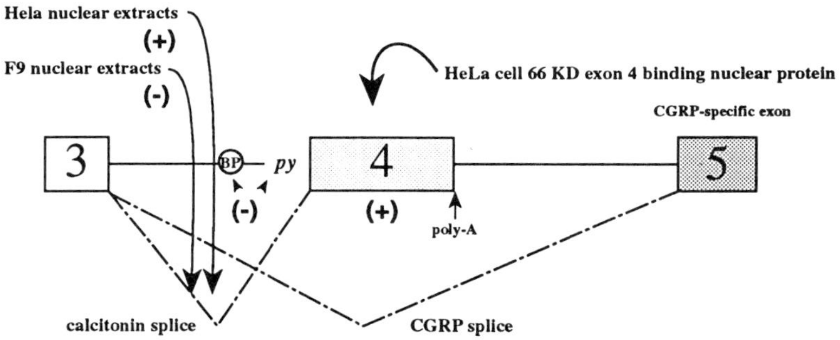

B. Model for Calcitonin/CGRP pre-mRNA processing

(A) Regulation of sxl *splicing in* Drosophila. Tra *and* tra-2 *are peptide factors produced in the female fly that promote the splice to the female-specific exon 4. In the absence of such factors an interrupted polypyrimidine stretch* (py) *is inefficiently recognized by the splicing machinery and exon 4 is skipped. (B) Model of calcitonin/ CGRP splicing parallels* sxl *splicing in* Drosophila. *The splice to the calcitonin exon is inhibited by a non-canonical branch point (BP) (U in human, C in rat), and a short polypyrimidine stretch* (py)*. Exon 4 sequences appear to mediate activation of the calcitonin splice and to specifically bind a 66 kDa protein in HeLa cell nuclear extracts. F9 embryonal carcinoma cells do not contain this protein and have an inhibitory effect on the calcitonin splice when mixed with HeLa extracts. (+) and (−) indicate regions of transcript or factors that activate or inhibit the splice to exon 4 respectively.*

against this possibility. Nuclear run-off analysis has shown that the site of transcriptional termination does not influence calcitonin/CGRP splicing (Amara *et al.*, 1984). Fusion constructs in which both polyadenylation sites are inserted in place of an SV40 poly-A site, yield transcripts polyadenylated at either site in tissue culture cells regardless of the splice choice normally associated with a particular cell line (Leff *et al.*, 1987). Also when CGRP and calcitonin polyadenylation sites are substituted for one another, splice choices remain unaltered in stably transfected cell lines (Emeson *et al.*, 1989).

Regulatory elements at splice junctions

Since alternative poly-A site usage does not appear to be the critical determinant of alternative splice choice for calcitonin/CGRP transcript processing, attention has been diverted to other possible *cis*-acting targets of the splicing machinery. In many instances where splicing is regulated, transcripts harbour consensus elements that are inherently 'weaker' resulting in slower intron excision kinetics both *in vivo* and *in vitro* (Bovenberg *et al.*, 1986; Cote *et al.*, 1991). Unusual consensus sequences, when demonstrated to affect splicing efficiency, may signify a requirement for an additional (possibly tissue-specific) factor or factors that may influence splice site choice or enhance the efficiency of intron excision. In fact, anomalies in splice site consensus elements have been noted in the *Drosophila dsx* transcript where the polypyrimidine stretch is interrupted (Burtis and Baker, 1989; Hedley and Maniatis, 1991) and in the preprotachykinin 5′ splice site of exon 4 which has a defective U1 snRNA consensus binding sequence (Nasim *et al.*, 1990; Kuo *et al.*, 1991) (see below).

Differential splicing of two highly similar human growth hormone gene transcripts is, in part, the result of competition between 3′ splice sites and the influence of their respective branch point nucleotides (Estes *et al.*, 1990). The human growth hormone (hGH)-N transcript uses an alternative splicing pathway not shared by the nearly identical hGH variant ChGH-V) (DeNoto *et al.*, 1981; Cooke *et al.*, 1988). hGH-N transcripts contain an alternative 3′ splice site within exon 3 which is used to generate a 20 kDa isoform. A canonical adenosine branch point within the exon, which allows the splice to occur, is not present in hGH-V exon 3 (Estes *et al.*, 1990). Mutation of the analogous hGH-V nucleotide to adenosine allows the use of the exon 3 internal splice acceptor as with hGH-N.

The spacing of consensus sequence elements may also be important. The neuron-specific exon N1 in c-*src* appears to be excluded from the *src* transcript because splicing factors that interact with the closely situated N1 splice acceptor and donor sites may be hindered sterically (Black, 1991, 1992).

Although the rat and human calcitonin genes contain the necessary consensus elements that define splice donor, acceptor and branch point sequences, there are a number of unique sequence features which may be relevant to alternative splicing. In the mapped branch point for the calcitonin exon (Adema *et al.*, 1988) a uridine serves as the branch acceptor at the intron 3,3′ splice site of the human gene while, in the rat gene, a cytidine residue is the acceptor. An adenosine residue generally serves as the branch point nucleotide for most introns. When uridine is changed to adenosine in human calcitonin/CGRP, splice enhancement has been demonstrated *in vitro* (Adema *et al.*, 1988, 1990; Cote *et al.*, 1990). This uridine residue therefore appears to be inhibitory for the exon 3–4 splice (Fig. 3). Similarly, the polypyrimidine sequence which is significantly shorter at the calcitonin 3′ splice site (Adema *et al.*, 1990; Bennett and Amara, 1991) may influence splicing by altering interactions between the spliceosome and the transcript.

Deletion analysis of splice junction sequences

Experiments introducing the gene and a variety of mutated gene constructs into different cell types have shown that different cell lines splice calcitonin/CGRP gene transcripts differently (Leff *et al.*, 1987). These cell line-specific differences might be due to the presence or absence of splice modulating factors which in turn interact with *cis* elements on the nascent transcript. Alternatively, these *cis* elements may themselves determine splicing efficiency by directly affecting spliceosome activity.

In cell lines that normally produce CGRP transcripts, experiments suggest that inhibitory factors block the splice to the calcitonin exon. In the F9 cell line, for example, no calcitonin product is made if F9 cells are transfected with a construct that has been mutated to block the splice to CGRP (lacking the intron 4,3′ splice site) (Emeson *et al.*, 1989). Since F9 cells normally make only CGRP, processing of the calcitonin/CGRP transcript to CGRP mRNA must not in and of itself block calcitonin mRNA production. Consistent with this view is the observation that F9 splice site usage can be made to favour calcitonin (82%) over CGRP (18%) by substitution of a constitutive 3′-splice site for the calcitonin intron 3,3′ splice site (Emeson *et al.*, 1989). These results suggest that the F9 cell line either contains an inhibitor and/or lacks an activator that regulates the calcitonin splice, and that the 3′ splice site is somehow involved.

Perturbation of calcitonin processing, however, has little or no direct effect on the capacity for the CGRP-specific splice to occur. In HeLa cells, which normally do not make CGRP, deletion of a 77 bp fragment containing the calcitonin exon 4 splice acceptor does not block the production of mature CGRP transcripts. A similar result is obtained in A20 lymphocytes when a construct harbouring a 350 bp deletion including the exon 4 splice acceptor sequences is expressed (Leff *et al.*, 1987). These results suggest that calcitonin is the regulated splice since a cell line normally capable of producing only calcitonin can still use the CGRP exon 5 splice acceptor when the calcitonin splice acceptor is deleted.

Cis-acting elements of the calcitonin/CGRP transcript have been further analysed using intron deletions that extend towards the 3′ splice site of the intron upstream of calcitonin exon 4 (Emeson *et al.*, 1989). These constructs were introduced into both F9 and HeLa cells. The two largest deletions -58/-27 and -58/-17 (origin at 5′ end of exon 4), did not significantly change the preference for the calcitonin splice choice in HeLa cells, but did alleviate the calcitonin block in F9 cells. A 4 to 5-fold increase in calcitonin splice site usage over wild type was observed when F9 cells were transfected with these deletion constructs. These results lend further support to the notion that a *cis* element causes repression of calcitonin exon 4 splicing and that the element is contained within the deleted sequences.

Direct evidence for the existence of an inhibitory factor that blocks the splice to exon 4 has been obtained *in vitro* (Cote *et al.*, 1991). In HeLa cell extracts capable of processing calcitonin/CGRP transcript to calcitonin, addition of F9 extract blocks splicing to calcitonin and results in spliced products containing the CGRP-specific exon 5. This inhibitory effect also occurs in spite of the presence of a mutated intron 3 branch point which appears to be required for the *in vitro* splice to exon 4 in HeLa cell extract alone.

Role of exon sequences

While intronic sequences in the vicinity of the calcitonin/CGRP splice sites have been demonstrated to play a role in determination of splice choice, there is also evidence that exonic sequences are important for calcitonin gene alternative splicing. In stably transfected cells harbouring a construct in which most of the calcitonin exon and its polyadenylation site are deleted, splicing to either the CGRP or calcitonin exon does not occur and only nuclear precursors accumulate, irrespective of cell type (Leff *et al.*, 1987). As discussed previously, when the poly-A site alone is mutated only the calcitonin splice is blocked. Furthermore, when exon 4 sequences near the intron 3,3′ splice site are deleted, transcripts are spliced to generate CGRP (Cote *et al.*, 1990). This suggests an important role for exon 4 sequences in the CGRP splicing machinery, perhaps in the formation of RNA secondary structure (Leff *et al.*, 1987).

The β-CGRP gene and human CGRP II genes are related to but distinct from calcitonin/CGRP. In both cases, RNA processing removes a truncated calcitonin pseudo-exon and the intron 3,3′ splice site (Steenbergh *et al.*, 1986). This pseudo-exon contains numerous in-frame stop codons and is missing a 3′ sequence and the polyadenylation signal that is normally present in calcitonin/CGRP. Consequently, neither gene encodes a functional calcitonin transcript. Calcitonin/CGRP gene constructs, in which these pseudo-exon sequences are substituted for exon 4, direct splicing to CGRP exon 5 exclusively. Splicing of this construct is identical in transfected HeLa cells, as well as in *in vitro* extracts (Cote *et al.*, 1992). Mutations introduced into an exon 4 repeat sequence GCGG, located 22 nucleotides from the intron 3,3′ splice site also appear to disrupt the splice to exon 4, although the inhibitory effects are weaker than those of the β-CGRP sequences.

U.v.-crosslinking experiments have identified a 66 kDa protein from HeLa nuclear extracts that specifically interacts with calcitonin exon 4 sequences (Cote *et al.*, 1992). F9 nuclear extracts do not contain this protein. It has been suggested that the calcitonin exon 4 factor may interact with the RNA in a manner analogous to that of the *tra*-2 gene product of *Drosophila* (Cote *et al.*, 1992). The *tra*-2 gene regulates *dsx* splicing through interactions localized to exon 4 of the *dsx* transcript (Fig. 3).

On the basis of the experiments described above, a possible model of calcitonin/CGRP splicing emerges (Fig. 3). In addition to the three usual intron consensus sequences, two regions appear crucial for the splice choice. The first of these two sequence elements resides near the 3′ end of intron 3. The second sequence element is within exon 4. In both cases, *in vitro* evidence suggests that these sequences are targets for the binding of splice modulating factors. In F9 cell lines factors are inhibitory and block processing to calcitonin mRNA. The inhibitory factor(s) binds directly or indirectly to the 3′ region of intron 3, upstream of calcitonin/CGRP exon 4, and blocks splicing to the calcitonin-specific transcript. Thus, by default, CGRP mRNA is synthesized in F9 cells. The factors interacting with exon 4, however, in some way facilitate splicing to calcitonin. These activating factors are present in HeLa cells but absent from F9 cells.

Preprotachykinin

Post-translational processing of the preprotachykinin (PPT) gene products gives rise to substance P, neurokinin A, neuropeptide K and neuropeptide γ, among others (Fig. 2) (Maggio, 1988). An additional level of regulatory complexity exists at the level of PPT pre-mRNA splicing. Transcript processing yields a total of three well-characterized splicing variants (Carter and Krause, 1990) as well as a fourth, more recently described, spliced product identified in dorsal root ganglia (DRG) (Fig. 2) (Harmar *et al.*, 1990). The PPT pre-mRNA contains seven exons. Differential splicing yields the predominant product γ-PPT, lacking exon 4 (which codes for a portion of neuropeptide K), β-PPT, the second most abundant product which contains all seven exons, δ-PPT, lacking both exons 4 and 6 (which code for portions of neuropeptide K and neurokinin A) and is as abundant as β-PPT in DRGs (Harmar *et al.*, 1990), and α-PPT which only lacks exon 6 (Carter and Krause, 1990; Bannon *et al.*, 1992) and is the least abundant ($<1\%$ in brain tissues). Variability is determined by the presence or absence of only two of the seven exons: exon 4 and exon 6 (Fig. 2).

In addition to differences in the abundance of specific, spliced transcripts, the distribution of these PPT slice variants can differ between different organisms. In bovine PPT, for example, splicing generates predominantly α- and β-PPT. β-PPT is expressed most highly in thyroid and intestine, while α-PPT is more highly represented in brain tissues (Nawa *et al.*, 1984). In contrast, in rat the ratios of γ:β:α mRNAs are relatively invariant (Carter and Krause, 1990).

A diversity of regulatory mechanisms must exist to achieve differences in the abundance and tissue-specific distribution of spliced transcripts. Intron excision may be modulated at the sequence level by affecting spliceosome assembly and/or activity or by incorporating cell-specific factors that modify the splicing pathway, as has been discussed for calcitonin/CGRP. Comparison of rat and bovine PPT RNA splice donor and acceptor sites demonstrate divergence from consensus sequences in the exon 1 splice donor which is constitutively spliced in both species, and in the exon 4 splice donor (Carter and Krause, 1990) which contains a functionally significant sequence anomaly. In the following section we will discuss how such unusual sequences affect processing, and particularly how altering the context in which an exon is found can redefine it as an intron.

The mechanism of exon 4 exclusion

Much attention has been directed toward understanding how exon 4 is excluded from the rat PPT transcript during pre-mRNA processing of the most abundant rat transcript for γ-PPT. RNAse protection and Northern blot analysis of *in vitro* PPT reaction products using intron- and exon-specific probes have failed to demonstrate the presence of RNA species in which intron 3 is removed before intron 4. This ordered removal of introns from the PPT transcript has been shown in HeLa cell extracts using a minigene containing a segment of the gene spanning exons 3, 4 and 5 (Nasim *et al.*, 1990). It appears that removal of intron 3 is a rate-limiting step that is contingent upon the prior removal of intron 4. Splicing experiments *in vitro* have also demonstrated that splicing of exon 3 to exon 4 appears to be inherently inefficient. Processing of a minigene construct comprised of only exons 3 and 4 joined by their intervening intron occurs at minute levels compared with that observed with analogous constructs containing exons 4 and 5 and exons 3 and 5 (Nasim *et al.*, 1990).

Sequences downstream are thought to have a role in the modulation of the inherently inefficient splicing of intron 3. In particular, the 5′ splice site of the intron is thought to be important for the determination of splice choice. Efficient splicing of intron 3 is activated in a minigene construct in which intron 4 is replaced with exon 5 and a portion of intron 5 (Nasim *et al.*, 1990). Significantly, exons 4 and 5 alone are not sufficient to activate intron 3 splicing. Rather, intronic sequences further downstream are required.

It has therefore been proposed that the inherent defect of intron 3 splicing is the result of a weak intron 4,5′ splice site. In the ordered intron removal of PPT splicing exon 4 inclusion is dependent on the replacement of the intron 4,5′ splice site by the intron 5,5′ splice site (Nasim *et al.*, 1990). This hypothesis is supported by experiments which show that mutation of the intron 4,5′ splice site, so that it matches the splice site consensus sequence (improving complementarity to U1 snRNA), results in more efficient splicing of intron 3 (Nasim *et al.*, 1990; Kuo *et al.*, 1991). These effects of a 5′ splice site on an upstream intron have been shown to result directly from increased U1 binding at the mutated (consensus) splice site with consequent acceleration of intron 3 lariat formation and splicing to exon 4 (Kuo *et al.*, 1991). Mutations that reduce 5′ splice-site consensus sequence homology and complementarity to U1 snRNA lead to exon 4 skipping as in the case of the wild-type gene.

How distinguishable are the U1 binding sites within the PPT transcript? The intron 4,5′ splice site differs by only one base relative to introns not associated with an alternatively spliced exon. It appears that a single nucleotide mismatch is sufficient to modify U1 binding and leads to exon 4 skipping (Kuo *et al.* 1991). Therefore, U1 snRNA–5′ splice site interaction appears to be critical for the regulation of exon recognition. It has also been shown that spliceosome assembly at a 3′ splice site is highly dependent on the presence of a functional downstream 5′ splice site. Together these interactions may serve to define an exon before alterations in spliceosome conformation lead to intron splicing (Robberson *et al.*, 1990). Such interactive events may be blocked in preprotachykinin RNA processing by the inefficient binding of U1 to the intron 4,5′ splice site resulting in the treatment of exon 4 sequences as intronic by the spliceosome machinery.

What mechanisms might exist to allow exon 4 inclusion in the mRNA transcript? It may be that a nuclear factor serves to facilitate recognition of the PPT intron 4,5′ splice site or that such a factor promotes excision of intron 4 so that the next downstream 5′ splice site that can efficiently bind U1 is utilized. In either case the presence of a stronger splice site activates intron 3 lariat and spliceosome formation and excision of intron 3. Such a factor may bind to the transcript directly, perhaps changing the context in which the exon is perceived by the processing machinery, or it may interact with the spliceosome itself.

Summary and conclusion

This review has focused on the mechanisms of mRNA processing of neuropeptides using two of the best understood examples of alternatively spliced neuropeptide RNAs: calcitonin/CGRP and PPT. On the background of work that has demonstrated the fundamental molecular operations of the spliceosome–RNA interaction in yeast and *Drosophila*, parallels can be drawn that help elucidate the factors contributing to RNA splicing of neuropeptides. In the case of the calcitonin/CGRP gene, unusual transcript splice sites appear to have an important role in the determination of whether or not a given exon is spliced in a manner analogous to pre-mRNA processing events in the *Drosophila* sex-determination pathway. Similarly, evidence now exists for the occurrence of tissue and cell line-specific nuclear proteins that are also involved in the regulation of calcitonin/CGRP splicing patterns.

In addition to demonstrating a role for U1–5′ splice site interactions in the determination of splice choice, the study of PPT transcript splicing has demonstrated the importance of splice site context in exon definition. In the splicing of transcripts including PPT exon 4, a downstream splice must first occur which removes the splice site anomaly. This results in the transposition of a sequence upstream that is recognized as a strong, constitutive 5′ splice site. It may also be proposed that a factor must be present to allow recognition of the weak intron 4,5′ splice site, or exon 4 sequences are viewed by the splicing machinery as intronic. These may represent general mechanisms by which transcripts choose between including or excluding a single exon, unlike the mechanism of calcitonin/CGRP splicing in which a choice is made between two exons which are alternatively spliced.

These findings have encouraged investigators to search for *cis*-acting sequences and *trans*-acting factors that can induce exon recognition by the splicing machinery in the non-optimal milieu of an unusual splice site. As in the case of *sxl*, protein factors that bind RNA transcripts in a cell-specific manner will no doubt be characterized in the near future as mediators of cell-specific or regulated splicing in neuropeptide gene expression.

References

Adam, S. A., Nakagawa, T., Swanson, M. S., Woodruff, T. K. and Dreyfuss, G. (1986) mRNA polyadenylate-binding protein: gene isolation and sequencing and identification of a ribonucleoprotein consensus sequence. Mol. Cell. Biol. **6**(8), 2932–2934

Adema, G. J., Bovenberg, R. A. L., Jansz, H. S. and Baas, P. D. (1988) Unusual branch point selection involved in splicing of the alternatively processed calcitonin/CGRP-I pre-mRNA. Nucleic Acids Res. **16**(20), 9513–9526

Adema, G. J., van Hulst, K. L. and Baas, P. D. (1990) Uridine branch acceptor is a *cis*-acting element involved in regulation of the alternative processing of calcitonin/CGRP-I pre-mRNA. Nucleic Acids Res. **18**(18), 5365–5373

Amara, S. G., Jonas, V., Rosenfeld, M. G., Ong, E. S. and Evans, R. M. (1982) Alternative RNA processing in calcitonin gene expression generates mRNAs encoding different polypeptide products. Nature (London) **298**, 240–244

Amara, S. G., Evans, R. M. and Rosenfeld, M. G. (1984) Calcitonin/calcitonin gene-related peptide transcription unit: tissue specific expression involves selective use of alternative polyadenylation sites. Mol. Cell. Biol. **4**(10), 2151–2160

Amara, S. G., Arriza, J. L., Leff, S. E., Swanson, L. W., Evans, R. M. and Rosenfeld, M. G. (1985) Expression in brain of a messenger RNA encoding a novel neuropeptide homologous to calcitonin gene-related peptide. Science **229**, 1094–1097

Amrein, H., Gorman, M. and Nothiger, R. (1988) The sex-determining gene *tra*-2 of *Drosophila* encodes a putative RNA binding protein. Cell **55**, 1025–1035

Baker, B. S. (1989) Sex in flies: the splice of life. Nature (London) **340**, 521–524

Bannon, M. J., Poosch, M. S., Haverstick, D. M., Mandal, A., Xue, I. C.-H., Shibata, K. and Dragovic, L. J. (1992) Preprotachykinin gene expression in the human basal ganglia: characterization of mRNAs and pre-mRNAs produced by alternate RNA splicing. Mol. Brain Res. **12**, 225–231

Barabino, S. M., Blencowe, B. J., Ryder, U., Sproat, B. S. and Lamond, A. I. (1990) Targeted snRNP depletion reveals an additional role for mammalian U1 snRNP in spliceosome assembly. Cell **63**, 293–302

Bennett, M. M. and Amara, S. G. (1991) Molecular mechanisms of cell-specific and regulated expression of the calcitonin/α-CGRP and β-CGRP genes. Ann. N.Y. Acad. Sci. **657**, 36–49

Berget, S. M., Moore, C. and Sharp, P. (1977) Spliced segments at the 5′ terminus of adenovirus 2 late mRNA. Proc. Natl. Acad. Sci. U.S.A. **74**(8), 3171–3175

Bindereif, A. and Green, M. R. (1987) An ordered pathway of snRNP binding during mammalian pre-mRNA splicing complex assembly. EMBO J. **6**(8), 2415–2424

Black, D. L. (1991) Does steric interference between splice sites block the splicing of a short *c-src* neuron-specific exon in non-neuronal cells? Genes Dev. **5**, 389–402

Black, D. L. (1992) Activation of *c-src* neuron-specific splicing by an unusual RNA element *in vivo* and *in vitro*. Cell **69**, 795–807

Boggs, R. T., Gregor, P., Idriss, S., Belote, J. M. and Mckeown, M. (1987) Regulation of sexual differentiation in *D. melanogaster* via alternative splicing of RNA from the transformer gene. Cell **50**, 739–747

Bovenberg, R. A. L., van de Meerendonk, W. P. M., Baas, P. D., Steenbergh, P. H., Lips, C. J. M. and Janz, H. S. (1986) Model for alternative RNA processing in human calcitonin gene expression. Nucleic Acids Res. **14**(22), 8785–8803

Breitbart, R. E., Andreadis, A. and Nadal-Ginard, B. (1987) Alternative splicing: a ubiquitous mechanism for the generation of multiple protein isoforms from single genes. Annu. Rev. Biochem. **56**, 467–495

Bringmann, P. and Luhrmann, R. (1986) Purification of the individual snRNPs U1, U2, U5, and U4/U6 from HeLa cells and characterization of their constituents. EMBO J. **5**(13), 3509–3516

Brow, D. A. and Guthrie, C. (1988) Spliceosomal RNA U6 is remarkably conserved from yeast to mammals. Nature (London) **334**, 213–218

Burgess, S., Couto, J. R. and Guthrie, C. (1990) A putative ATP binding protein influences the fidelity of branchpoint recognition in yeast splicing. Cell **60**, 705–717

Burtis, K. C. and Baker, S. B. (1989) *Drosophila* doublesex gene controls somatic sexual differentiation by producing alternatively spliced mRNAs encoding related sex-specific polypeptides. Cell **56**, 997–1010

Carter, M. S. and Krause, J. E. (1990) Structure, expression and some regulatory mechanisms of the rat preprotachykinin gene encoding substance P, neurokinin A, neuropeptide K, and neuropeptide γ. J. Neurosci. **10**(7), 2203–2214

Cech, T. R. (1990) Self-splicing of group I introns. Annu. Rev. Biochem. **59**, 543–568

Cheng, S.-C. and Abelson, J. (1987) Spliceosome assembly in yeast. Genes Dev. **1**, 1014–1027

Chou, T.-B., Zachar, Z. and Bingham, P. M. (1987) Developmental expression of a regulatory gene is programmed at the level of splicing. EMBO J. **6**(13), 4095–4104

Chow, L. T., Gelinas, R. E., Broker, T. R. and Roberts, R. J. (1977) An amazing sequence arrangement at the 5′ ends of adenovirus 2 messenger RNA. Cell **12**, 1–8

Company, M., Arenas, J. and Abelson, J. (1991) Requirement of the RNA helicase-like protein PRP22 for release of messenger RNA from spliceosomes. Nature (London) **349**, 487–493

Cooke, N. E., Ray, J., Watson, M. A., Estes, P. A., Kuo, B. A. and Liebhaber, S. A. (1988) Human growth hormone gene and the highly homologous growth hormone variant gene display different splicing patterns. J. Clin. Invest. **82**, 270–275

Cote, G. J., Nguyen, I. N., Berget, S. M. and Gagel, R. F. (1990) Calcitonin exon sequences influence alternative RNA processing. Mol. Endocrinol. **4**, 1744–1749

Cote, G. J. Nguyen, I. N., Lips, C. J. M., Berget, M. B. S. and Gagel, R. F. (1991) Validation of an *in vitro* RNA processing system for CT/CGRP precursor mRNA. Nucleic Acids Res. **19**(13), 3601–3606

Cote, G. J., Stolow, D. T., Peleg, S., Berget, S. M. and Gagel, R. F. (1992) Identification of exon sequences and an exon binding protein involved in alternative RNA splicing of calcitonin/CGRP. Nucleic Acids Res. **20**(9), 2361–2366

Crenshaw, III, E. B., Russo, A. F., Swanson, L. W. and Rosenfeld, M. G. (1987) Neuron-specific alternative RNA processing in transgenic mice expressing a metallothionein–calcitonin fusion gene. Cell **49**, 389–398

Dalbadie-McFarland, G. and Abelson, J. (1990) PRP5: a helicase-like protein required for mRNA splicing in yeast. Proc. Natl. Acad. Sci. U.S.A. **87**, 4236–4240

Dal Toso, R., Sommer, B., Ewert, M., Herb, A., Pritchett, D. B., Bach, A., Shivers, B. D. and Seeburg, P. H. (1989) The dopamine D2 receptor: two molecular forms generated by alternative splicing. EMBO J. **8**, 4025–4034

de Ferra, F., Engh, H., Hudson, L., Kamholz, J., Puckett, C., Molineaux, S. and Lazzarini, R. A. (1985) Alternative splicing accounts for the four forms of myelin basic protein. Cell **43**, 721–727

Delsert, C. D. and Rosenfeld, M. G. (1992) A tissue-specific small nuclear ribonucleoprotein and the regulated splicing of the calcitonin/calcitonin gene-related protein transcript. J. Biol. Chem. **267**, 14573–14579

DeNoto, F. M., Moore, D. D. and Goodman, H. M. (1981) Human growth hormone DNA sequence and mRNA structure: possible alternative splicing. Nucleic Acids Res. **9**, 3719–3730

Edwards, R. H., Selby, M. J. and Rutter, W. J. (1986) Differential RNA splicing predicts two distinct nerve growth factor precursors. Nature (London) **319**, 784–787

Emeson, R. B., Hedjran, F., Yeakley, J. M., Guise, J. W. and Rosenfeld, M. G. (1989) Alternative production of calcitonin and CGRP mRNA is regulated at the calcitonin-specific splice acceptor. Nature (London) **341**, 76–80

Estes, P. A., Cooke, N. E. and Liebhaber, S. A. (1990) A difference in the splicing patterns of the closely related normal and variant human growth hormone gene transcripts is determined by a minimal sequence divergence between two potential splice-acceptor sites. J. Biol. Chem. **265(32)**, 19863–19870

Fu, X.-Y., Ge, H. and Manley, J. L. (1988) The role of the polypyrimidine stretch at the SV40 early pre-mRNA 3′ splice site in alternative splicing. EMBO J. **7**, 809–817

Galli, G., Guise, J. W., Tucker, P. W. and Nevins, J. R. (1988) Poly(A) site choice rather than splice site choice governs the regulated production of IgM heavy-chain RNAs. Proc. Natl. Acad. Sci. U.S.A. **85**, 2439–2443

Ge, H. and Manley, J. L. (1990) A protein factor, ASF, controls cell-specific alternative splicing of SV40 early pre-mRNA *in vitro*. Cell **62**, 25–34

Ge, H., Ping, Z. and Manley, J. L. (1991) Primary structure of the human splicing factor ASF reveals similarities with *Drosophila* regulators. Cell **66**, 373–382

Giros, B., Sokoloff, P., Martres, M. P., Riou, J. F., Emorine, L. J. and Schwartz, J. C. (1989) Alternative splicing directs the expression of two D2 dopamine receptor isoforms. Nature (London) **342**, 923–926

Goldman, D. L., Deneris, E., Luyten, W., Kochar, A., Patrick, J. and Heinemann, S. (1987) Members of a nicotinic acetylcholine receptor gene family are expressed in different regions of the mammalian central nervous system. Cell **48**, 965–973

Goridis, C., Hirn, M., Santoni, M.-J., Gennarini, G., Deagostini-Bazin, H., Jordan, B. R., Kiefer, M. and Steinmetz, M. (1985) Isolation of mouse N-CAM-related cDNA: detection and cloning using monoclonal antibodies. EMBO J. **4**, 631–635

Gough, N. (1987) Putting a stop to an immunoglobulin message. Trends Genet. **3(9)**, 238–239

Grima, B., Lamouroux, A., Boni, C., Julien, J.-F., Javoy-Agid, F. and Mallet, J. (1987) A single human gene encoding multiple tyrosine hydroxylases with different predicted functional characteristics. Nature (London) **326**, 707–711

Guthrie, C. (1991) Messenger RNA splicing in yeast: clues to why the spliceosome is a ribonucleoprotein. Science **253**, 157–163

Guthrie, C. and Patterson, B. (1988) Spliceosomal snRNAs. Annu. Rev. Genet. **22**, 387–419

Harmar, A. J., Hyde, V. and Chapman, K. (1990) Identification and cDNA sequence of δ-preprotachykinin, a fourth splicing variant of the rat substance P precursor. FEBS Lett. **275(1)**, 22–24

Hashimoto, C. and Steitz, J. A. (1984) U4 and U6 RNAs coexist in a single small nuclear ribonucleoprotein particle. Nucleic Acids Res. **12(7)**, 3283–3293

Hedley, M. L. and Maniatis, T. (1991) Sex-specific splicing and polyadenylation of *dsx* pre-mRNA requires a sequence that binds specifically to *tra*-2 protein *in vitro*. Cell **65**, 579–586

Hoshijima, K., Inoue, K., Higuchi, I., Sakamoto, H. and Shimura, Y. (1991) Control of *doublesex* alternative splicing by *transformer* and *transformer*-2 in *Drosophila*. Science **252**, 833–836

Kitaguchi, N., Takahashi, Y., Tokushima, Y., Shiojiri, S. and Ito, H. (1988) Novel precursor of Alzheimer's disease amyloid protein shows protease inhibitory activity. Nature (London) **331**, 530–532

Klessig, D. F. (1977) Two adenovirus mRNAs have a common 5′ terminal leader sequence encoded at least 10 kb upstream from their main coding regions. Cell **12**, 9–21

Krainer, A. R., Conway, G. C. and Kozak, D. (1990*a*) The essential pre-mRNA splicing factor SF2 influences 5′ splice site selection by activating proximal sites. Cell **62**, 35–42

Krainer, A. R., Conway, G. C. and Kozak, D. (1990*b*) Purification and characterization of pre-mRNA splicing factor SF2 from HeLa cells. Genes Dev. **4**, 1158–1171

Krainer, A. R., Mayeda, A., Kozak, D. and Binns, G. (1991) Functional expression of cloned human splicing factor SF2: homology to RNA-binding proteins U1 70K, and *Drosophila* splicing regulators. Cell **66**, 383–394

Kuo, H.-C., Nasim, F.-U. H. and Grabowski, P. J. (1991) Control of alternative splicing by the differential binding of U1 small nuclear ribonucleoprotein particle. Science **251**, 1045–1050

Lamond, A. I., Konarska, M. M., Grabowski, P. J. and Sharp, P. A. (1988) Spliceosome assembly involves the binding and release of U4 small nuclear ribonucleoprotein. Proc. Natl. Acad. Sci. U.S.A. **85**, 411–415

Leff, S. E., Rosenfeld, M. G. and Evans, R. M. (1986) Complex transcriptional units: diversity in gene expression by alternative RNA processing. Annu. Rev. Biochem. **55**, 1091–1117

Leff, S. E., Evans, R. M. and Rosenfeld, M. G. (1987) Splice commitment dictates neuron-specific alternative RNA processing in calcitonin/CGRP gene expression. Cell **48**, 517–524

Legrain, P., Seraphin, B. and Rosbash, M. (1988) Early commitment of yeast pre-mRNA to the spliceosome pathway. Mol. Cell. Biol. **8**(9), 3755–3760

Li, H. and Bingham, P. M. (1991) Arginine/serine-rich domains of the *su(w^a)* and *tra* RNA processing regulators target proteins to a subnuclear compartment implicated in splicing. Cell **67**, 335–342

Maggio, J. E. (ed.) (1988) Tachykinins. Annu. Rev. Neurosci. **11**, 13–28

Maniatis, T. (1991) Mechanisms of alternative pre-mRNA splicing. Science **251**, 33–34

Martinez, R., Mathey-Prevot, B., Bernards, A. and Baltimore, D. (1987) Neuronal pp60$^{c\text{-}src}$ contains a six amino acid insertion relative to its non-neuronal counterpart. Science **237**, 411–416

Mather, E. L., Nelson, K. J., Haimovich, J. and Perry, R. P. (1984) Mode of regulation of immunoglobulin μ and δ-chain expression varies during B-lymphocyte maturation. Cell **36**, 329–338

McAllister, G., Amara, S. G. and Lerner, M. R. (1988) Tissue-specific expression and cDNA cloning of small nuclear ribonucleoprotein-associated polypeptide N. Proc. Natl. Acad. Sci. U.S.A. **85**, 5296–5300

Minvielle, S., Giscard-Dartevelle, S., Cohen, R., Taboulet, J., Labye, F., Jullienne, A., Rivaille, P., Milhaud, G., Moukhtar, M. S. and Lasmoles, F. (1991) A novel calcitonin carboxyl-terminal peptide produced in medullary thyroid carcinoma by alternative RNA processing of the calcitonin/calcitonin gene-related peptide gene. J. Biol. Chem. **266**(36), 24627–24631

Monsma, F. J., Mcvittie, L. D., Gerfen, C. R., Mahan, L. C. and Sibly, D. R. (1989) Multiple D2 dopamine receptors produced by alternative RNA splicing. Nature (London) **342**, 926–929

Mount, S. M. (1982) A catalogue of splice junction sequences. Nucleic Acids Res. **10**, 459

Mount, S. M., Pettersson, I., Hinterberger, M., Aavo, K. and Steitz, J. (1983) The U1 small nuclear RNA–protein complex selectively binds a 5′ splice site *in vitro*. Cell **33**, 509–518

Murray, B. A., Hemperly, J. J., Gallin, W. J., MacGregor, J. S., Edelman, G. M. and Cunningham, B. A. (1984) Isolation of cDNA clones for the chicken neural cell adhesion molecule (N-CAM). Proc. Natl. Acad. Sci. U.S.A. **81**, 5584–5588

Nagoshi, R. N. and Baker, B. (1990) Regulation of sex specific RNA splicing at the *Drosophila doublesex* gene: *cis*-acting mutations in exon sequences alter sex specific RNA splicing patterns. Genes Dev. **4**, 789–805

Nasim, F.-U., Spears, P. A., Hoffman, H. M., Kuo, H.-C. and Grabowski, P. J. (1990) A sequential splicing mechanism promotes selection of an optional exon by repositioning a downstream 5′ splice site in preprotachykinin pre-mRNA. Genes Dev. **4**, 1172–1184

Nawa, H., Kotani, H. and Nakanishi, S. (1984) Tissue-specific generation of two preprotachykinin mRNAs from one gene by alternative RNA splicing. Nature (London) **312**(20), 729–734

Nelson, K. K. and Green, M. R. (1989) Mammalian U2 snRNP has a sequence-specific RNA-binding activity. Genes Dev. **3**, 1562–1571

Newman, A. and Norman, C. (1991) Mutations in yeast U5 snRNA alter the specificity of 5′ splice-site cleavage. Cell **65**, 115–123

Newman, A. J. and Norman, C. (1992) U5 snRNA interacts with exon sequences at 5′ and 3′ splice sites. Cell **68**, 743–754

Padgett, R. A., Grabowski, P. J., Konarska, M. M., Seiler, S. and Sharp, P. A. (1986) Splicing of messenger RNA precursors. Annu. Rev. Biochem. **55**, 1119–1150

Parker, R., Siliciano, P. G. and Guthrie, C. (1987) Recognition of the TACTAAC box during mRNA splicing in yeast involves base pairing to the U2-like snRNA. Cell **49**, 229–239

Peterson, M. L. and Perry, R. P. (1989) The regulated production of mm and ms mRNA is dependent on the relative efficiencies of the ms poly(A) site usage and the c m4- to-M1 splice. Mol. Cell. Biol. **9**(2), 726–738

Pikielny, C. W., Rymond, B. C. and Rosbash, M. (1986) Electrophoresis of ribonucleoproteins reveals an ordered assembly pathway of yeast splicing complexes. Nature (London) **324**, 341–345

Ponte P., Gonzalez-DeWhitt, J., Schilling, J., Miller, J., Hsu, D., Greenberg, B., Davis, K., Wallace, W., Lieberburg, I., Fuller, F. and Cordell, B. (1988) A new A4 amyloid mRNA contains a domain homologous to serine proteinase inhibitors. Nature (London) **331**, 525–527

Query, C. C., Bentley, R. C. and Keene, J. D. (1989) A common RNA recognition motif identified within a defined U1 RNA binding domain of the 70K U1 snRNP protein. Cell **57**, 89–101

Reich, C. I., Vanhoy, R. W., Porter, G. L. and Wise, J. A. (1992) Mutations at the 3′ splice site can be suppressed by compensatory base changes in U1 snRNA in fission yeast. Cell **69**, 1159–1169

Robberson, B. L., Cote, G. J. and Berget, S. M. (1990) Exon definition may facilitate splice site selection in RNAs with multiple exons. Mol. Cell. Biol. **10**(1), 84–94

Rosenfeld, M. G., Mermod, J.-J., Amara, S. G., Swanson, L. W., Sawchenko, P. E., Rivier, J., Vale, W. W. and Evans, R. M. (1983) Production of a novel neuropeptide encoded by the calcitonin gene via tissue specific RNA processing. Nature (London) **304**, 129–135

Ruby, S. W. and Abelson, J. (1988) An early hierarchic role of U1 small nuclear ribonucleoprotein in spliceosome assembly. Science **242**, 1028–1035

Ruskin, B. and Green, M. R. (1985) Role of the 3′ splice site consensus sequence in mammalian pre-mRNA splicing. Nature (London) **317**, 732–734

Ruskin, B., Krainer, A. R., Maniatis, T. and Green, M. R. (1984) Excision of an intact intron as a novel lariat structure during pre-mRNA splicing *in vitro*. Cell **38**, 317–331

Ruskin, B., Zamore, P. D. and Green, M. R. (1988) A factor, U2AF, is required for U2 snRNP binding and splicing complex assembly. Cell **52**, 207–219

Sachs, A. B., Bond, M. W. and Kornberg, R. D. (1986) A single gene from yeast for both nuclear and cytoplasmic polyadenylate-binding proteins: domain structure and expression. Cell **45**, 827–835

Schmauss, C., Brines, M. L. and Lerner, M. R. (1992) The gene encoding the small nuclear ribonucleoprotein-associated protein N is expressed at high levels in neurons. J. Biol. Chem. **267(12)**, 8521–8529

Schwarz, T. L., Tempel, B. L., Papazian, D. M., Jan, Y. N. and Jan, L. Y. (1988) Multiple potassium-channel components are produced by alternative splicing at the *Shaker* locus in *Drosophila*. Nature (London) **331**, 137–142

Schwer, B. and Guthrie, C. (1991) PRP16 is an RNA-dependent ATPase that interacts transiently with the spliceosome. Nature (London) **349**, 494–499

Seraphin, B. and Rosbash, M. (1989) Identification of functional U1 snRNA–pre-mRNA complexes committed to spliceosome assembly and splicing. Cell **59**, 349–358

Siliciano, P. G., Brow, D. A., Roiha, H. and Guthrie, C. (1987) An essential snRNA from *S. cerevisiae* has properties predicted for U4, including interaction with a U6-like snRNA. Cell **50**, 585–592

Steenbergh, P. H., Hoppener, J. W. M., Zandberg, J., Visser, A., Lips, C. J. M. and Jansz, H. S. (1986) Structure and expression of the human calcitonin/CGRP genes. FEBS Lett. **209(1)**, 97–103

Steitz, J. A. (1992) Splicing takes a holiday. Science **257**, 888–889

Tani, T. and Ohshima, Y. (1989) The gene for the U6 small nuclear RNA in fission yeast has an intron. Nature (London) **337**, 87–90

Tanzi, R. E., McClatchey, A. I., Lamperti, E. D., Villa-Komaroff, L., Gusella, J. F. and Neve, R. L. (1988) Protease inhibitor domain encoded by an amyloid protein precursor mRNA associated with Alzheimer's disease. Nature (London) **331**, 527–530

Tian, M. and Maniatis, T. (1992) Positive control of pre-mRNA splicing *in vitro*. Science **256**, 237–240

Wassarman, D. A. and Steitz, J. A. (1991) Alive with DEAD proteins. Nature (London) **349**, 463–464

Wieringa, B., Hofer, E. and Weissman, C. (1984) A minimal intron length but no specific internal sequence is required for splicing the large rabbit β-globin intron. Cell **37**, 915–925

Zamore, P. D. and Green, M. R. (1989) Identification, purification, and biochemical characterization of U2 small nuclear ribonucleoprotein auxiliary factor. Proc. Natl. Acad. Sci. U.S.A. **86**, 9243–9247

Zamore, P. D. and Green, M. R. (1991) Biochemical characterization of U2 snRNP auxiliary factor: an essential pre-mRNA splicing factor with a novel intra nuclear distribution. EMBO J. **10(1)**, 207–214

Zhuang, Y. and Weiner, A. M. (1986) A compensatory base change in U1 snRNA suppresses a 5′ splice site mutation. Cell **46**, 827–835

Zhuang, Y. and Weiner, A. M. (1989) A compensatory base change in human U2 snRNA can suppress a branch site mutation. Genes Dev. **3**, 1545–1552

2

Post-transcriptional regulation of calcitonin and calcitonin gene-related peptide mRNA expression

Ronald B. Emeson and Jonathan T. Lu

Department of Pharmacology, Vanderbilt University, Nashville, TN 37232-6600, U.S.A.

Introduction

The production of peptide neurotransmitters requires the activities of a number of complex cellular processes involving the transcription and processing of RNA transcripts, the translation and modification of prohormone precursors and the regulated secretion of bioactive peptide products (Sossin *et al.*, 1989; Jung and Scheller, 1991). Considerable progress has been made during the last decade regarding the identification of primary amino acid sequences for many peptide prohormones and the factors regulating their expression including transcription factors, enzymes involved in post-translational processing as well as agents modulating peptide hormone release at synaptic terminals. RNA processing has been shown to play a central role in the control of gene expression in higher eukaryotes and the regulated use of alternative splice sites or polyadenylation signals can lead to the cell-specific production of distinct protein products from a single primary RNA transcript (Leff *et al.*, 1986; Breitbart *et al.*, 1987; Smith *et al.*, 1989). Alternative splicing can affect almost all aspects of protein function, ranging from determination of cellular and subcellular localization to modulations of enzymic activity. The biological role of such changes can be dramatically amplified when the proteins produced by such post-transcriptional processes are themselves important regulatory molecules including transcription factors, biosynthetic enzymes, peptide prohormones, cell-surface receptors and ion channels. Recent studies have indicated that alternative splicing is responsible for the generation of a number of proteins involved in chemical synaptic transmission in the central and peripheral nervous systems including peptide neurotransmitters (substance P/K, calcitonin/calcitonin gene-related peptide (CGRP); Amara *et al.* 1982*b*; Nawa *et al.*, 1984), biogenic amine receptors (D2 dopamine receptor; O'Malley *et al.*, 1990), ion channels (glutamate receptor, *Drosophila* voltage-dependent potassium channel; Schwarz *et al.*, 1988; Sommer *et al.*, 1990) and growth factors (Edwards *et al.*, 1986).

The calcitonin/CGRP gene is responsible for the generation of a number of distinct peptide hormones and peptide neurotransmitters mediating various physiological processes ranging from proliferation of bone osteoblasts (Burns *et al.*, 1989*a*, *b*, 1992) to regulation of acetylcholine receptor synthesis (New and Mudge, 1986; Fontaine *et al.*, 1987; Miles *et al.*, 1989). The starting point for the study of this transcription unit began with the isolation and characterization of clones complementary to calcitonin mRNA from a rat medullary thyroid carcinoma (MTC) cDNA library (Amara *et al.*, 1980). Subsequent analysis of calcitonin gene expression demonstrated the production of a novel calcitonin-related RNA species referred to as CGRP mRNA (Amara *et al.*, 1982*b*). Isolation and sequence analyses of the calcitonin genomic DNA, as well as calcitonin and CGRP cDNAs confirmed that both calcitonin and CGRP are generated by differential RNA splicing from a single genomic locus (Rosenfeld *et al.*, 1981, 1982, 1983; Amara *et al.*, 1982, 1984).

CGRP was one of the first examples of a neuropeptide that was initially predicted, and later identified, as a result of molecular cloning, without the biological basis for chemical characterization that existed for many other bioactive peptides. The discovery of a second gene encoding a CGRP-like molecule (Amara *et al.*, 1985) further complicated the biological role of CGRP as both of these peptides were shown to be expressed in the central and peripheral nervous systems, to act at multiple CGRP receptors in widespread effector systems and to possess potent bioactivities in the regulation of diverse physiological processes. The calcitonin/CGRP gene and its resulting protein products have been studied extensively and used as an important model for examination of tissue-specific gene expression, regulation of cell-specific RNA processing events, post-translational processing of peptide prohormones and the biological activity of multiple peptide products.

Calcitonin

The existence of a calcium-regulating hormone with a biological activity antagonistic to that of parathyroid hormone was first demonstrated by Copp *et al.* (1962). This hormone, referred to as calcitonin for its ability to control the level or 'tone' of calcium in the blood, was soon isolated and subjected to amino acid sequence analysis (Potts *et al.*, 1968; Niall *et al.*, 1969). Calcitonin was shown to be a 32 amino acid amidated peptide hormone that is synthesized by the parafollicular C-cells of the thyroid gland and the ultimobranchial gland in mammals and non-mammalian vertebrates, respectively (Kumar *et al.*, 1963; Hirsch *et al.*, 1964; Minvielle *et al.*, 1987). The primary activity of calcitonin is to decrease circulating levels of calcium and phosphate by inhibiting bone resorption (Hirsch *et al.*, 1964; Reynolds, 1968; Lenz *et al.*, 1985). Calcitonin inhibits such resorption by a cyclic AMP-dependent osteoclast inactivation, by decreasing tubular reabsorption of both calcium and phosphate in the proximal straight tubule of the kidney and by inhibiting the actions of vitamin D 1-α-hydroxylase, thus favouring formation of the less active 24,25(OH)-derivative of vitamin D (Breimer *et al.*, 1988; Boden and Kaplan, 1990).

The role of calcitonin as a hypocalcaemic hormone was questioned during the 1970s when it became apparent that chronically elevated or reduced levels of calcitonin were only occasionally associated with significant alterations in circulating plasma calcium concentrations (Copp and Kline, 1989). It was suggested that the role of calcitonin may be limited to redressing the hypercalcemia resulting from post-prandial calcium resorption (Munson and Gray, 1970). Calcitonin has also been shown to protect bone from demineralization by inhibition of osteoclastic activity and osteocytic osteolysis during periods of potential calcium shortage such as growth, pregnancy and lactation (Lewis *et al.*, 1971; Stevenson *et al.*, 1979). Although the precise physiological role(s) of calcitonin remain controversial, the pharmacological actions of calcitonin regarding inhibition of bone resorption have demonstrated significant therapeutic value in pathophysiological disorders such as Paget's disease and severe osteoporosis.

Analyses of DNA complementary to rat calcitonin mRNA predicted the structure of a 17.5 kDa polypeptide precursor for calcitonin, from which this 32 amino acid peptide is post-translationally excised and modified (Fig. 1). Calcitonin messenger RNA is approximately 1 kb in length, encoding a predicted protein of 136 or 141 amino acids in rat and human, respectively (Amara *et al.*, 1980; Allison *et al.*, 1981; Jacobs *et al.*, 1981). The amino terminus of this prohormone contains a 25 amino acid hydrophobic signal peptide characteristic of secreted proteins, the calcitonin peptide itself, as well as amino- and carboxy-terminal flanking peptides. These latter peptides are separated from calcitonin by a pair of basic amino acid residues (Lys-Arg) at the amino terminus and Gly-Lys-Arg-Arg at the carboxy terminus. These basic residues represent predicted sites of post-translational proteolytic processing (Andrews *et al.*, 1989; Darby and Smyth, 1990; Neurath, 1991), while the glycine residue at the C-terminus of calcitonin in the prohormone serves as an amino donor for the amide nitrogen found in the final bioactive peptide product (Bradbury and Smyth, 1991; Eipper *et al.*, 1992).

The 57 amino acid N-terminal peptide derived from the calcitonin prohormone (N-proCT) has been identified in extracts of rat and human C-cells and this peptide has been shown to be secreted in equimolar amounts with calcitonin (Burns *et al.*, 1989*a*). N-proCT has been highly conserved

Fig. 1. Cellular processes involved in the tissue-specific production of calcitonin and calcitonin gene-related peptide

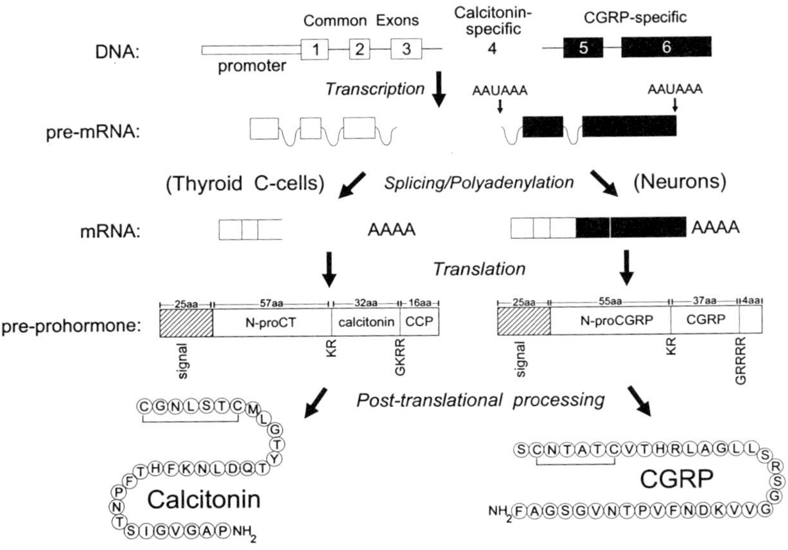

A schematic representation of the rat calcitonin/CGRP gene, primary RNA transcript, mature mRNA species, peptide prohormone and biologically active peptide products is presented, demonstrating differential patterns of mRNA and peptide expression in thyroid C-cells and neurons. The location of polyadenylation signals (AAUAAA) and the sizes of predicted peptide products are indicated; sites of post-translational proteolytic processing are designated with the one-letter amino acid code.

during vertebrate evolution, suggesting that this peptide may serve an important physiological function. Recent studies have demonstrated that N-proCT acts as a cellular mitogen, causing the proliferation of bone osteoblasts and their progenitors (Burns *et al.*, 1989*a*, *b*, 1992; Guenther and Fleisch, 1991). Despite the frequent reports of a correlation between changes in thyroid C-cell activity and bone mass (Parthemore and Deftos, 1979; Shamonki *et al.*, 1980), it has been difficult to prove that calcitonin is an effective bone-forming agent, much less a critical physiological regulator of bone formation (Zaidi *et al.*, 1988, 1990). It has been suggested that N-proCT may possess a number of the anabolic osteoblastic effects originally attributed to the calcitonin peptide itself.

The C-terminal flanking peptide of the calcitonin prohormone has been referred to as calcitonin C-terminal cleavage peptide (CCP) or katacalcin and ranges from 16 (rat) to 21 (human, bovine, chicken, salmon) amino acids in length (Amara *et al.*, 1980; Hillyard *et al.*, 1983; Lasmoles *et al.*, 1985; Poschl *et al.*, 1987; Collyear *et al.*, 1991). While the existence of CCP was originally based upon cDNA sequence analyses alone, specific antisera prepared against a synthetic hexadecapeptide corresponding to the predicted amino acid sequence has allowed radioimmunochemical detection of CCP in both normal rat thyroid glands and medullary thyroid carcinomas (Amara *et al.*, 1982*a*). In the human thyroid, CCP has also been shown to be co-secreted with calcitonin (Hillyard *et al.*, 1983). Although there is significant sequence conservation between rat (44%) and bovine (52%) CCP when compared with the human sequence, the biological role of CCP remains unknown.

Calcitonin gene-related peptide

The production of multiple mRNA transcripts from the single rat calcitonin gene was first described during studies of calcitonin gene expression in serially passaged MTCs (Rosenfeld *et al.*, 1981). Thyroid C-cell hyperplasia and the ultimate development of MTC are common physiological events in rodents occurring in 10 to 40% of older animals (Boorman *et al.*, 1972). Production of calcitonin by these tumours decreases spontaneously during a single transplantation, or gradually during successive transplantations, from the usual 1.5 to 4.5% of total cellular protein to less than 0.3% of total protein (Roos *et al.*, 1979). Immunohistochemical analyses with calcitonin-specific antisera demonstrated that 95% of the cells in high calcitonin-producing tumours contained calcitonin-related immunoreactive material in contrast to <5% of the cells in low calcitonin-producing tumours. While originally considered to be a consequence of altered transcriptional regulation (Rosenfeld *et al.*, 1981), the 'switching' of medullary thyroid carcinomas from states of high to low calcitonin production was shown to result from changes in RNA processing events generating a novel 1.2 kb RNA species, referred to as CGRP mRNA (Rosenfeld *et al.*, 1981). Sequence analyses of rat CGRP cDNA predicted a 16 kDa prohormone precursor which could be subsequently modified by post-translational proteolytic cleavage to produce a 37 amino acid amidated peptide hormone (CGRP), as well as N- and C-terminal flanking peptides analogous to those derived from the calcitonin prohormone (Fig. 1).

Isolation and sequence analyses of the calcitonin genomic DNA, as well as calcitonin and CGRP cDNAs confirmed that both calcitonin and CGRP are generated by differential RNA splicing from a single genomic locus (Rosenfeld *et al.*, 1981, 1982, 1983; Amara *et al.*, 1982*b*, 1984). Examinations of the sequence for the two gene products revealed that CGRP and calcitonin share sequence identity through nucleotide 227 of the coding region, predicting that the initial 76 amino acids of each prohormone are identical. The nucleotide sequences then diverge completely, encoding unique C-terminal domains (Amara *et al.*, 1982*b*). Protein-processing signals within the C-terminal region of CGRP predict the post-translational excision of a 37 amino acid peptide containing an α-amidated phenylalanine residue at its C-terminus (Amara *et al.*, 1982). Based upon the structure of the calcitonin/CGRP gene and the resultant mRNAs, production of mature calcitonin transcripts involves splicing of the first three exons, which are common to both mRNAs, to the calcitonin-specific exon (exon 4) which encodes the entire sequence for calcitonin, CCP and 3′-untranslated information. Alternative splicing of the first three exons to the fifth and sixth exons, which contain the entire CGRP coding sequence and 3′-non-coding sequences, respectively, results in the production of CGRP mRNA. The 55 residue N-terminal flanking peptide (N-proCGRP) is 90% homologous to the 57 residue peptide derived from the calcitonin prohormone. Their first 51 amino acids are identical, reflecting the fact that the first three exons in the calcitonin/CGRP gene are utilized in both calcitonin and CGRP mRNA transcripts. The predicted peptide flanking the C-terminus of CGRP is only four amino acids in length.

Using CGRP-specific antisera, CGRP-related immunoreactive material was shown to have a unique anatomical distribution in the central nervous system, distinct from that of any known neuropeptide (Fig. 2; Kawai *et al.*, 1985; Skofitsch and Jacobowitz, 1985*a*, *b*; Kresse *et al.*, 1992). CGRP-immunoreactive cell bodies are present in several regions of the hypothalamus including the preoptic and medial preoptic areas, the periventricular and anterior hypothalamic nuclei, the perifornical area and in the lateral hypothalamus/medial forebrain bundle area. CGRP-positive cells are also found in the medial amygdaloid nucleus, the hippocampus, the ventromedial nucleus of the thalamus, in the periventricular grey and in the fasciculus retroflexus extending laterally over the lemniscus medialis. In the mesencephalon, CGRP is found in the superior colliculus, parabigeminal nucleus, and in the peripeduncular area ventral to the medial geniculate body. In the hindbrain, cells are found in the parabrachial nucleus, ventral tegmental nucleus, lateral lemniscus, superior olive, lateral cuneate nucleus and the nucleus of the solitary tract. CGRP-immunoreactive cell bodies are present in all cranial nuclei including the trochlear nerve (IV), the trigeminal nerve (V), the facial nerve (VII), the hypoglossal nerve (XII), the oculomotor nerve (III) and the nucleus ambiguus. Central CGRP-stained pathways include projections from the thalamus to insular cortex (Shimada *et al.*, 1985*b*), from the hypothalamus to the lateral septal area (Sakanaka *et al.*, 1985) and from the parabrachial nucleus to the visual sensory thalamus and cortex (Yasui *et al.*, 1989). Immunoreactive afferents to the amygdala, bed nucleus of the stria terminalis and certain hypothalamic nuclei also arise from the parabrachial nucleus (Shimada *et al.*, 1985*a*; Schwaber *et al.*, 1988; Yamano *et al.*, 1988).

Although the physiological function(s) of CGRP have not been precisely defined, based upon its anatomical distribution, CGRP is thought to play an important role in autonomic, somatosensory, integrative and

Fig. 2. Schematic representation of a sagittal view of the adult rat brain demonstrating a summary of the major CGRP-stained cell groups (●) and pathways (arrows)

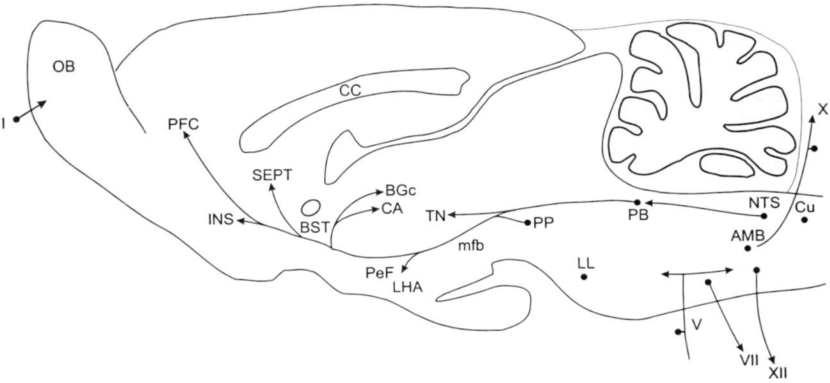

*For a more complete description of CGRP-containing pathways in the central nervous system, see accompanying text or Skofitsch and Jacobowitz (1985*a,b*). Anatomical locations are abbreviated as follows: nucleus of the solitary tract (NTS); cuneate nucleus (Cu); nucleus ambiguus (AMB); parabrachial nucleus (PB); lateral lemniscus (LL); peripeduncular nucleus (PP); thalamic taste nucleus (TN); medial forebrain bundle (mfb); perifornical hypothalamic nucleus (PeF); lateral hypothalamic area (LHA); central amygdaloid nucleus (CA); caudal regions of the caudaoputamen and globis pallidus (BGc); bed nucleus of the stria terminalis (BST); lateral septal nucleus (SEPT); infralimbic prefrontal area (PFC); insular area of the cerebral cortex (INS); olfactory bulb (OB); primary olfactory fibres (I); corpus callosum (CC); trigeminal nerve (V); facial nerve (VII); vagus nerve (X); hypoglossal nerve (XII); adapted from Rosenfeld* et al. *(1983).*

motor functions (Rosenfeld *et al.*, 1983; Kawai *et al.*, 1985; Skofitsch and Jacobowitz, 1985*a*, *b*). More recent studies have identified CGRP throughout the mammalian gastrointestinal tract, within intrinsic endocrine cells, arising from a set of neurons in the myenteric and submucosal plexuses and originating from sensory neurons whose cell bodies are located in the nodose and dorsal root ganglia (Mulderry *et al.*, 1985; Feher *et al.*, 1986). Many studies have suggested that CGRP may serve not only as a neurotransmitter or neuromodulator in the enteric nervous system, but may also participate in local neurohormonal (paracrine) regulation of gastrointestinal effector systems (Beglinger *et al.*, 1988; Holzer and Guth 1991; Raybould, 1992). CGRP has also been identified in cardiac C-fibre afferents where it has demonstrated a potent vasodilatory activity on coronary arteries, in the endocrine pancreas where this peptide may play an important role in the regulation of carbohydrate metabolism (Leighton *et al.*, 1989; Molina *et al.*, 1990; Franco-Cereceda, 1991), and in the mouse olfactory lobe where it has been postulated to serve as a trophic factor in neuronal development (Denis-Donini, 1989, 1992). Additionally, CGRP has been co-localized with acetylcholine in presynaptic terminals of the neuromuscular junction where it stimulates both acetylcholine receptor synthesis and the rate of receptor desensitization (New and Mudge, 1986; Miles *et al.*, 1989; Osterlund *et al.*, 1989).

A second CGRP-encoding gene (β-CGRP)

In addition to the gene encoding both calcitonin and CGRP, referred to as the calcitonin/α-CGRP gene, subsequent studies have uncovered the expression of a second CGRP-containing gene that encodes a peptide differing from α-CGRP by a single amino acid (Amara *et al.*, 1985; Steenbergh *et al.*, 1985, 1986); this peptide is referred to as β-CGRP (Fig. 3). The sequence of the mRNA encoded by the rat gene reveals a 394 nucleotide open reading frame in which the first 256 nucleotides encode a 25 amino acid signal peptide and a 57 amino acid N-terminal flanking peptide; this flanking peptide has a 70% amino acid sequence similarity with the N-terminal peptides derived from the calcitonin and α-CGRP prohormones. There are two sets of dibasic amino acid pairs (Arg-Lys) within this region, suggesting that three distinct peptide products could be post-translationally generated from this region of the prohormone. In contrast, there are fewer than 5% base substitutions in the next 114 nucleotides of the β-CGRP mRNA, constituting the β-CGRP-encoding domain of the RNA transcript. This sequence predicts the excision of a 37 amino acid peptide containing a C-terminal phenylalanyl-amide residue, differing by only a single amino acid (lysine^{35} for glutamic acid^{35}) from the primary sequence of α-CGRP. The β-CGRP gene is very similar in overall structure to the α-CGRP gene (Fig. 3), but the former does not contain an exon encoding a β-calcitonin-like peptide. Rather, the β-CGRP gene contains a non-functional exon which has significantly diverged from its α-calcitonin-specific (exon 4) counterpart and lacks a detectable polyadenylation signal. Splicing to this 'vestigial' exon is never observed *in vivo*, even when using a highly sensitive polymerase chain reaction (PCR)-based assay (Bennett and Amara, 1992).

Analyses of RNA from various regions of rat brain have demonstrated an overlapping but non-identical pattern of expression for these genes throughout the central nervous system. While the relative distribution of α- and β-CGRP is quite similar, the level of expression of β-CGRP is less than 20% of that found for α-CGRP in many brain regions, including the peripeduncular and parabrachial nuclei, superior olive and trigeminal ganglion. The expression of β-CGRP exceeds that for α-CGRP in the oculomotor (III), trochlear (IV) and trigeminal (V) cranial nerves and is roughly

Fig. 3. Schematic diagram of the rat α-and β-CGRP genes and their encoded peptide pre-prohormones

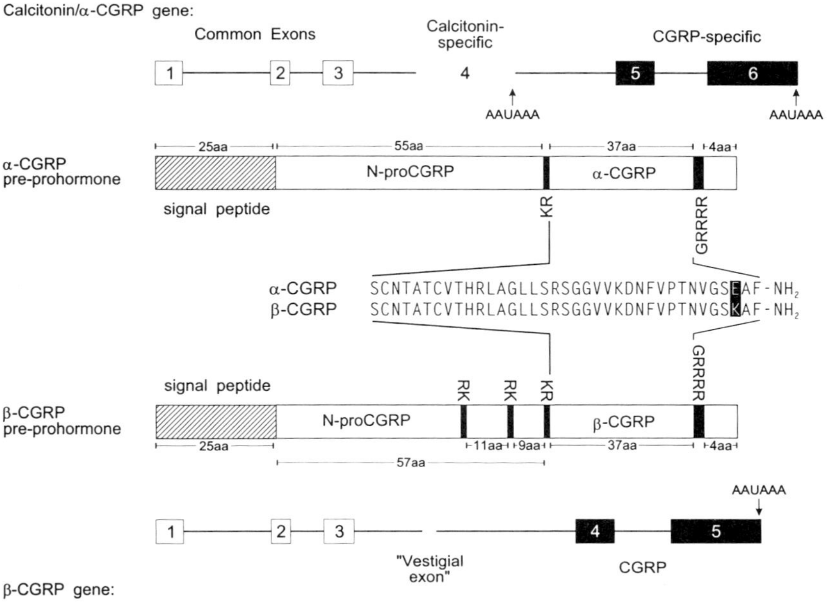

The genomic structure of the rat α-and β-CGRP genes are presented indicating the common (open rectangle), calcitonin-specific (shaded rectangle) and CGRP-specific (black rectangle) exons. The amino acid sequences of α- and β-CGRP are presented showing a single amino acid difference ($K^{35} \rightarrow E^{35}$) with the one-letter amino acid code. The location of polyadenylation signals (AAUAAA), the sites of predicted post-translational proteolytic cleavage at pairs of basic amino acids and the sizes of predicted peptide products are indicated (Amara et al.*, 1985; Bennett and Amara, 1992).*

equivalent in the hypoglossal nerve (XII) and the nucleus ambiguus (Amara *et al.*, 1985). The primary amino acid sequences of human α- and β-CGRP have also been determined and demonstrate highly conserved structures, differing from the rat peptides by only five amino acids (Steenbergh *et al.*, 1986; Petermann *et al.*, 1987). Pharmacological comparisons of human α- and β-CGRP have indicated that these peptides exhibit qualitatively similar biological properties but are not necessarily equipotent (Holman *et al.*, 1986; Williams *et al.*, 1988).

The close structural homology of α- and β-CGRP raises the possibility that not all of the CGRP-like immunoreactive material detected by antisera raised against α-CGRP is necessarily the product of α-CGRP gene expression. Preliminary studies have suggested that sensory neurons preferentially express α-CGRP while enteric neurons express β-CGRP (Mulderry *et al.*, 1988). In organs where CGRP occurs predominantly in sensory fibres, the concentration of α-CGRP was found to be 3–6-times higher than that of β-CGRP. In the intestine, the concentration of β-CGRP was up to 7-times higher than the product of the α-gene. The bulk of α-CGRP-containing fibres within the gastrointestinal tract have been shown to be eliminated by either extrinsic denervation or by degeneration of primary afferents utilizing capsaicin (Lee *et al.*, 1987; Mulderry *et al.*, 1988; Sternini, 1992). Recently, monoclonal antibodies have been raised that distinguish between human α- and β-CGRP (Andrew *et al.*, 1990); however, their usefulness in discriminating between extremely similar peptides in nervous tissue has yet to be examined.

Recent studies have identified the presence of an additional calcitonin-encoding transcript derived from the calcitonin/α-CGRP gene in a human MTC (Minvielle *et al.*, 1991; Adema and Baas, 1992; Cohen *et al.*, 1992). Sequence analyses of PCR-amplified cDNA products have revealed that this transcript encodes a calcitonin prohormone containing a unique CCP (CCP II). Both CCP and CCP II are identical for their first 13 amino acids, diverging completely for the final eight amino acid residues. Nucleotide sequence comparisons between this novel PCR product and the human calcitonin/CGRP gene have indicated that this RNA transcript results from the use of a cryptic splice donor site 172 bp downstream from the beginning of the calcitonin-specific exon; this cryptic donor site is then spliced to the authentic CGRP-specific 3′-splice junction, thereby changing the open reading frame of exon 5. Immunocytochemical analyses, utilizing an antiserum directed against the unique C-terminus of CCP II, have detected the presence of this peptide in human MTCs and normal thyroid tissue (Minvielle *et al.*, 1991; Cohen *et al.*, 1992). This alternative RNA transcript may be restricted to humans, however, as nucleotide sequence comparisons of rat, rabbit, chicken and salmon cDNAs do not appear to contain a cryptic splice donor site in this region (Amara *et al.*, 1980; Lasmoles *et al.*, 1985; Martial *et al.*, 1990; Jansz and Zandberg, 1992).

Mechanisms of cell-specific alternative RNA processing

The human and rat calcitonin/CGRP genes are transcribed as large precursor RNA molecules containing six exons (Allison *et al.*, 1981; Jacobs *et al.*, 1983; Amara *et al.*, 1984); calcitonin mRNA is produced by splicing of the first three exons to the fourth exon, accompanied by endonucleolytic cleavage and polyadenylation at the 3′-end of the fourth exon (Fig. 1). CGRP mRNA production results from the splicing of the first three exons to the fifth and sixth exons and the use of a distal polyadenylation signal at the 3′-end of the sixth exon. While calcitonin mRNA production is largely limited to the

C-cells of the thyroid gland, CGRP mRNA is widely expressed in discrete cell types throughout the central and peripheral nervous systems.

S1 nuclease analyses have indicated that both calcitonin and CGRP mRNA are produced in the thyroid gland in a ratio of approximately 95–98:1 (Rosenfeld *et al.*, 1983, 1984; Tschopp *et al.*, 1984; Sabate *et al.*, 1985). Immunohistochemical and radioimmunoassay studies have further confirmed the presence of both calcitonin- and CGRP-immunoreactive material in parafollicular C-cells and have demonstrated that most of these cells, in both rat and human tissues, react with both calcitonin and CGRP antisera (Sabate *et al.*, 1985). These results have indicated that the ability to co-produce both calcitonin and CGRP peptide products is a feature of normal as well as transformed cells (Sabate *et al.*, 1985). By contrast, little or no detectable calcitonin mRNA has been observed in a number of neuronal structures throughout the brain utilizing S1 nuclease protection, RNA blotting and immunohistochemical analyses (Rosenfeld *et al.*, 1983; Tschopp *et al.*, 1984; Sabate *et al.*, 1985). This differential production of calcitonin and CGRP mRNAs has raised a number of fundamental questions regarding the molecular mechanisms by which a single messenger RNA precursor can be post-transcriptionally processed, in a tissue-specific manner, to yield two distinct mRNA products.

Cell-specific poly(A)-site selection

To examine the molecular mechanisms underlying the tissue-specific production of calcitonin and CGRP mRNAs, two distinct models have been proposed for the regulation of alternative splice site use. One model involves sequence or site-specific poly(A)-site selection, in which the production of alternative mRNAs is dependent upon a tissue-specific utilization of either the proximal or distal polyadenylation signals (AAUAAA) at the 3′-end of the fourth (calcitonin) and sixth (CGRP) exons, respectively. Two simple models have been postulated for the tissue-specific polyadenylation of transcripts at these two sites within the calcitonin/CGRP primary RNA transcript.

(1) In tissues which produce mature calcitonin transcripts, transcription of the calcitonin/CGRP pre-mRNA could terminate at, or adjacent to, the first polyadenylation site. Premature transcriptional termination would result in the formation of a calcitonin-specific primary RNA transcript lacking the CGRP-specific exons and containing a single polyadenylation signal. Readthrough transcription to the distal polyadenylation site would give rise to a CGRP pre-mRNA species, thereby resulting in the formation of mature CGRP transcripts. This model would further suggest that the CGRP-specific poly(A) signal is preferentially utilized in CGRP-producing cells when both polyadenylation signals are present.

(2) The second model suggests that transcription could always terminate at, or beyond, the second polyadenylation site and that alternative

RNA production could be determined by cell-specific poly(A) addition or 3′-splice site selection or both. The first model predicts that the calcitonin-specific exon will be transcribed at a substantially higher rate than the CGRP-specific exons in calcitonin-producing tissues, while the second model predicts that both regions will be transcribed at similar rates, because termination or attenuation immediately after the calcitonin exon is not critical for the use of the proximal polyadenylation signal.

To distinguish between these two possibilities, nuclear run-on transcription experiments were performed using nuclei isolated from calcitonin- and CGRP-producing MTCs (Amara *et al.*, 1984). For these studies, nuclei were prepared from a calcitonin-producing MTC tumour in which calcitonin mRNA was the predominant (>98%) product of calcitonin gene expression, and nascent nuclear RNA transcripts were elongated in the presence of [α-^{32}P]UTP; nascent radiolabelled RNA transcripts were then hybridized to immobilized plasmids containing inserts corresponding to unique sequence genomic fragments. Results from such experiments indicated that the apparent transcription rate across all regions of the calcitonin/CGRP gene from the transcriptional start (cap) site through a region 640 bp downstream of the CGRP polyadenylation site was essentially equivalent (Fig. 4).

As expected, the transcription rate through a region corresponding to a genomic fragment 1.2 kb upstream from the cap site was almost undetectable, whereas the rate through a 900 bp fragment beginning >0.6 kb downstream from the CGRP polyadenylation site was reduced 30–40% from that observed across the gene itself. These results indicated that the rate of transcription across the entire calcitonin/CGRP gene (from the cap site to the CGRP polyadenylation signal) is roughly equivalent with significant transcriptional attenuation beginning only downstream from the distal polyadenylation signal. An identical pattern of expression was observed in CGRP-producing MTCs (Fig. 4; Amara *et al.*, 1984). Additionally, RNA-blotting (Northern) analyses of poly(A^{+}) RNA from both calcitonin- and CGRP-producing MTC tumours detected the presence of an RNA species corresponding to the predicted size for the primary calcitonin/CGRP transcript in various MTC tumours, irrespective of the mature transcript generated (Rosenfeld *et al.*, 1981; Amara *et al.*, 1984; Bovenberg *et al.*, 1986). These data indicate that transcriptional termination at the 3′-end of the calcitonin-specific exon (exon 4) does not represent the regulatory mechanism responsible for differential calcitonin/CGRP mRNA production. These data also support a model in which poly(A)-site selection precedes RNA splicing events producing a single primary RNA transcript which is subsequently processed by site-specific poly(A) site utilization or cell-specific patterns of RNA splicing to yield calcitonin or CGRP mRNA.

An additional strategy that has been widely utilized to examine the post-transcriptional events involved in calcitonin/CGRP mRNA production has taken advantage of tissue culture model systems which mimic the RNA

Fig. 4. Alternative production of calcitonin and CGRP mRNAs is not due to cell-specific transcriptional termination

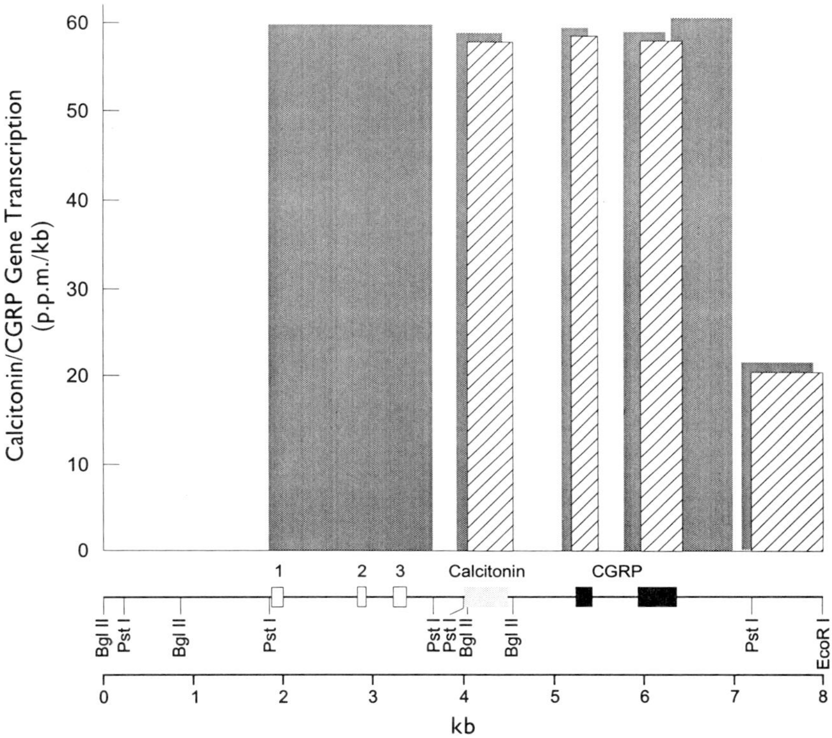

Nuclear run-on transcription analyses of radiolabelled, nascent calcitonin/CGRP transcripts were quantitated under conditions of DNA excess hybridization using immobilized plasmids containing the indicated genomic regions (Amara et al., *1984). Results are expressed as p.p.m. per kb of genomic probe. These studies were performed using rat medullary thyroid carcinomas producing predominantly calcitonin mRNA (> 95%; shaded bars) or CGRP transcripts (> 92%; striped bars). The genomic structure of the rat calcitonin/CGRP gene is schematically presented including an abbreviated restriction enzyme map.*

processing choices of tissues observed *in vivo* (Leff *et al.*, 1987; Emeson *et al.*, 1989; Adema *et al.*, 1990; Adema and Baas, 1991). For these studies, wild-type or mutant calcitonin/CGRP transcription units have been subcloned into several eukaryotic expression vectors and either permanently or transiently transfected into a variety of mammalian cell lines. The RNA processing choice made by each cell line was determined by RNA-blotting (Northern), ribonuclease protection or PCR-based analyses. Transfected lymphoid (A20), human embryonic kidney (293) and human epithelial

(HeLa) cell lines produced almost exclusively mature calcitonin transcripts, while rat pheochromocytoma (PC12) and mouse teratocarcinoma (F9) cell lines produced >90% CGRP mRNA.

To further examine the role of poly(A) site selection in the regulation of alternative mRNA production, mutant transcription units were devised in which the entire fourth, fifth and sixth exons from the calcitonin/CGRP gene were inserted in the place of the SV40 poly(A) site in pSV$_2$*gpt* and pSV$_2$*neo* plasmids (Mulligan and Berg, 1980; Southern and Berg, 1982) as indicated in Fig. 5A. These expression plasmids were then stably transfected into A20 and F9 cell lines to determine the site(s) of polyadenylation in the resulting gpt- and neo-fusion transcripts (Leff *et al.*, 1987). Analyses of mRNA isolated from permanently transfected cell lines indicated that mature fusion transcripts terminating at either poly(A) site could be found in both lymphocyte and F9 cells expressing these fusion genes (Leff *et al.*, 1987); in both cell types, however, mRNAs terminating at the calcitonin polyadenylation signal represented the predominant mature RNA transcript.

In a second series of experiments, genomic segments of 450–550 bp containing the polyadenylation sites for calcitonin and CGRP were inserted in tandem array at the 3′-end of the *Eco-gpt* gene in each of the two possible orders such that these two poly(A) sites substituted for the SV40 poly(A) site normally found in the pSV$_2$*gpt* vector (Fig. 5B; Leff *et al.*, 1987). Analyses of RNA isolated from several clonal lymphocyte lines expressing the tandem poly(A) constructions demonstrated stochastic use of the two polyadenylation signals, with a preference for the proximal poly(A) site in each of the two possible orientations. These results have also suggested that cell-specific utilization of polyadenylation signals within the calcitonin/CGRP primary RNA transcript is not a key regulatory event dictating alternative patterns of mRNA expression.

The mutant transcription units designed by Leff *et al.* clearly demonstrate that cell lines making discrete RNA processing choices do not have an apparent preference for the calcitonin or CGRP-specific polyadenylation signals, yet it should be noted that these mutant transcription units effectively separate the processes of alternative splicing from 3′-end formation. If the process of 3′-splice site selection is mechanistically linked to the choice of polyadenylation site, mutant transcription units in which these processes have been uncoupled may give rise to aberrant results not reflecting the normal regulatory mechanisms observed *in vivo*. To further address the role of cell-specific poly(A) site selection, a calcitonin/CGRP transcription unit was constructed in which the calcitonin (exon 4) poly(A) site was replaced by sequence information from the 3′-end of the sixth exon containing the CGRP polyadenylation signal (Fig. 5C; Emeson *et al.*, 1989). If production of mature calcitonin transcripts is solely dependent upon efficient utilization of the calcitonin-specific polyadenylation signal, CGRP-producing cells may not have the appropriate cellular machinery to either recognize or cleave at the 3′-end of the fourth exon, thereby resulting in the

Fig. 5. Mutant calcitonin/CGRP transcription units for determining the role of cell-specific poly(A)-site utilization in alternative mRNA production

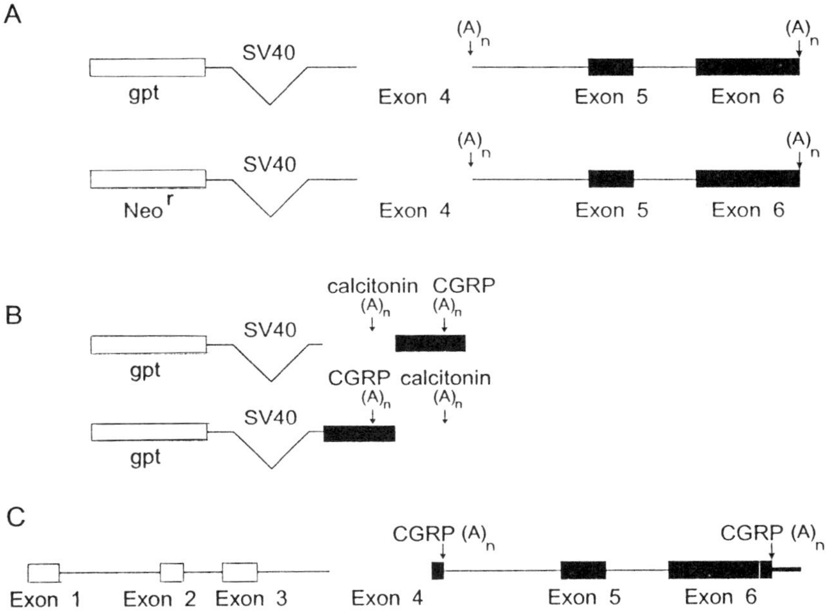

(A) Fusion genes containing exons 4 to 6 of the calcitonin/CGRP gene were inserted in place of the SV40 poly(A) addition site in pSV$_2$gpt and pSV$_2$neo as described (Leff et al., *1987) for transfection into cell lines demonstrating selective calcitonin/CGRP RNA processing choices. Sites of polyadenylation [(A)$_n$] are indicated. (B) Mutant transcription units containing 450–550 bp genomic fragments containing the calcitonin and CGRP polyadenylation signals were fused to the* Escherichia coli *guanosine phosphoribosyl transferase* (gpt) *gene and placed in tandem array as indicated (Leff* et al., *1987). (C) A mutant calcitonin/CGRP transcription unit in which the calcitonin-specific polyadenylation signal was deleted and replaced with sequence information from the 3′-end of the sixth exon is represented. The relative positions of the duplicated CGRP polyadenylation signal are indicated (Emeson* et al., *1989).*

formation of mature CGRP transcripts. Replacement of the exon 4 poly(A) site with a signal that can be utilized by CGRP-producing cells should result in efficient cleavage/polyadenylation at the proximal poly(A) signal resulting in an increased expression of calcitonin transcripts in the CGRP-producing (F9) cell line.

Analyses of mRNA from cell lines transfected with the 'poly(A) replacement' mutation revealed cell-specific RNA processing patterns similar to those elaborated by HeLa and F9 cells transfected with the wild-type gene. These experiments strongly suggest that differential poly(A) site utilization is not the key regulatory event involved in calcitonin/CGRP alternative RNA processing, suggesting that the tissue-specific production of mRNA transcripts from this gene is regulated by alternative splice site selection.

Cell-specific *trans*-acting regulatory factors

For many genes, alternative splicing may simply reflect a kinetic process in which the ratio of splice site selection is solely determined by a hierarchy of relative 'intrinsic strengths' among *cis*-active elements at competing splice sites (Dominski and Kole, 1991; Guo *et al.*, 1991; Kuo *et al.*, 1991; Mullen *et al.*, 1991). A number of features in pre-mRNA transcripts have been implicated in alternative splice site selection including intron size (Fu and Manley, 1987), the pyrimidine content of the 3′-splice site (Freyer *et al.*, 1989; Reed, 1989; Mullen *et al.*, 1991), the location and sequence composition of the branch point (Reed and Maniatis, 1988), RNA secondary structure (Watakabe *et al.*, 1989; Libri *et al.*, 1991), exon sequences (Reed and Maniatis, 1986; Black, 1991; Cote *et al.*, 1992) as well as proximity to other *cis*-active elements. This model suggests that splice site utilization is regulated exclusively by *cis*-active elements, yet the 'strength' of competing splice sites must reflect the interactions between *cis*-active elements and the constitutive splicing apparatus in specifying the nature of the final spliced product.

A second model of alternative splicing suggests that differential splice site utilization is mediated by *trans*-acting factors which affect differential RNA processing by binding to specific *cis*-active pre-mRNA sequences, thereby promoting or repressing the selection of alternative splice sites. Recently, several factors have been identified that regulate the alternative splicing patterns of specific RNA transcripts. These factors include gene products of the *Drosophila* sex-determination pathway such as *sex-lethal*, *transformer* and *transformer-2* which have been shown to bind to sequence-specific regulatory elements, thereby mediating the production of female-specific *sex-lethal*, *transformer* and *doublesex* mRNA transcripts (Sosnowski *et al.*, 1989; Inoue *et al.*, 1990; Ryner and Baker, 1991; Tian and Maniatis, 1992).

To determine whether the generation of mature calcitonin and CGRP transcripts requires the actions of cell-specific machinery, a strategy was employed to permit expression of the rat calcitonin/CGRP gene in tissues in a wide variety of neuronal and non-neuronal tissues (Crenshaw *et al.*, 1987). A eukaryotic expression vector containing the inducible mouse metallothionein-I promoter fused to the rat calcitonin/CGRP gene, was injected into the pronucleus of fertilized mouse oocytes for the generation of transgenic mice. Calcitonin/CGRP gene expression was then induced by zinc treatment of three mouse pedigrees expressing the transgene; the RNA processing choice made in a variety of transgenic tissues was extensively analysed. In all pedigrees, the calcitonin/CGRP transcription unit was expressed in a large number of tissues as indicated in Fig. 6; in general, these tissues represented regions in which the endogenous gene is not normally expressed. In tissues other than the heart and brain, calcitonin mRNA represented >90% of the mature transcripts derived from the transgene (Fig. 6). In the brain, the transgene was expressed in several novel regions and Northern blotting analyses of whole brain mRNA indicated approximately equal levels of calcitonin and CGRP transcripts.

Fig. 6. Summary of calcitonin/CGRP mRNA production from transgenic mice expressing a metallothionein–calcitonin/CGRP fusion gene

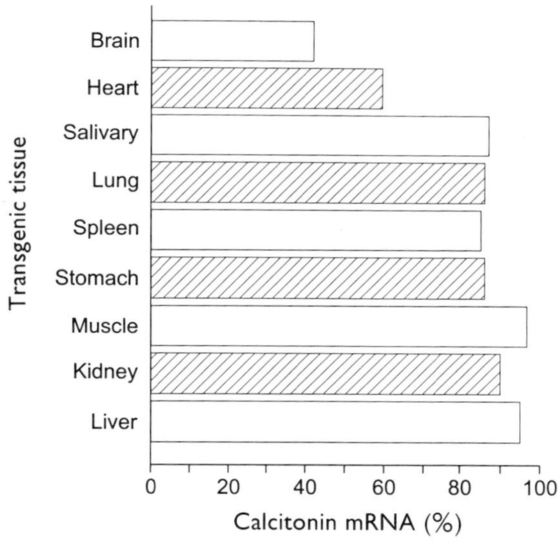

The RNA processing choice in a variety of transgenic tissues was determined by hybridization analyses to size-fractionated RNA from zinc-treated mice (Crenshaw et al., *1987); relative levels of calcitonin/CGRP mRNA expression utilizing calcitonin- and CGRP-specific probes were determined by densitometric quantitation of autoradiograms. The numerical values presented represent the proportion of calcitonin mRNA produced in each tissue relative to the total mature mRNA transcript derived from the transgene.*

Preliminary interpretations, based upon these RNA-blotting analyses, suggested that cells within the brain were either incapable of making a selective RNA processing choice with pre-mRNA transcripts derived from the transgene, or that distinct regions within the brain were selectively producing one mRNA species, while other regions generated the alternative transcript. To distinguish between these possibilities, hybridization histochemistry utilizing [^{35}S]cRNA probes specific for calcitonin and CGRP mRNAs revealed that calcitonin mRNA was primarily expressed in non-neuronal tissues including the pia mater, ependyma and the choroid plexus. In contrast, CGRP mRNA was largely associated with neuronal structures including the hippocampus (CA3 pyramidal cells and dentate gyrus), reticular nucleus, ventrobasal complex of the thalamus and layer V of the neocortex. Since mature calcitonin mRNA transcripts represented the predominant RNA species in almost all tissues examined, even in tissues in which the calcitonin/CGRP transcription unit was not normally expressed, the simplest interpretation of these results has suggested that calcitonin mRNA production is the unregulated or 'default' RNA processing pattern, relying solely upon constitutive splicing components for its synthesis. The ability to splice the calcitonin/CGRP primary transcript to produce mature CGRP mRNA must therefore be dependent upon a neuron-specific splicing machinery. Alternatively, production of calcitonin-specific transcripts could require a tissue-specific factor which is widely distributed, but absent from neuronal tissues.

Cis-active regulatory elements

The observation that alternative splicing of calcitonin/CGRP transcripts occurs in a tissue-specific manner suggests that cell-specific *trans*-acting factors must interact with the pre-mRNA. *Cis*-active elements within the transcript itself, however, must ultimately determine the response to *trans*-acting factors in specifying the nature of the final spliced product. If the premise is accepted that calcitonin mRNA production represents the unregulated RNA processing choice, at least two distinct models may be proposed by which a neuron-specific cellular machinery might promote the formation of mature CGRP transcripts. In the case of two competing 3′-splice sites, cell-specific regulation could result from the repression of the calcitonin-specific splicing choice (negative model), leading to an increase in utilization of the CGRP splice junction. Regulation could also result from a direct activation of the CGRP splice acceptor (positive model), without requiring inhibition of the calcitonin splice acceptor site. Similar positive and negative models can be proposed for a widely distributed factor promoting calcitonin production which either promotes utilization of the calcitonin-specific 3′-splice junction or inhibits use of the CGRP splice acceptor.

Transfection analyses

Tissue culture model systems mimicking the RNA processing choices of thyroid C-cells and neurons have been extensively utilized to identify potential *cis*-active regulatory elements involved in calcitonin/CGRP post-transcriptional processing. The RNA processing choice made by cell lines transfected with mutant calcitonin/CGRP transcription units represents a common experimental strategy for determining the effects of such mutations upon alternative 3′-splice site selection. When transfected with the wild-type calcitonin/CGRP gene, lymphoid (A20), human embryonic kidney (293) and human epithelial (HeLa) cell lines produce almost exclusively mature calcitonin transcripts, while rat pheochromocytoma (PC12), rat glioma (C6) and mouse teratocarcinoma (F9) cell lines produce >90% CGRP mRNA (Leff *et al.*, 1987; Emeson *et al.*, 1989; Adema *et al.*, 1990; Adema and Baas, 1991; Bennett and Amara, 1992).

To examine putative *cis*-active elements and/or RNA secondary structure involved in differential mRNA production, preliminary experiments have been directed at mutating sequences surrounding the calcitonin-

Fig. 7. Regulation of calcitonin/CGRP alternative RNA processing is at the calcitonin-specific splice acceptor

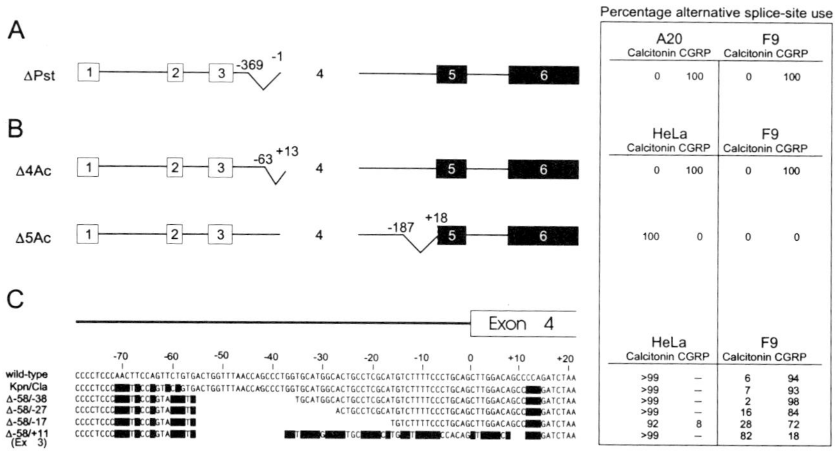

(A) Schematic diagram of a mutant calcitonin/CGRP gene in which a 360 bp PstI restriction fragment was deleted from the transcription unit (Leff et al., *1987). (B) Schematic diagrams of mutant calcitonin/CGRP transcription units in which either the calcitonin- (Δ4Ac) or the CGRP-specific (Δ5Ac) splice acceptor regions have been deleted (Emeson* et al., *1989). The coordinates of the deletions, relative to the beginning of either the fourth or fifth exons, are indicated. (C) Nucleotide sequences of the wild-type and mutant calcitonin-specific splice acceptors (Emeson* et al., *1989). Alterations from the wild-type sequences are indicated by the boxed letters; coordinates are relative to the beginning of the fourth exon. The numerical values represent the relative values of calcitonin- and CGRP-specific splice acceptor use in permanently transfected A20, HeLa and F9 cell lines.*

and CGRP-specific splice junctions (Leff *et al.*, 1987; Emeson *et al.*, 1989). Deletion of a 380 bp or 590 bp fragment from the third and fourth introns, respectively, had no effect upon RNA processing patterns in permanently transfected HeLa or F9 cells indicating that the RNA sequence information within these regions was not essential for alternative splice site selection (Emeson *et al.*, 1989). Deletion of a 369 bp Pstl fragment from the third intervening sequence extending from −369 and −1, relative to the beginning of exon 4 (ΔPst; Fig. 7A), resulted in significant levels of CGRP mRNA production in a stably transfected mouse lymphocyte cell line (A20) with no mature calcitonin transcripts detected (Leff *et al.*, 1987). Since the 3′-border of this Pstl deletion coincides exactly with the calcitonin-specific splice acceptor site (CTGCAG) and a potential branch point sequence (CTGAY) is found 38 nucleotides upstream from the reconstructed Pstl site, there is some question regarding whether or not the calcitonin-specific 3′-splice junction had been inactivated in this reconstructed transcription unit. Nonetheless, the observation that calcitonin-producing cells can effectively utilize the CGRP-specific splice acceptor, has suggested that there is no absolute requirement for a positive cell-specific machinery to promote the splicing of exon 3 to exon 5. To address this question and further localize *cis*-active regulatory information, a 77 bp fragment containing the calcitonin-specific 3′-splice junction was deleted from the calcitonin/CGRP transcription unit (Δ4Ac; Fig. 7B). Transfection of this mutant gene into HeLa cells resulted in efficient utilization of the CGRP-specific splice acceptor, yielding an RNA processing pattern virtually identical to A20 cells transfected with the ΔPst gene or F9 cells transfected with either the wild-type or Δ4Ac transcription units (Emeson *et al.*, 1989).

The conclusion that a cell-specific machinery is not required to promote directly the splicing of exon 3 to exon 5 has suggested that factors mediating differential mRNA production may indirectly promote the expression of mature CGRP transcripts by repressing the formation of calcitonin mRNA. To investigate the potential role of an inhibitory factor acting at the calcitonin-specific 3′-splice junction, the ability of CGRP-producing cells to alter their splice-site selection in favour of calcitonin mRNA production was examined by removal of the CGRP-specific splice acceptor site (Δ5Ac; Fig. 7B; Emeson *et al.*, 1989). Transfection of this mutant transcription unit into HeLa cells resulted in the normal production of mature calcitonin mRNA; however, F9 cells produced only unprocessed RNA species. These results suggested that, in the absence of a functional CGRP-specific splice acceptor, CGRP-producing cells were incapable of altering their differential splice site selection to produce calcitonin mRNA, consistent with a cell-specific inhibition of the calcitonin splice acceptor site. It should be noted, however, that the RNA splicing patterns observed upon transfection of the ΔPst, Δ4Ac and Δ5Ac transcription units into HeLa, A20 and F9 cells could also be explained by the presence of a cell-specific factor which is required for calcitonin mRNA production, and yet absent from the CGRP-producing (F9) cell line.

To further identify potential regulatory sequences near the calcitonin-specific splice acceptor that might interact with either a calcitonin-promoting or inhibitory machinery, a series of mutant calcitonin/CGRP transcription units were constructed containing nested deletions as schematized in Fig. 7C (Emeson *et al.*, 1989). In control studies, the introduction of *Kpn*I and *Cla*I restriction sites, required for construction of these nested deletions, had little effect upon cell-specific alternative splice site selection. Deletion of a region between −58 and −38 (Δ−58/−38) also had no apparent effect on alternative splicing. Further deletion from −58 to −27 (Δ−58/−27) resulted in continued use of the calcitonin acceptor in HeLa cells and an apparent increase in the ratio of calcitonin to CGRP mRNA in transfected F9 cells. Additional deletion from −58 to −17 (Δ−58/−17) further increased the percentage of calcitonin-specific splice acceptor use in the CGRP-producing cell line. These results were interpreted to suggest that *cis*-active element(s) near the calcitonin acceptor inhibited the formation of calcitonin transcripts in F9 cells and that deletion of this region permitted the increased production of calcitonin mRNA in the CGRP-producing cell line. These results are not consistent with a model requiring a calcitonin mRNA promoting factor, as deletion of regulatory sequences near the calcitonin splice acceptor site should not lead to an increased production of calcitonin transcripts in F9 cells.

Further nested deletions near the calcitonin-specific splice acceptor were not feasible, owing to the presence of constitutive regulatory elements [$(Py)_{6-8}$GCAG] defining the exon 4 acceptor itself. Deletion of constitutive *cis*-active elements would have resulted in an inactivation of this 3′-splice junction, yielding results similar to those previously obtained with the exon 4 splice acceptor deletion (Δ4Ac; Fig. 7B). To further define *cis*-active elements in this region, sequence information surrounding the calcitonin-specific splice acceptor, corresponding to the coordinates −58 to +11, was removed and replaced with a heterologous splice acceptor that could be utilized by both cell types [Δ−58/+11 (Ex3); Fig. 7C]; a region of the exon 3 splice acceptor, from −39 to +8 relative to the beginning of the third exon, was chosen for these studies. When transfected into HeLa cells, this mutant transcription unit resulted in the efficient production of calcitonin transcripts, while F9 cells now produced >80% calcitonin mRNA (Fig. 7C). These results demonstrated that removal of sequence information from the calcitonin-specific 3′-splice junction and its replacement with an acceptor that could be utilized by F9 cells results in the efficient production of mature calcitonin transcripts in a CGRP-producing cell line.

Splicing analyses *in vitro* of the human calcitonin/CGRP gene have indicated that the branch point consensus sequence (YNYURAY) preceding the calcitonin-specific exon contains a unique uridine residue as the branch point nucleotide rather than the adenine moiety which normally occurs in this position (Adema *et al.*, 1988). In the rat gene, a cytidine nucleotide at the same position (−23 relative to the beginning of the fourth exon) is surrounded by an otherwise perfect match for the loosely defined metazoan branch point consensus. To examine the role of a non-standard branch point nucleotide in the regulation of alternative RNA processing, the branch point uridine or cytidine residues were mutated to the canonical adenine nucleotide in the human and rat calcitonin/CGRP genes, respectively (Adema *et al.*, 1990; Bennett and Amara, 1992). Transfection of the mutant transcription unit into embryonic kidney, glioma, pheochromocytoma and teratocarcinoma cells resulted in the predominant or exclusive production of mature calcitonin transcripts, regardless of the RNA processing phenotype observed with the wild-type calcitonin/CGRP gene (Adema *et al.*, 1990; Adema and Baas, 1991; Bennett and Amara, 1992). These results have been interpreted to suggest that the branch point nucleotide is a specific *cis*-active regulatory element dictating cell-specific RNA processing patterns. Alternatively, changing a competing 3′-splice junction to a stronger site by introduction of a perfect branch point consensus may merely overcome the existing regulation even if such sequences are not directly involved in interactions with a cell-specific regulatory machinery. Indeed, replacement of the calcitonin-specific splice acceptor with sequences derived from the exon 3 acceptor site [Δ−58/+11 (Ex3)] transferred a potential branch point sequence (TGCUAGAC) which may have increased calcitonin mRNA production in F9 cells by increasing the relative strength of the exon 4 splice acceptor site.

Splicing analyses *in vitro*

The development of systems *in vitro* that accurately splice exogenous pre-mRNA substrates has allowed the dissection of a number of biochemical steps involved in the constitutive splicing reaction that are not technically feasible using splicing models *in vivo* (Grabowski *et al.*, 1984; Hardy *et al.*, 1984; Krainer *et al.*, 1984; Padgett *et al.*, 1984; Ruskin *et al.*, 1984). These cell-free splicing systems have taken advantage of whole cell or nuclear extracts of cultured cell lines containing components that can be solubilized by moderate salt extraction of lysed cells or crude nuclei, respectively (Manley *et al.*, 1980; Dignam *et al.*, 1983). Despite these advances, the development of splicing systems *in vitro* to demonstrate the alternative production of calcitonin and CGRP mRNAs has been both problematic and controversial. To analyse the molecular events involved in the alternative production of human calcitonin and CGRP mRNA transcripts, Bovenberg *et al.* (1988) constructed a plasmid coding for a model pre-mRNA substrate extending from exon 3 to exon 5 (Fig. 8A); this region of the pre-mRNA was hypothesized to contain all of the *cis*-active sequences required for differential mRNA production. Pre-mRNA transcripts synthesized *in vitro*, utilizing bacteriophage SP6 RNA polymerase, were incubated with crude nuclear extracts isolated from either HeLa, PC12 or human sarcoma (Ewing 1B) cell lines. Results from these experiments indicated that nuclear extracts from all three cell lines were able to direct the splicing of exon 3 to exon 5, albeit at an extremely low level of efficiency. Deletion of intervening sequence information (Fig. 8B), thought not to be involved in alternative RNA processing, enhanced the efficiency of the splicing reaction *in vitro* (Bovenberg *et al.*, 1988). Based on these results it was hypothesized that efficient calcitonin splicing requires inactivation of the dominant CGRP-specific 3′-splice junction, a process which could be achieved by a cell-specific cleavage/polyadenylation at the 3′-end of the fourth exon (Adema *et al.*, 1988; Bovenberg *et al.*, 1988).

Fig. 8. Pre-mRNA substrates for the splicing of calcitonin/CGRP transcripts *in vitro*

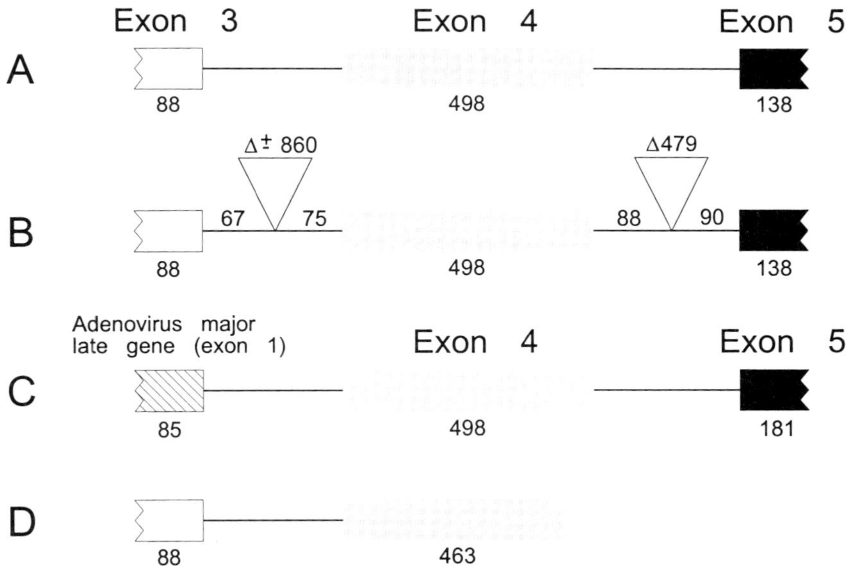

A schematic representation of several 'minigene' constructs is presented which have been used to assess RNA processing patterns in vitro *utilizing crude nuclear extracts from mammalian cell lines. (A) A wild-type minigene transcription unit extending from exon 3 to exon 5 and (B) deletion of intervening sequence information (Bovenberg* et al.*, 1988). (C) Replacement of the third exon with sequences derived from the adenovirus major late exon 1 (Cote* et al.*, 1990). (D) Truncation of the calcitonin/CGRP transcription unit within the calcitonin-specific (fourth) exon (Adema* et al.*, 1988). The sizes of exons, introns and deletions are indicated.*

In a separate series of experiments, Cote *et al.* (1990) synthesized a calcitonin/CGRP pre-mRNA substrate extending from exon 3 through exon 5 in which the third exon had been replaced with sequences from the adenovirus major late gene exon 1 (Fig. 8C). When this pre-mRNA was incubated with a crude nuclear extract isolated from HeLa cells, inefficient utilization of the calcitonin-specific splice acceptor site and the exon 4 polyadenylation signal was observed; CGRP-specific splicing was not detected. A series of increasingly large deletions, removing sequence information from both the fourth intervening sequence and exon 4, resulted in an increase in the use of the CGRP-specific 3′-splice junction. These studies have been interpreted to suggest that specific sequence information within the first 45 nucleotides of the fourth exon was responsible for suppressing CGRP mRNA formation with a HeLa cell nuclear extract.

To enhance the production of mature calcitonin transcripts, pre-mRNA substrates terminating within the fourth exon were synthesized for use in the splicing system *in vitro* (Fig. 8D; Adema *et al.*, 1988). Incubation of this truncated pre-mRNA transcript with HeLa cell nuclear extracts resulted in a low level of exon 3–exon 4 (calcitonin-specific) splicing. Biochemical analyses of splicing intermediates generated in this reaction *in vitro* have indicated that a uridine residue, 23 nucleotides upstream from the calcitonin-specific exon, represents the major branch point nucleotide (Adema *et al.*, 1988), in contrast to an adenosine residue which represents the branch point nucleotide for almost all other genes examined. This observation has led to the hypothesis that the poor efficiency of exon 3–exon 4 splicing elaborated *in vitro*, and the apparent dominance of the CGRP-specific splice junction, may result from the presence of a non-standard branch point nucleotide which serves to decrease the efficiency of the calcitonin-specific splicing reaction. Only in those instances in which the CGRP-specific acceptor site is removed from the pre-mRNA substrate, either by truncation for the reaction *in vitro* (Fig. 8D) or by cleavage/polyadenylation in the situation *in vivo*, will the calcitonin-specific splicing reaction occur (Adema *et al.*, 1988).

To further enhance the efficiency of the splicing reaction *in vitro*, a variety of mutations were introduced into various calcitonin/CGRP 'mini-genes'. These modifications included further deletion of intronic sequence information (Adema *et al.*, 1988; Cote *et al.*, 1990, 1991), replacement of the polypyrimidine tract at the calcitonin acceptor with the corresponding region from human β-globin exon 2, and replacement of the uridine branch point nucleotide with an adenosine residue (Adema *et al.*, 1990; Cote *et al.*, 1990, 1991; Adema and Baas, 1991).

Although substitution of the polypyrimidine tract did not increase the efficiency of calcitonin-specific splicing, introduction of an adenosine residue within the branch point consensus did result in a significant increase in the splicing of exon 3 to exon 4; this result was observed whether utilizing a pre-mRNA transcript truncated at the end of the fourth exon or with pre-mRNA substrates extending through exon 5 (Adema *et al.*, 1990; Adema and Baas, 1991). Alteration of a single nucleotide residue within the exon 4 branch point consensus sequence (U → A) could alter the RNA processing pattern *in vitro* demonstrated by HeLa cell nuclear extracts in favour of calcitonin mRNA production, suggesting that the uridine branch point nucleotide represents a major *cis*-active element involved in calcitonin/CGRP differential RNA processing.

While many minigene constructs have been utilized for the synthesis of pre-mRNA substrates for splicing reactions *in vitro*, in few instances have the mutations introduced into these transcription units been assessed for their effects upon differential splice site selection in either *in vivo* or transfected tissue culture cell model systems. Appropriate interpretation of *in vitro* splicing results is dependent upon the hypothesis that the various pre-mRNA substrates utilized in this experimental paradigm are capable of directing cell-specific RNA processing patterns. For example, deletion of exon 4 sequence information to less than 45 nucleotides results in an increased production of CGRP-specific splicing in HeLa cell nuclear extracts (Cote *et al.*, 1990). Although these results have been interpreted to suggest that a critical *cis*-active element within this region is responsible for the suppression of CGRP mRNA production, it is equally likely that this pre-mRNA substrate does not contain sufficient sequence information for efficient use of the calcitonin-specific splice junction, owing to the extremely short size to which the fourth exon has been reduced.

Trans-acting factors mediating calcitonin/CGRP alternative splice site selection

A number of *trans*-acting factors regulating the alternative splice site selection of *Drosophila* P-transposase (Siebel *et al.*, 1992) and genes involved in the *Drosophila* sex-determination pathway have recently been identified (Sosnowski *et al.*, 1989; Inoue *et al.*, 1990; Ryner and Baker, 1991; Tian and Maniatis, 1992). These studies have further suggested that such *trans*-acting factors directly interact with *cis*-active regulatory sequences present in the pre-mRNA. Such regulatory factors may affect differential RNA processing by binding to specific *cis*-active pre-mRNA sequences, thereby promoting or repressing the selection of alternative splice sites (Inoue *et al.*, 1990; Valcarcel *et al.*, 1993). In mammalian systems, at least one cellular factor, termed alternative splicing factor (ASF) or splicing factor 2 (SF2), has been found to alter 5′-splice site selection *in vitro* using either model pre-mRNA substrates derived from the human β-globin gene (Krainer *et al.*, 1990*a*) or a pre-mRNA transcript derived from the early region of SV30 encoding large T and small t antigens (Ge and Manley, 1990). This protein factor is required for early steps in spliceosome assembly and for the subsequent first cleavage–ligation reaction (Krainer *et al.*, 1990*a*, *b*). More recently, two members of a conserved family of nuclear phosphoproteins (SRp33 and SRp55), containing both an RNA recognition motif and arginine/serine-rich domain, have been shown to functionally replace human SF2 in constitutive and alternative splicing reactions *in vitro* (Mayeda *et al.*, 1992). Although SF2 is required for the general splicing process, differences in the relative concentrations or activities of factors involved in constitutive splicing could play a role in mediating alternative RNA processing events. Indeed, the relative concentration of SF2 and the heterogeneous ribonucleoprotein (hnRNP) A1 has been shown to modulate alternative 5′-splice site selection of model β-globin RNA substrates spliced *in vitro* (Mayeda and Krainer, 1992). The development of systems *in vitro* that accurately splice calcitonin/CGRP pre-mRNA substrates has allowed the biochemical identification of at least two polypeptides that may be involved in the differential production of calcitonin and CGRP mRNAs.

An intron binding protein

To identify *trans*-acting factors controlling the cell-specific production of calcitonin and CGRP mRNAs, Roesser *et al.* (1993) examined the *cis*-active elements surrounding the calcitonin-specific splice acceptor. For these studies, a variety of minigenes were constructed containing the first two exons of human β-globin gene (βg1–2), a genomic fragment extending from exon 3 to exon 4 of the rat calcitonin/CGRP gene (cal3–4) or a chimaeric molecule containing the first exon of the human β-globin gene and the calcitonin-specific exon (βg1–cal4) as schematized in Fig. 9A. Stable and transient transfection into HeLa cells resulted in the production of accurately and efficiently spliced transcripts for each of the minigene transcription units. F9 cells, however, efficiently spliced the β-globin pre-mRNA transcript, but produced very little mature cal3–4 or βg1–cal4 mRNA. Mutation of the putative branch point cytidine residue to an adenosine in the βg1–cal4 transcription unit (βg1–cal4 BP) resulted in the efficient utilization of the calcitonin-specific 3′-splice junction in F9 cells as had been previously demonstrated for the entire human and rat calcitonin/CGRP genes (Adema *et al.*, 1990; Bennett and Amara, 1992). These results were used to support the conclusion of Emeson *et al.* (1989), that suggests a negative regulation of the calcitonin splice acceptor site in F9 cells. Further, these data indicated that all of the information required for the cell-specific repression of exon 4 acceptor use was contained within the βg1–cal4 minigene (Roesser *et al.*, 1993).

To identify and isolate *trans*-acting factors from neuronal tissue that inhibit utilization of the calcitonin-specific splice acceptor, an *in vitro* system was developed by which to assess the activity of such factor(s) (Roesser *et al.*, 1993). In this assay, βg1–2 and βg1–cal4 RNAs were synthesized *in vitro* and incubated with HeLa cell nuclear extracts; the formation of splicing products and intermediates was then examined. βg1–2 and β1–cal4 RNAs were accurately spliced in an ATP-dependent manner when incubated with

Fig. 9. Calcitonin/CGRP minigene constructs for transfection and splicing analyses *in vitro*

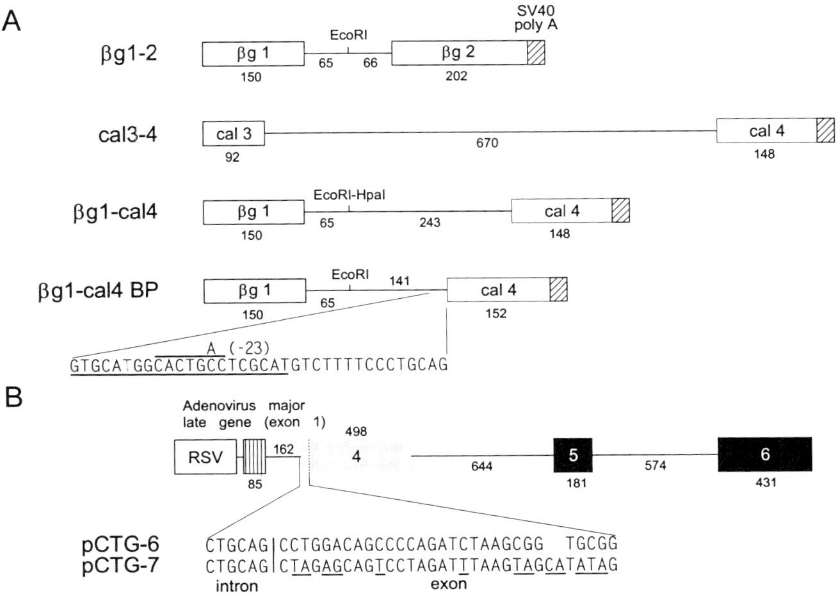

(A) A series of transcription units were placed under control of the SRα promoter for stable transfection into HeLa and F9 cells to investigate calcitonin splice acceptor use; alternatively, these constructs served as templates for RNA synthesis in vitro *with T3 RNA polymerase, and the subsequent RNA products were incubated with HeLa cell nuclear extracts for examination of splicing products and intermediates. The control βg1−2 construct consists of the human β-globin first exon, intervening sequence and the first 202 nucleotides of exon 2. An* Eco*RI restriction site, 65 nucleotides downstream from the exon 1 5′-splice junction, was introduced for the creation of β-globin/calcitonin chimeras and is schematically presented. The diagonally hatched rectangles represent the SV40 polyadenylation signal present in the eukaryotic expression vector. The cytosine to adenosine mutation in βg1−cal4 BP is indicated by an 'A' at position −23 and the seven nucleotide sequence under the bar indicates the putative branch point sequence. The 21 nucleotide target sequence (underlined) has been used for u.v. cross-linking and competition analyses with HeLa cell nuclear extracts and partially purified preparations from rat brain (Roesser* et al.*, 1993). (B) A schematic diagram of the 'wild-type' pCTG-6 minigene for transfection and splicing analyses* in vitro *is depicted (Cote* et al.*, 1992); the positions of the Rous sarcoma virus promoter and adenovirus major late exon 1 are depicted. A comparison of the nucleotide sequences surrounding the calcitonin-specific 3′-splice junction in the pCTG-6 transcription unit and a mutant gene in which the first 30 nucleotides of the calcitonin-specific exon has been replaced with the corresponding region from the human β-CGRP gene (pCTG-7) units is presented; mutated nucleotides are underlined. Exon and intron sizes are indicated.*

HeLa cell nuclear extracts. The addition of nuclear extracts from rat brain and F9 cells inhibited βg1–cal4 splicing but had no effect upon βg1–2 mRNA production. Calcitonin-splice acceptor utilization was apparently blocked at an early step in the splicing reaction, as no intermediate products were shown to accumulate. Thus, use of the calcitonin splice acceptor was preferentially inhibited *in vitro* by the addition of nuclear extracts from CGRP-producing cells, consistent with the presence of a regulator inhibiting calcitonin-specific 3′-splice site selection. Nuclear extracts prepared from rat spleen did not inhibit calcitonin splice acceptor use *in vitro*, demonstrating the tissue-specific nature of this inhibitory effect.

Inhibition of βg1–cal4 and βg1–cal4 BP splicing *in vitro* was then used as an assay to follow the fractionation of rat brain nuclear extract using gel-filtration (Sephadex G-150) and ion-exchange (DEAE-cellulose) chromatography. Partially purified extracts from rat brain were used to assess the RNA binding specificity of this inhibitory activity by examining the effect of adding unlabelled competitor RNAs to the βg1–cal4 BP splicing reaction. Because of the observations of Emeson *et al.* (1989) that intron sequences near the calcitonin 3′-splice junction might mediate the inhibition of calcitonin splice acceptor use in F9 cells (Fig. 7C), the ability of unlabelled RNA containing putative regulatory sequences extending from −16 to −36 (relative to the beginning of the fourth exon) was examined for its ability to protect βg1–cal4 BP from inhibition by the active rat brain fraction from Sephadex G150. Addition of a 40-fold molar excess of the target sequence, but not tRNA or an unrelated sequence derived from the pBluescript polylinker, almost completely protected βg1–cal4 BP splicing from inhibition; no effect upon βg1–2 splicing was observed. The ability of added target RNA to specifically protect βg1–cal4 and βg1–cal4 BP splicing *in vitro* implies that the rat brain activity-inhibiting calcitonin splice acceptor use acts by binding directly to specific RNA sequences. Further support in favour of this hypothesis comes from mobility-shift and u.v.-cross-linking analyses in which partially purified extracts from rat brain were shown to directly interact with the 21 nucleotide target sequence revealing the presence of two brain-specific polypeptides (approximately 43 and 41 kDa) which preferentially bind RNA containing the calcitonin splice acceptor (Roesser *et al.*, 1993).

An exon binding protein

Splicing analyses *in vitro* of mutant calcitonin/CGRP pre-mRNA substrates have suggested that a *cis*-active regulatory element within the first 45 nucleotides of the calcitonin-specific exon is required to promote utilization of the calcitonin-specific 3′-splice junction (Cote *et al.*, 1990). To further localize regulatory sequences within this region, both transfection and splicing analyses *in vitro* were employed (Cote *et al.*, 1992). These studies involved substitution of the first 30 nucleotides of the calcitonin-specific exon with the corresponding region fom the human β-CGRP (CALC II) gene (Fig. 9B). The β-CGRP gene contains a non-functional exon (exon 4) which has significantly diverged from its α-calcitonin-specific counterpart, and β-calcitonin-specific splicing is never observed *in vivo* (Bennett and Amara, 1992). The human β-CGRP gene contains 13 nucleotide differences and a two nucleotide insertion when compared with the calcitonin-α-CGRP gene (Steenbergh *et al.*, 1986). HeLa cells transiently transfected with a 'wild-type' minigene transcription unit (pCTG-6) produced predominantly calcitonin mRNA with little or no detectable inclusion of the CGRP-specific exon.

Introduction of β-CGRP exon 4 sequences into the calcitonin-specific exon (pCTG-7) resulted in a complete shift in splicing phenotype as HeLa cells produced exclusively CGRP mRNA transcripts. Similar results were obtained from splicing analyses *in vitro* with these minigene transcription units. These results indicate that sequences within the first 30 nucleotides of the calcitonin-specific exon are required for utilization of the calcitonin splice acceptor site; substitution of these sequences with the closely related region from the β-CGRP gene, in which calcitonin-splicing is never observed, results in the exclusive production of CGRP mRNA. These data are consistent with a model in which HeLa cells contain *trans*-acting regulatory factor(s) which interact with sequences in the calcitonin-specific exon and are required for inclusion of the fourth exon into mature mRNA transcripts.

U.v.-cross-linking analyses were then performed to determine if RNA-binding proteins could be identified that interact with this region of the calcitonin-specific exon. U.v.-cross-linking to short RNA probes (55 and 57 nucleotides in length), containing only exon 4 sequences, revealed the presence of a 66 kDa protein which could be cross-linked to the wild-type (pCTG-6), but not to the mutant (pCTG-7) exon sequences. Significant cross-linking of this 66 kDa protein to either RNA substrate was not observed in nuclear extracts derived from F9 cells. Furthermore, cross-linking experiments using a mixture of HeLa and F9 nuclear extracts also resulted in the absence of a 66 kDa protein binding. An absence of protein cross-linking in the presence of both F9 and HeLa cell nuclear extracts has led to the hypothesis that a dominant F9 factor inhibits recognition of exon 4 by inhibiting the binding of a 66 kDa polypeptide to calcitonin-specific exon sequences (Cote *et al.*, 1992).

The role of SmN in CGRP-specific mRNA production

The Sm class of small nuclear ribonucleoproteins (snRNPs) is composed of a group of particles located in the nucleus of all higher eukaryotes (Luhrmann *et al.*, 1990; Zieve and Sauterer, 1990). These snRNP particles have previously been shown to be involved in the splicing of mRNA precursors and are composed of uridine-rich small nuclear RNAs complexed with several polypeptides (Maniatis and Reed, 1987; Luhrmann, 1990). The polypeptide components of these snRNPs are recognized by antisera from patients with systemic lupus erythematosus and other related autoimmune disorders (Craft, 1992). Utilizing anti-Sm sera, a novel 28 kDa polypeptide, referred to as 'N' or SmN, was cloned and shown to demonstrate a high level of expression in the brain, with lesser levels in the heart, and no detectable expression in other tissues examined (McAllister *et al.*, 1988, 1989).

Expression of the calcitonin/CGRP gene in transgenic mice, under control of the mouse metallothionein promoter, revealed that although most tissues made predominantly calcitonin trancripts, the heart and brain contained significant levels of transgene-derived CGRP mRNA (Fig. 6) (Crenshaw *et al.*, 1987). The pattern of CGRP expression in these transgenic studies correlated precisely with the reported pattern of SmN gene expression suggesting that the SmN polypeptide itself may be directly involved in the regulation of CGRP-specific mRNA processing. Further support for the regulatory role of SmN protein was provided by analyses of cell lines derived from rat or human MTC tumours demonstrating that such cell lines contain high levels of SmN mRNA and produce predominantly CGRP transcripts (Delsert and Rosenfeld, 1992). Additionally, transfected F9 cells, stably expressing the calcitonin/CGRP transcription unit, produced $>95\%$ CGRP transcripts and contained high levels of SmN mRNA while similarly transfected HeLa and mouse lymphocyte (A20) cell lines produced $>99\%$ calcitonin transcripts and contained no detectable level of the SmN factor (Delsert and Rosenfeld, 1992).

Despite these correlations, SmN mRNA was not detected in human embryonic kidney (293) or rat fibroblast (Rat-1) cell lines even though such lines produced mature CGRP transcripts when transiently transfected with the rat calcitonin/CGRP gene (Delsert and Rosenfeld, 1992). Exogenous expression of the SmN factor in a calcitonin-producing cell line (HeLa) expressing the calcitonin/CGRP gene was then used as an experimental paradigm to examine the role of SmN in regulating the tissue-specific production of CGRP transcripts (Delsert and Rosenfeld, 1992). Stable transfection of a plasmid containing the SmN transcription unit and a selectable marker gene (dihydrofolate reductase) did not result in an increased production of CGRP mRNA in the model cell line. An unambiguous demonstration of increased SmN protein in these HeLa cells, however, was precluded due to the presence of SmB′ protein which was recognized by Sm antisera and comigrated with SmN on both one- and two-dimensional gel electrophoresis systems. Hybridization analyses of rat tissues *in situ* utilizing an SmN-specific probe demonstrated that although SmN mRNA expression was limited to discrete regions of the brain and heart, significant expression could be detected in the inferior colliculus, a region of the brain which demonstrated calcitonin-specific mRNA production in transgenic mice (Crenshaw *et al.*, 1987; Li *et al.*, 1989). Whether the specific cells expressing SmN were identical to those expressing the calcitonin/CGRP transgene was not determined. While these studies strongly suggest that SmN is neither sufficient nor required for the tissue-specific production of CGRP mRNA, additional studies will be necessary to exclude SmN as a *trans*-acting factor modulating calcitonin/CGRP 3′-splice site selection.

Conclusion

Transfectional analyses of HeLa and F9 cells with mutant calcitonin/CGRP transcription units have indicated that the tissue-specific production of calcitonin and CGRP mRNAs is the result of alternative splice site selection primarily regulated by *cis*-active sequences near the calcitonin-specific 3′-splice junction (Leff *et al.*, 1987; Emeson *et al.*, 1989; Adema *et al.*, 1991; Cote *et al.*, 1992). These data in conjunction with transgenic mouse analyses (Crenshaw *et al.*, 1987) have suggested a model for the post-transcriptional regulation of calcitonin/CGRP differential RNA processing in which calcitonin mRNA represents the non-regulated RNA processing choice in non-neuronal cells, relying solely upon the constitutive splicing apparatus for its synthesis. According to this model, the production of mature CGRP transcripts requires the actions of a putative neuron-specific negative regulatory machinery which serves to indirectly promote the expression of CGRP mRNA by repressing utilization of the calcitonin-specific 3′-splice junction. More recent studies, taking advantage of splicing systems *in vitro* which accurately splice calcitonin/CGRP pre-mRNA substrates, have demonstrated the presence of such neuron-specific factor(s) which appear to specifically inhibit utilization of the calcitonin 3′-splice junction by binding to intronic sequences preceding the fourth exon (Roesser *et al.*, 1993). In contrast to these observations, however, Cote *et al.* (1992) have identified a 66 kDa protein which appears to bind to regulatory elements within the first 55 nucleotides of the calcitonin-specific exon and is required for production of mature calcitonin mRNA. Additionally, the presence of dominant factors derived from F9 nuclear extracts appears to inhibit the recognition of exon 4 by inhibiting binding of the required 66 kDa polypeptide (Cote *et al.*, 1992).

It is possible that the rat brain inhibitory activity identified by Roesser *et al.* (1993) corresponds to the dominant inhibitory activity isolated from F9 cells as described by Cote *et al.*, (1992). Although inhibition of calcitonin splice acceptor use involves a 21 nucleotide intronic sequence (Roesser *et al.*, 1993), activation of the calcitonin 3′-splice junction requires a 55 nucleotide region of exonic information. These data, implicating a widely distributed RNA-binding factor which is required for calcitonin mRNA production and a neuron-specific RNA-binding protein which opposes calcitonin splice site activation, emphasize the extreme complexity which may underlie calcitonin/CGRP RNA processing. Further definition of the *cis*-active regulatory sequences involved in the production of calcitonin and CGRP mRNAs, as well as a refinement of functional assays for such cell-specific machineries, may facilitate further isolation and characterization of the tissue-specific *trans*-acting factor(s) involved in the differential RNA processing of this complex neuroendocrine gene.

References

Adema, G. J. and Baas, P. D. (1991) Deregulation of alternative processing of calcitonin/CGRP-pre-mRNA by a single point mutation. Biochem. Biophys. Res. Commun. **178**, 985–992

Adema, G. J. and Baas, P. D. (1992) A novel calcitonin-encoding mRNA is produced by alternative processing of calcitonin/calcitonin gene-related peptide-I pre-mRNA. J. Biol. Chem. **267**, 7943–7948

Adema, G. J., Bovenberg, R. A., Jansz, H. S. and Baas, P. D. (1988) Unusual branch point selection involved in splicing of the alternatively processed calcitonin/CGRP-I pre-mRNA. Nucleic Acids Res. **16**, 9513–9526

Adema, G. J., van-Hulst, K. L. and Baas, P. D. (1990) Uridine branch acceptor is a *cis*-acting element involved in regulation of the alternative processing of calcitonin/CGRP-I pre-mRNA. Nucleic Acids Res. **18**, 5365–5373

Allison, J., Hall, L., MacIntyre, I. and Craig, R. K. (1981) The construction and partial characterization of plasmids containing complementary DNA sequences to human calcitonin precursor polyprotein. Biochem. J. **199**, 725–731

Amara, S. G., David, D. N., Rosenfeld, M. G., Roos, B. A. and Evans, R. M. (1980) Characterization of rat calcitonin mRNA. Proc. Natl. Acad. Sci. U.S.A. **77**, 4444–4448

Amara, S. G., Jonas, V., O'Neil, J. A., Vale, W., Rivier, J., Roos, B. A., Evans, R. M. and Rosenfeld, M. G. (1982*a*) Calcitonin COOH-terminal cleavage peptide as a model for identification of novel neuropeptides predicted by recombinant DNA analysis. J. Biol. Chem. **257**, 2129–2132

Amara, S. G., Jonas, V., Rosenfeld, M. G., Ong, E. S. and Evans, R. M. (1982*b*) Alternative RNA processing in calcitonin gene expression generates mRNAs encoding different polypeptide products. Nature (London) **298**, 240–244

Amara, S. G., Evans, R. M. and Rosenfeld, M. G. (1984) Calcitonin/calcitonin gene-related peptide transcription unit: tissue-specific expression involves selective use of alternative polyadenylation sites. Mol. Cell. Biol. **4**, 2151–2160

Amara, S. G., Arriza, J. L., Leff. S. E., Swanson, L. W., Evans, R. M. and Rosenfeld, M. G. (1985) Expression in brain of a messenger RNA encoding a novel neuropeptide homologous to calcitonin gene-related peptide. Science **229**, 1094–1097

Andrew, D. P., Bidgood, T. D., Bose, C., Galfre, G. and Sherwood, M. (1990) Monoclonal antibodies distinguishing α and β forms of calcitonin gene-related peptide. J. Immunol. Methods **154**, 87–94

Andrews, P. C., Brayton, K. A. and Dixon, J. E. (1989) Post-translational proteolytic processing of precursors to regulatory peptides. Experientia (Suppl.) **56**, 192–209

Beglinger, C., Born, W., Hildebrand, P., Ensinck, J. W., Burkhardt, F., Fischer, J. A. and Gyr, K. (1988) Calcitonin gene-related peptides I and II and calcitonin: distinct effects on gastric acid secretion in humans. Gastroenterology **95**, 958–965

Bennett, M. M. and Amara, S. G. (1992) Molecular mechanisms of cell-specific and regulated expression of the calcitonin/alpha-CGRP and beta-CGRP genes. Ann. N.Y. Acad. Sci. **657**, 36–49

Black, D. L. (1991) Does steric interference between splice sites block the splicing of a short c-*src* neuron-specific exon in non-neuronal cells? Genes Dev. **5**, 389–402

Boden, S. D. and Kaplan, F. S. (1990) Calcium homeostasis. Orthop. Clin. North Am. **21**, 31–42

Boorman, G. A., Noord, M. V. and Hollander, C. F. (1972) Naturally occurring medullary thyroid carcinoma in the rat. Arch. Pathol. **94**, 35–41

Bovenberg, R. A., van-de-Meerendonk, W. P., Baas, P. D., Steenbergh, P. H., Lips, C. J. and Jansz, H. S. (1986) Model for alternative RNA processing in human calcitonin gene expression. Nucleic Acids Res. **14**, 8785–8803

Bovenberg, R. A., Adema, G. J., Jansz, H. S. and Baas, P. D. (1988) Model for tissue specific calcitonin/CGRP-I RNA processing from *in vitro* experiments. Nucleic Acids Res. **16**, 7867–7883

Bradbury, A. F. and Smyth, D. G. (1991) Peptide amidation. Trends Biochem. Sci. **16**, 112–115

Breimer, L. H., MacIntyre, I. and Zaidi, M. (1988) Peptides from the calcitonin genes: molecular genetics, structure and function. Biochem. J. **255**, 377–390

Breitbart, R. E., Andreadis, A. and Nadal-Ginard, B. (1987) Alternative splicing: a ubiquitous mechanism for the generation of multiple protein isoforms from single genes. Annu. Rev. Biochem. **56**, 467–495

Burns, D. M., Birnbaum, R. S. and Roos, B. A. (1989*a*) A neuroendocrine peptide derived from the amino-terminal half of rat procalcitonin. Mol. Endocrinol. **3**, 140–147

Burns, D. M., Forstrom, J. M., Friday, K. E., Howard, G. A. and Roos, B. A. (1989*b*) Procalcitonin's amino-terminal cleavage peptide is a bone-cell mitogen. Proc. Natl. Acad. Sci. U.S.A. **86**, 9519–9523

Burns, D. M., Howard, G. A. and Roos, B. A. (1992) An assessment of the anabolic skeletal actions of the common-region peptides derived from the CGRP and calcitonin prohormones. Ann. N.Y. Acad. Sci. **657**, 50–62

Cohen, R., Delehaye, M. C., Minvielle, S., Segond, N., Taboulet, J., Noel, M., Guliana, J. M., Calmettes, C., Milhaud, G. and Lasmoles, F. (1992) CCP II: a novel calcitonin carboxy terminal peptide is expressed in normal thyroid tissue. Biochem. Biophys. Res. Commun. **185**, 330–334

Collyear, K., Girgis, S. I., Saunders, G., MacIntyre, I. and Holt, G. (1991) Predicted structure of the bovine calcitonin gene-related peptide and the carboxy-terminal flanking peptide of bovine calcitonin precursor. J. Mol. Endocrinol. **6**, 147–152

Copp, D. H. and Kline, L. (1989) Vertebrate Endocrinology: Fundamentals and Biomedical Implications, p. 79, Academic Press, San Diego

Copp, D. H., Cameron, E. C., Cheney, B., Davidson, A. G. F. and Henze, K. (1962) Evidence for calcitonin — a new hormone from the parathyroid that lowers blood calcium. Endocrinology, **70**, 638–649

Cote, G. J., Nguyen, I. N., Berget, S. M. and Gagel, R. F. (1990) Calcitonin exon sequences influence alternative RNA processing. Mol. Endocrinol. **4**, 1744–1749

Cote, G. J., Nguyen, I. N., Lips, C. J., Berget, S. M. and Gagel, R. F. (1991) Validation of an *in vitro* RNA processing system for CT/CGRP precursor mRNA. Nucleic Acids Res. **19**, 3601–3606

Cote, G. J., Stolow, D. T., Peleg, S., Berget, S. M. and Gagel, R. F. (1992) Identification of exon sequences and an exon binding protein involved in alternative RNA splicing of calcitonin/CGRP. Nucleic Acids Res. **20**, 2361–2366

Craft, J. (1992) Antibodies to snRNPs in systemic lupus erythematosus. Rheum. Dis. Clin. North Am. **18**, 311–335

Crenshaw, E. B., III, Russo, A. F., Swanson, L. W. and Rosenfeld, M. G. (1987) Neuron-specific alternative RNA processing in transgenic mice expressing a metallothionein–calcitonin fusion gene. Cell **49**, 389–398

Darby, N. J. and Smyth, D. G. (1990) Endopeptidases and prohormone processing. Biosci. Rep. **10**, 1–13

Delsert, C. D. and Rosenfeld, M. G. (1992) A tissue-specific small nuclear ribonucleoprotein and the regulated splicing of the calcitonin/calcitonin gene-related protein transcript. J. Biol. Chem. **267**, 14573–14579

Denis-Donini, S. (1989) Expression of dopaminergic phenotypes in the mouse olfactory bulb induced by the calcitonin gene-related peptide. Nature (London) **339**, 701–703

Denis-Donini, S. (1992) Calcitonin gene-related peptide influence on central nervous system differentiation. Ann. N.Y. Acad. Sci. **657**, 344–350

Dignam, J. D., Lebovitz, R. M. and Roeder, R. G. (1983) Accurate transcription initiation by RNA polymerase II in a soluble extract from isolated mammalian nuclei. Nucleic Acids Res. **11**, 1475–1489

Dominski, Z. and Kole, R. (1991) Selection of splice sites in pre-mRNAs with short internal exons. Mol. Cell. Biol. **11**, 6075–6083

Edwards, R. H., Selby, M. J. and Rutter, W. J. (1986) Differential RNA splicing predicts two distinct nerve growth factor precursors. Nature (London) **319**, 784–787

Eipper, B. A., Stoffers, D. A. and Mains, R. E. (1992) The biosynthesis of neuropeptides: peptide alpha-amidation. Annu. Rev. Neurosci. **15**, 57–85

Emeson, R. B., Hedjran, F., Yeakley, J. M., Guise, J. W. and Rosenfeld, M. G. (1989) Alternative production of calcitonin and CGRP mRNA is regulated at the calcitonin-specific splice acceptor. Nature (London) **341**, 76–80

Feher, E., Burnstock, G., Varndell, I. M. and Polak, J. M. (1986) Calcitonin gene-related peptide-immunoreactive nerve fibres in the small intestine of the guinea-pig: electron-microscopic immunocytochemistry. Cell Tissue Res. **245**, 353–358

Fontaine, B., Klarsfeld, A. and Changeux, J. P. (1987) Calcitonin gene-related peptide and muscle activity regulate acetylcholine reeptor alpha-subunit mRNA levels by distinct intracellular pathways. J. Cell Biol. **105**, 1337–1342

Franco-Cereceda, A. (1991) Calcitonin gene-related peptide and human epicardial coronary arteries: presence, release and vasodilator effects. Br. J. Pharmacol. **102**, 506–510

Freyer, G. A., O'Brien, J. P. and Hurwitz, J. (1989) Alterations in the polypyrimidine sequence affect the *in vitro* splicing reactions catalysed by HeLa cell-free preparations. J. Biol. Chem. **264**, 14631–14637

Fu, X. Y. and Manley, J. L. (1987) Factors influencing alternative splice site utilization *in vivo*. Mol. Cell. Biol. **7**, 738–748

Ge, H. and Manley, J. L. (1990) A protein factor, ASF, controls cell-specific alternative splicing of SV40 early pre-mRNA *in vitro*. Cell **62**, 25–34

Grabowski, P. J., Padgett, R. A. and Sharp, P. A. (1984) Messenger RNA splicing *in vitro*: an excised intervening sequence and a potential intermediate. Cell **37**, 415–427

Guenther, H. L. and Fleisch, H. (1991) The procalcitonin amino-terminal cleavage peptide (N-proCT) lacks biological activity on normal clonal rat osteoblastic and preosteoblastic cells *in vitro*. Calcif. Tissue Int. **49**, 138–140

Guo, W., Mulligan, G. J., Wormsley, S. and Helfman, D. M. (1991) Alternative splicing of β-tropomyosin pre-mRNA: *cis*-acting elements and cellular factors that block the use of a skeletal muscle exon in nonmuscle cells. Genes Dev. **5**, 2096–2107

Hardy, S. F., Grabowski, P. J., Padgett, R. A. and Sharp, P. A. (1984) Cofactor requirements of splicing of purified messenger RNA precursors. Nature (London) **308**, 375–377

Hillyard, C. J., Myers, C., Abeyasekera, G., Stevvensvenson, J. C., Craig, R. K. and MacIntyre, I. (1983) Katacalcin: a new plasma calcium-lowering hormone. Lancet **1**, 846–848

Hirsch, P. F., Voelkel, E. F. and Munson, P. L. (1964) Thyrocalcitonin hypocalcemic hypophosphatemic principle of the thyroid gland. Science **146**, 412–413

Holman, J. J., Craig, R. K. and Marshall, I. (1986) Human alpha-and beta-CGRP and rat alpha-CGRP are coronary vasodilators in the rat. Peptides **7**, 231–235

Holzer, P. and Guth, P. H. (1991) Neuropeptide control of rat gastric mucosal blood flow. Increase by calcitonin gene-related peptide and vasoactive intestinal polypeptide, but not substance P and neurokinin A. Circ. Res. **68**, 100–105

Inoue, K., Hoshijima, K., Sakamoto, H. and Shimura, Y. (1990) Binding of the *Drosophila sex-lethal* gene to the alternative splice site of *transformer* primary transcript. Nature (London) **344**, 461–463

Jacobs, J. W., Goodman, R. H., Chin, W. W., Dee, P. C., Habener, J. F., Bell, N. H. and Potts, J. T., Jr. (1981) Calcitonin messenger RNA encodes multiple polypeptides in a single precursor. Science **213**, 457–459

Jacobs, J. W., Simpson, E., Penschow, J., Hudson, P., Coghlan, J. and Niall, H. (1983) Characterization and localization of calcitonin messenger ribonucleic acid in rat thyroid. Endocrinology **113**, 1616–1622

Jansz, H. S. and Zandberg, J. (1992) Identification and partial characterization of the salmon calcitonin/CGRP gene by polymerase chain reaction. Ann. N.Y. Acad. Sci. **657**, 63–69

Jung, L. J. and Scheller, R. H. (1991) Peptide processing and targeting in the neuronal secretory pathway. Science **251**, 1330–1335

Kawai, Y., Takami, K., Shiosaka, S., Emson, P. C., Hillyard, C. J., Girgis, S., MacIntyre, I. and Tohyama, M. (1985) Topographic localization of calcitonin gene-related peptide in the rat brain: an immunohistochemical analysis. Neuroscience **15**, 747–763

Krainer, A. R., Maniatis, T., Ruskin, B. and Green, M. R. (1984) Normal and mutant human beta-globin pre-mRNAs are faithfully and efficiently spliced *in vitro*. Cell **36**, 993–1005

Krainer, A. R., Conway, G. C. and Kozak, D. (1990*a*) The essential pre-mRNA splicing factor SF2 influences 5′ splice site selection by activating proximal sites. Cell **62**, 35–42

Krainer, A. R., Conway, G. C. and Kozak, D. (1990*b*) Purification and characterization of pre-mRNA splicing factor SF2 from HeLa cells. Genes Dev. **4**, 1158–1171

Kresse, A., Jacobowitz, D. M. and Skofitsch, G. (1992) Distribution of calcitonin gene-related peptide in the central nervous system of the rat by immunocytochemistry and *in situ* hybridization histochemistry. Ann. N.Y. Acad. Sci. **657**, 455–457

Kumar, M. A., Foster, G. V. and MacIntyre, I. (1963) Further evidence for calcitonin — a rapid acting hormone which lowers plasma calcium. Lancet **2**, 480–492

Kuo, H. C., Nasim, F. H. and Grabowski, P. J. (1991) Control of alternative splicing by the differential binding of U1 small nuclear ribonucleoprotein particle. Science **251**, 1045–1050

Lasmoles, F., Jullienne, A., Day, F., Minvielle, S., Mihaud, G. and Moukhtar, M. S. (1985) Elucidation of the nucleotide sequence of chicken calcitonin mRNA: direct evidence for the expression of a lower vertebrate calcitonin-like gene in man and rat. EMBO J. **4**, 2603–2607

Lee, Y., Hayashi, N., Hillyard, C. J., Girgis, S. I., MacIntyre, I., Emson, P. C. and Tohyama, M. (1987) Calcitonin gene-related peptide-like immunoreactive sensory fibers form synaptic contact with sympathetic neurons in the rat celiac ganglion. Brain Res. **407**, 149–151

Leff, S. E., Rosenfeld, M. G. and Evans, R. M. (1986) Complex transcriptional units: diversity in gene expression by alternative RNA processing. Annu. Rev. Biochem. **55**, 1091–1117

Leff, S. E., Evans, R. M. and Rosenfeld, M. G. (1987) Splice commitment dictates neuron-specific alternative RNA processing in calcitonin/CGRP gene expression. Cell **48**, 517–524

Leighton, B., Foot, E. A., Cooper, G. G. and King, J. M. (1989) Calcitonin gene-related peptide-1 (CGRP-1) is a potent regulator of glycogen metabolism in rat skeletal muscle. FEBS Lett. **249**, 357–361

Lenz, H. J., Rivier, J. E. and Brown, M. R. (1985) Biological actions of human and rat calcitonin and calcitonin gene-related peptide. Regul. Pept. **12**, 81–89

Lewis, P. B., Rafferty, M., Shelley, M. and Robinson, C. J. (1971) A suggested physiological role of calcitonin: the protection of the skeleton during pregnancy and lactation. J. Endocrinol. **49**, ix–x

Li, S., Klein, E. S., Russo, A. F., Simmons, D. M. and Rosenfeld, M. G. (1989) Isolation of cDNA clones encoding small nuclear ribonucleoparticle-associated proteins with different tissue specificities. Proc. Natl. Acad. Sci. U.S.A. **86**, 9778–9782

Libri, D., Piseri, A. and Fiszman, M. Y. (1991) Tissue-specific splicing *in vivo* of the beta-tropomyosin gene: dependence on an RNA secondary structure. Science **252**, 1842–1845

Luhrmann, R. (1990) Functions of U-snRNPs. Mol. Biol. Rep. **14**, 183–192

Luhrmann, R., Kastner, B. and Bach, M. (1990) Structure of spliceosomal snRNPs and their role in pre-mRNA splicing. Biochim. Biophys. Acta **1087**, 265–292

Maniatis, T. and Reed, R. (1987) The role of small nuclear ribonucleoprotein particles in pre-mRNA splicing. Nature (London) **325**, 673–678

Manley, J. L., Fire, A., Cano, A., Sharp, P. A. and Gefter, M. L. (1980) DNA-dependent transcription of adenovirus genes in a soluble whole-cell extract. Proc. Natl. Acad. Sci. U.S.A. **77**, 3855–3859

Martial, K., Minvielle, S., Jullienne, A., Segond, N., Milhaud, G. and Lasmoles, F. (1990) Predicted structure of rabbit N-terminal, calcitonin and katacalcin peptides. Biochem. Biophys. Res. Commun. **171**, 1111–1114

Mayeda, A. and Krainer, A. R. (1992) Regulation of alternative pre-mRNA splicing by hnRNP A1 and splicing factor SF2. Cell **68**, 365–375

Mayeda, A., Zahler, A. M., Krainer, A. R. and Roth, M. B. (1992) Two members of a conserved family of nuclear phosphoproteins are involved in pre-mRNA splicing. Proc. Natl. Acad. Sci. U.S.A. **89**, 1301–1304

McAllister, G., Amara, S. G. and Lerner, M. R. (1988) Tissue-specific expression and cDNA cloning of small nuclear ribonucleoprotein-associated polypeptide N. Proc. Natl. Acad. Sci. U.S.A. **85**, 5296–5300

McAllister, G., Roby-Shemkovitz, A., Amara, S. G. and Lerner, M. R. (1989) cDNA sequence of the rat U snRNP-associated protein N: description of a potential Sm epitope. EMBO J. **8**, 1177–1181

Miles, K., Greengard, P. and Huganir, R. L. (1989) Calcitonin gene-related peptide regulates phosphorylation of the nicotinic acetylcholine receptor in rat myotubes. Neuron **2**, 1517–1524

Minvielle, S., Cressent, M., Delehaye, M. C., Segond, N., Milhaud, G., Jullienne, A., Moukhtar, M. S. and Lasmoles, F. (1987) Sequence and expression of the chicken calcitonin gene. FEBS Lett. **223**, 63–68

Minvielle, S., Giscard-Dartevelle, S., Cohen, R., Taboulet, J., Labye, F., Jullienne, A., Rivaille, P., Milhaud, G., Moukhtar, M. S. and Lasmoles, F. (1991) A novel calcitonin carboxyl-terminal peptide produced in medullary thyroid carcinoma by alternative RNA processing of the calcitonin/calcitonin gene-related peptide gene. J. Biol. Chem. **266**, 24627–24631

Molina, J. M., Cooper, G. J., Leighton, B. and Olefsky, J. M. (1990) Induction of insulin resistance *in vivo* by amylin and calcitonin gene-related peptide. Diabetes **39**, 260–265

Mulderry, P. K., Ghatei, M. A., Bishop, A. E., Allen, Y. S., Polak, J. M. and Bloom, S. R. (1985) Distribution and chromatographic characterisation of CGRP-like immunoreactivity in the brain and gut of the rat. Regul. Pept. **12**, 133–143

Mulderry, P. K., Ghatei, M. A., Spokes, R. A., Jones, P. M., Pierson, A. M., Hamid, Q. A., Kanse, S., Amara, S. G., Burrin, J. M., Legon, S., *et al.* (1988) Differential expression of alpha-CGRP and beta-CGRP by primary sensory neurons and enteric autonomic neurons of the rat. Neuroscience **25**, 195–205

Mullen, M. P., Smith, C. W. J., Patton, J. G. and Nidal-Ginard, B. (1991) α-Tropomyosin mutually exclusive exon selection: competition between branchpoint/polypyrimidine tract determines default exon choice. Genes Dev. **5**, 642–655

Mulligan, R. C. and Berg, P. (1980) Expression of a bacterial gene in mammalian cells. Science **209**, 1422–1427

Munson, P. L. and Gray, T. K. (1970) Function of thyrocalcitonin in normal physiology. Fed. Proc. **29**, 1206–1208

Nawa, H., Kotani, H. and Nakanishi, S. (1984) Tissue-specific generation of two preprotachykinin mRNAs from one gene by alternative RNA splicing. Nature (London) **312**, 729–734

Neurath, H. (1991) Proteolytic processing and regulation. Enzyme **45**, 239–243

New, H. V. and Mudge, A. W. (1986) Calcitonin gene-related peptide regulates muscle acetylcholine receptor synthesis. Nature (London) **323**, 809–811

Niall, H. D., Keutmann, H. T., Copp, D. H. and Potts, J. T., Jr. (1969) Amino acid sequence of salmon ultimobranchial calcitonin. Proc. Natl. Acad. Sci. U.S.A. **64**, 771–778

O'Malley, K. L., Mack, K. J., Gandelman, K. Y. and Todd, R. D. (1990) Organization and expression of the rat D2A receptor gene: identification of alternative transcripts and a variant donor splice site. Biochemistry **29**, 1367–1371

Osterlund, M., Fontaine, B., Devillers-Thiery, A., Geoffroy, B. and Changeux, J. P. (1989) Acetylcholine receptor expression in primary cultures of embryonic chick myotubes. I. Discoordinate regulation of alpha-, gamma- and delta-subunit gene expression by calcitonin gene-related peptide and by muscle electrical activity. Neuroscience **32**, 279–287

Padgett, R. A., Konarska, M. M., Grabowski, P. J., Hardy, S. F. and Sharp, P. A. (1984) Lariat RNAs as intermediates and products in the splicing of messenger RNA precursors. Science **225**, 898–903

Parthemore, J. G. and Deftos, L. J. (1979) Calcitonin secretion in primary hyperparathyroidism. J. Clin. Endocrinol. Metab. **49**, 223–226

Petermann, J. B., Born, W., Chang, J. Y. and Fischer, J. A. (1987) Identification in the human central nervous system, pituitary, and thyroid of a novel calcitonin gene-related peptide, and partial amino acid sequence in the spinal cord. J. Biol. Chem. **262**, 542–545

Poschl, E., Lindley, I., Hofer, E., Seifert, J. M., Brunowsky, W. and Besemer, J. (1987) The structure of procalcitonin of the salmon as deduced from its cDNA sequence. FEBS Lett. **226**, 96–100

Potts, J. T., Jr., Niall, H. D., Keutmann, H. T., Brewer, H. B., Jr. and Deftos, L. J. (1968) The amino acid sequence of porcine thyrocalcitonin. Proc. Natl. Acad. Sci. U.S.A. **59**, 1321–1328

Raybould, H. E. (1992) Inhibitory effects of calcitonin gene-related peptide on gastrointestinal motility. Ann. N.Y. Acad. Sci. **657**, 248–257

Reed, R. (1989) The organization of 3′ splice-site sequences in mammalian introns. Genes Dev. **3**, 2113–2123

Reed, R. and Maniatis, T. (1986) A role for exon sequences and splice-site proximity in splice-site selection. Cell **46**, 681–690

Reed, R. and Maniatis, T. (1988) The role of the mammalian branchpoint sequence in pre-mRNA splicing. Genes Dev. **2**, 1268–1276

Reynolds, J. J. (1968) Inhibition by calcitonin of bone resorption induced *in vitro* by vitamin A. Proc. R. Soc. London, Ser. B **170**, 61–69

Roesser, J. R., Littschwager, K. and Leff, S. E. (1993) Regulation of tissue-specific splicing of the calcitonin/calcitonin gene-related peptide gene by RNA-binding proteins. J. Biol. Chem. **268**, 8366–8375

Roos, B. A., Yoon, M. J., Frelinger, A. L., Pensky, A. E., Birnbaum, R. S. and Lambert, P. W. (1979) Tumor growth and calcitonin during serial transplantation of rat medullary thyroid carcinoma. Endocrinology **105**, 27–32

Rosenfeld, M. G., Amara, S. G., Roos, B. A., Ong, E. S. and Evans, R. M. (1981) Altered expression of the calcitonin gene associated with RNA polymorphism. Nature (London) **290**, 63–65

Rosenfeld, R. M., Lin, C. R., Amara, S. G., Stolarsky, L., Roos, B. A., Ong, E. S. and Evans, R. M. (1982) Calcitonin mRNA polymorphism: peptide switching associated with alternative RNA splicing events. Proc. Natl. Acad. Sci. U.S.A. **79**, 1717–1721

Rosenfeld, M. G., Mermod, J. J., Amara, S. G., Swanson, L. W., Sawchenko, P. E., Rivier, J., Vale, W. W. and Evans, R. M. (1983) Production of a novel neuropeptide encoded by the calcitonin gene via tissue-specific RNA processing. Nature (London) **304**, 129–135

Rosenfeld, M. G., Amara, S. G. and Evans, R. M. (1984) Alternative RNA processing: determining neuronal phenotype. Science **225**, 1315–1320

Ruskin, B., Krainer, A. R., Maniatis, T. and Green, M. R. (1984) Excision of an intact intron as a novel lariat structure during pre-mRNA splicing *in vitro*. Cell **38**, 317–331

Ryner, L. C. and Baker, B. S. (1991) Regulation of *doublesex* pre-mRNA processing occurs by 3′-splice site activation. Genes Dev. **5**, 2071–2085

Sabate, M. I., Stolarsky, L. S., Polak, J. M., Bloom, S. R., Varndell, I. M., Ghatei, M. A., Evans, R. M. and Rosenfeld, M. G. (1985) Regulation of neuroendocrine gene expression by alternative RNA processing. Colocalization of calcitonin and calcitonin gene-related peptide in thyroid C-cells. J. Biol. Chem. **260**, 2589–2592

Sakanaka, M., Magari, S., Emson, P. C., Hillyard, C. J., Girgis, S. I., MacIntyre, I. and Tohyama, M. (1985) The calcitonin gene-related peptide-containing fiber projection from the hypothalamus to the lateral septal area including its fine structures. Brain Res. **344**, 196–199

Schwaber, J. S., Sternini, C., Brecha, N. C., Rogers, W. T. and Card, J. P. (1988) Neurons containing calcitonin gene-related peptide in the parabrachial nucleus project to the central nucleus of the amygdala. J. Comp. Neurol. **270**, 416–426

Schwarz, T. L., Tempel, B. L., Papazian, D. M., Jan, Y. N. and Jan, L. Y. (1988) Multiple potassium-channel components are produced by alternative splicing at the Shaker locus in *Drosophila*. Nature (London) **331**, 137–142; **332**, 740

Shamonki, I. M., Frumar, A. M., Tataryn, I. V., Meldrum, D. R., Davidson, B. H., Parthemore, J. G., Judd, H. L. and Deftos, L. J. (1980) Age-related changes of calcitonin secretion in females. J. Clin. Endocrinol. Metab. **50**, 437–439

Shimada, S., Shiosaka, S., Emson, P. C., Hillyard, C. J., Girgis, S., MacIntyre, I. and Tohyama, M. (1985 *a*) Calcitonin gene-related peptidergic projection from the parabrachial area to the forebrain and diencephalon in the rat: an immunohistochemical analysis. Neuroscience **16**, 607–616

Shimada, S., Shiosaka, S., Hillyard, C. J., Girgis, S. I., MacIntyre, I., Emson, P. C. and Tohyama, M. (1985 *b*) Calcitonin gene-related peptide projection from the ventromedial thalamic nucleus to the insular cortex: a combined retrograde transport and immunocytochemical study. Brain Res. **344**, 200–204

Siebel, C. W., Fresco, L. D. and Rio, D. C. (1992) The mechanism of somatic inhibition of *Drosophila* P-element pre-mRNA splicing: multiprotein complexes at an exon pseudo-5′ splice site control U1 snRNP binding. Genes Dev. **6**, 1386–1401

Skofitsch, G. and Jacobowitz, D. M. (1985 *a*) Calcitonin gene-related peptide: detailed immunohistochemical distribution in the central nervous system. Peptides **6**, 721–745

Skofitsch, G. and Jacobowitz, D. M. (1985 *b*) Quantitative distribution of calcitonin gene-related peptide in the rat central nervous system. Peptides **6**, 1069–1073

Smith, C. W., Patton, J. G. and Nadal-Ginard, B. (1989) Alternative splicing in the control of gene expression. Annu. Rev. Genet. **23**, 527–577

Sommer, B., Keinänen, K., Verdoorn, T. A., Wisden, W., Burnashev, N., Herb, A., Köhler, M., Takagi, T., Sakmann, B. and Seeburg, P. H. (1990) Flip and flop: a cell-specific functional switch in glutamate-operated channels of the CNS. Nature (London) **249**, 1580–1585

Sosnowski, B. A., Belote, J. M. and McKeown, M. (1989) Sex-specific alternative splicing of RNA from the *transformer* gene results from sequence-dependent splice-site blockage. Cell **58**, 449–459

Sossin, W. S., Fisher, J. M. and Scheller, R. H. (1989) Cellular and molecular biology of neuropeptide processing and packaging. Neuron **2**, 1407–1417

Southern, P. J. and Berg, P. (1982) Transformation of mammalian cells with a bacterial gene under control of the SV40 early region promoter. J. Mol. Appl. Genet. **1**, 327–341

Steenbergh, P. H., Hoppener, J. W. Zandberg, J., Lips, C. J. and Jansz, H. S. (1985) A second human calcitonin/CGRP gene. FEBS Lett. **183**, 403–407

Steenbergh, P. H., Hoppener, J. W., Zandberg, J., Visser, A., Lips, C. J. and Jansz, H. S. (1986) Structure and expression of the human calcitonin/CGRP genes. FEBS Lett. **209**, 97–103

Sternini, C. (1992) Enteric and visceral afferent CGRP neurons. Targets of innervation and differential expression patterns. Ann. N.Y. Acad. Sci. **657**, 170–186

Stevenson, J. C., Hillyard, C. J., MacIntyre, I., Cooper, H. and Whitehead, M. I. (1979) A physiological role for calcitonin: protection of the maternal skeleton. Lancet **2**, 769–770

Tian, M. and Maniatis, T. (1992) Positive control of pre-mRNA splicing *in vitro*. Science **256**, 237–240

Tschopp, F. A., Tobler, P. H. and Fischer, J. A. (1984) Calcitonin gene-related peptide in the human thyroid, pituitary and brain. Mol. Cell. Endocrinol. **36**, 53–57

Valcarcel, J., Singh, R., Zamore, P. D. and Green, M. R. (1993) The protein *sex-lethal* antagonizes the splicing factor U2AF to regulate alternative splicing of transformer pre-mRNA. Nature (London) **362**, 171–175

Watakabe, A., Inoue, K., Sakamoto, H. and Shimura, Y. (1989) A secondary structure at the 3′ splice site affects the *in vitro* splicing reaction of mouse immunoglobulin mu chain pre-mRNAs. Nucleic Acids Res. **17**, 8159–8169

Williams, G., Cardoso, H., Ball, J. A., Mulderry, P. K., Cooke, E. and Bloom, S. R. (1988) Potent and comparable vasodilator actions of A and B calcitonin gene-related peptides on the superficial subcutaneous vasculature of man. Clin. Sci. **75**, 309–313

Yamano, M., Hillyard, C. J., Girgis, S., MacIntyre, I., Emson, P. C. and Tohyama, M. (1988) Presence of a substance P-like immunoreactive neurone system from the parabrachial area to the central amygdaloid nucleus of the rat with reference to coexistence with calcitonin gene-related peptide. Brain Res. **451**, 179–188

Yasui, Y., Saper, C. B. and Cechetto, D. F. (1989) Calcitonin gene-related peptide immunoreactivity in the visceral sensory cortex, thalamus, and related pathways in the rat. J. Comp. Neurol. **290**, 487–501

Zaidi, M., Chambers, T. J., Bevis, P. J., Beacham, J. L., Gaines-Das, R. E. and MacIntyre, I. (1988) Effects of peptides from the calcitonin genes on bone and bone cells. Q. J. Exp. Physiol. **73**, 471–485

Zaidi, M., Moonga, B. S., Bevis, P. J., Bascal, Z. A. and Breimer, L. H. (1990) The calcitonin gene peptides: biology and clinical relevance. Crit. Rev. Clin. Lab. Sci. **28**, 109–174

Zieve, G. W. and Sauterer, R. A. (1990) Cell biology of the snRNP particles. Crit. Rev. Biochem. Mol. Biol. **25**, 1–46

3

Expression of the oxytocin and vasopressin genes

W. Scott Young, III

Laboratory of Cell Biology, National Institute of Mental Health, Bethesda, MD 20892, U.S.A.

Introduction

The nonapeptides oxytocin (OT) and vasopressin (VP) have held prominent roles in the fields of endocrinology and neuroscience. By the early 1900s, the effects of neurohypophysial substances on blood pressure, rapidity of labour, and milk secretory activity were described (Oliver and Schäfer, 1895; Dale, 1909; Ott and Scott, 1911; Schäfer and Mackenzie, 1911). The concept of neurosecretion arose from the studies of these peptides (Bargmann and Scharrer, 1951). The early determination of their amino acid sequences (Fig. 1) (Du Vigneaud, 1954) permitted more refined studies of their biological effects and distributions. VP and OT were both determined to exist in the magnocellular neurons of the paraventricular nuclei (PVN) and supraoptic nuclei (SON) of the hypothalamus (Swaab *et al.*, 1975; Vandesande and Dierickx, 1975) and processed from their precursor forms along the axonal projection to the neural lobe (Brownstein *et al.*, 1980). The orders of the nonapeptide, neurophysin and glycopeptide (VP precursor only) components in the preprohormones were subsequently determined (Russell *et al.*, 1981). The functions of the neurophysins and glycopeptide remain essentially unknown, although the neurophysins can bind their respective nonapeptides during axonal transport. Excellent discussions of the physiological roles of VP and OT are found elsewhere (Leng and Bicknell, 1986; Jard and Jamison, 1991; Reeves and Andreoli, 1992) and a recent review discusses the physiology of magnocellular neurons of the PVN and SON (Renaud and Bourque, 1991).

The extensive distributions of VP and OT fibres and receptors in the central nervous system (CNS) (Sofroniew, 1985; Tribollet *et al.*, 1988; Snijdewint *et al.*, 1989; Berecek and Swords, 1990) implied that, in addition to the well-known antidiuretic effect of VP and smooth muscle contracting effect of OT in the periphery, these peptides would have central (or centrally originating) effects as well. Perhaps a role for central VP in cardiovascular regulation is best accepted (Berecek and Swords, 1990). VP in parvocellular

Fig. 1. OT and VP are nonapeptides with amidated C-terminals

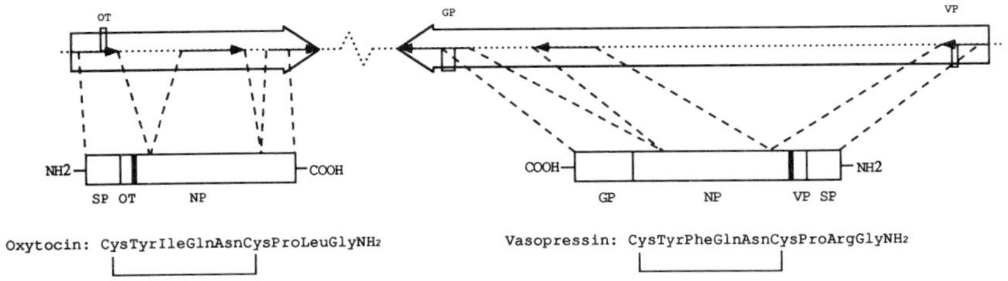

They are encoded by their respective genes which are present on the same chromosome (approximately 3.5–12 kb apart) and are transcribed towards each other. Exons are depicted by solid arrows within the nuclear RNAs (open arrows). The sequence encoding OT neurophysin is located 3′ to that encoding OT and the sequence encoding VP neurophysin is located between those encoding VP and the glycopeptide (GP). SP is signal peptide.

neurons of the PVN, suprachiasmatic nucleus (SCN), and bed nucleus of the stria terminalis, for example, is likely to be important for regulation of glucocorticoid synthesis, circadian rhythm effects and sexual behaviour (see below). Even 'higher' CNS processes such as memory appear to be regulated by OT and VP (e.g. de Wied, 1971). Remarkable changes in levels of hypothalamic OT and its receptors caused by changes in sexual behaviour and hormonal environment also signify their CNS importance (de Kloet *et al.*, 1986; Schumacher *et al.*, 1989; Jirikowski *et al.*, 1991; Johnson *et al.*, 1991).

Our knowledge of the regulation of expression of the genes encoding VP and OT greatly expanded after Richter and co-workers determined the cDNA sequences for the bovine preprohormones (Land *et al.*, 1982, 1983). The rest of this chapter will be devoted to those studies that have examined the structure and component functions of the VP and OT genes as well as the distribution and regulation of VP and OT transcripts. Post-translational processing and secretion are reviewed nicely elsewhere (Gainer *et al.*, 1988).

Structures of the VP and OT genes

The cloning of the VP and OT cDNAs confirmed the order of the nonapeptide, neurophysin and glycopeptide (also known as copeptin) components described above. Subsequent analysis of the corresponding bovine genes revealed a three exon structure (Fig. 1). A signal peptide, the nonapeptide, and a portion of the neurophysin are located in the first exon. The second exon codes for the bulk of the neurophysin. The third exon codes for the remainder of the neurophysin and the 39 amino acid copeptin in the case of VP. This exon/intron arrangement has also been demonstrated for rat (Ivell and Richter, 1984*b*), mouse (Hara *et al.*, 1990), humans (Sausville *et al.*, 1985) and teleost fish (Morley *et al.*, 1990). Sausville *et al.* (1985) initially demonstrated that the human VP and OT genes are both located on chromosome 20 linked to dynorphin (which is co-expressed with VP in the PVN and SON — see below) (Summar *et al.*, 1990) and postulated a gene duplication and inversion to explain their inverted arrangement. This arrangement has also been demonstrated in mouse (Hara *et al.*, 1990), rat (Mohr *et al.*, 1988) and cattle (Ang *et al.*, 1991) chromosomes.

Related members of the VP/OT family have been identified in vertebrates and invertebrates (Fig. 2). Intriguingly, whereas vasotocins-1 and -2 of a teleost fish have the same exon/intron arrangement as found in mammals (Morley *et al.*, 1990), the isotocins-1 and -2 of the fish do not have introns within the coding region (Figueroa *et al.*, 1989). Both isotocins and vasotocins contain neurophysin extensions that contain a leucine-rich core segment similar to that found in copeptin. It appears as if the copeptin-related segment was lost during the evolution of the ancestral gene to OT in mammals. Analysis of the toad vasotocin and mesotocin cDNA sequences (Nojiri *et al.*, 1987) suggests that the former is a VP precursor and the latter an OT homologue. Further aspects of and speculations about this family's evolution (Fig. 3) are presented in the above references and elsewhere (Sherwood and Parker, 1990; Urano *et al.*, 1992). As more species are examined, of course, questions about the evolutionary pathways should be answered.

Analyses of VP sequences have revealed the nature of several pathophysiological states. The Brattleboro rat's diabetes insipidus is caused by the lack of a single residue in the second exon that leads to a frameshift and a new precursor with a highly changed poly-lysine carboxy terminus (Schmale and Richter, 1984). This mutated precursor is not processed to release VP (Schmale *et al.*, 1984); instead, the mutant protein does not move beyond the endoplasmic reticulum and *trans*-Golgi (Ivell *et al.*, 1986) to appear in the secretory granules where proteolytic processing would normally occur (Brownstein *et al.*, 1980). Of potential fundamental importance is the observation that as homozygous Brattleboro rats age, VP appears with greater frequency in neurons of the PVN and SON, suggesting some sort of active gene modification is occurring in non-dividing cells (van Leeuwen *et al.*, 1989). Initial studies have shown that various insertions which correct the frameshift are occurring (Evans *et al.*, 1991), but more analysis is necessary for confirmation.

The human VP gene has also been examined for mutations to explain autosomal dominant, familial central diabetes insipidus (FCDI). Bahnsen *et al.* (1992) demonstrated that a G→T conversion changed the neurophysin Gly-17 to Val and that the mutation segregated with the pheno-

Fig. 2. Alignment of preprohormone amino acid sequence of members of the OT and VP family

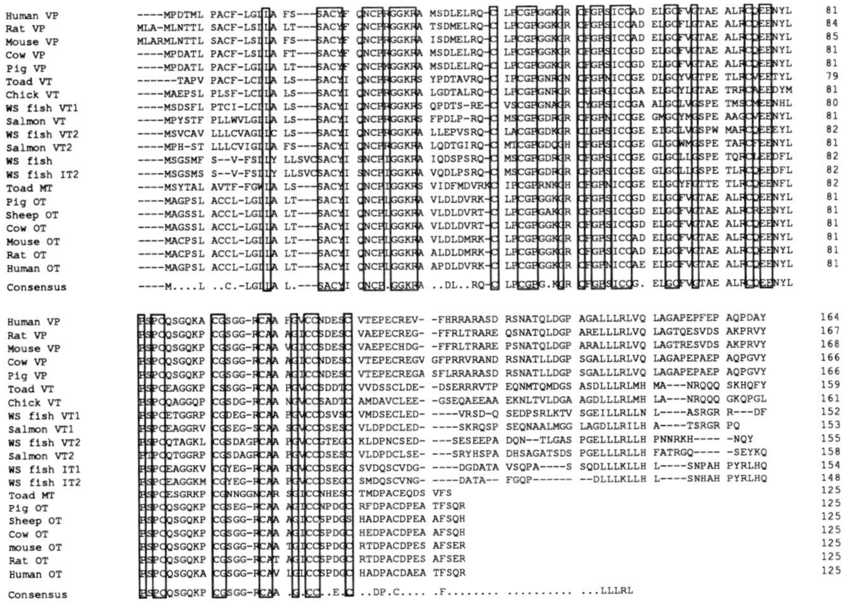

Boxed columns indicate 100% identity and the consensus row indicates > 50% identity. Alignments were performed by the GeneWorks program (IntelliGenetics, Mountain View, CA, U.S.A.) that scores mismatches and gaps rather than matches. Abbreviations: WS, white sucker; VT, vasotocin; IT, isotocin; MT, mesotocin. References: Human VP (Sausville et al.*, 1985) and OT (Rehbein* et al.*, 1986), rat VP and OT (Schmitz* et al.*, 1991), cow VP (Land* et al.*, 1982) and OT (Land* et al.*, 1983), pig VP (Rehbein* et al.*, 1986) and OT (Rehbein and Richter, 1990), toad VT and MT (Nojiri* et al.*, 1987), chick VT (GenBank Accession # X55130), WS fish VT1 and VT2 (Morley* et al.*, 1990) and IT1 and IT2 (Figueroa* et al.*, 1989), and salmon VT and VT2 (Hyodo* et al.*, 1991).*

type. In a second study of FCDI, two patients within a pedigree had a single base substitution (G→A) on one allele resulting in a Gly for Ser substitution at amino acid 57 in the neurophysin moiety (Ito *et al.*, 1991). None of 10 patients with idiopathic central diabetes insipidus or five normal patients had a mutation. It is not known how this dominant negative mutation leads to lower VP levels. Bahnsen *et al.* (1992) suggest that the Gly-17 to Val mutation may effect the secondary structure just enough to permit transport beyond the endoplasmic reticulum (unlike the Brattleboro rat above) yet perturb the interaction of the normal precursors, perhaps upsetting the formation of tetrameric structures suggested by a crystallographic method

Fig. 3. Hypothetical scheme of the evolution of some members of the neurohypophysial hormone precursors as proposed by Urano *et al.* (1992)

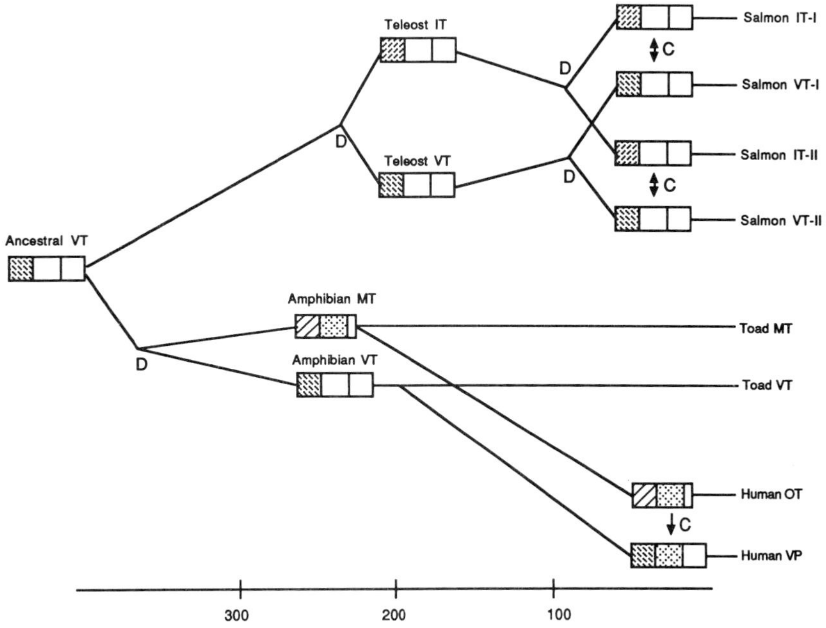

VT, vasotocin; IT, isotocin; MT, mesotocin. Axis is in millions of years.

(Chen *et al.*, 1991). One would also be hard pressed to supply an explanation if members of these pedigrees also have reduced numbers of neurons in the PVN and SON as occasionally reported for FCDI (see Braverman *et al.*, 1965; Treip, 1984 and references within). In another study, the genetic locus was closely linked with the VP gene on chromosome 20 (Repaske *et al.*, 1990).

A number of features are apparent upon examination of this VP/OT gene locus. As noted above, the two genes are linked within 3.5–12 kb of each other, transcription occurring towards the other gene. The two genes in the rat are linked by a long interspersed repeated DNA element (LINE) which contains seven open reading frames between 99 and 556 amino acids in length (Schmitz *et al.*, 1991). These elements contain open reading frames (ORFs) with homology to reverse transcriptases of retroviruses (Xiong and Eickbush, 1990). ORF3 from the rat LINE (Schmitz *et al.*, 1991) has extensive homology (85% amino acid identify in a stretch of 442) with, for example, the reverse transcriptase domain of a human transposable element

(Dombrowski *et al.*, 1991). Specific RNA probes (in combination with RNase) have not been used for *in situ* hybridization localization of expression. The significance of this rat LINE is unknown, but it is probably unrelated to OT and VP regulation. This LINE is not necessary for rat OT transgene expression (see below) and certainly is not present in its entirety between the mouse OT and VP genes.

The sequences of the second exons of OT and VP are highly conserved with respect to each other within a given species, more so than either the OT or VP second exons across species (Fig. 4). Although it has not been demonstrated, it is tempting to speculate that this conservation (probably through gene conversion; Lewin, 1990) is preserved because it is used to co-ordinately regulate gene expression. Perhaps tissue-specific factors bind to exon B to allow one gene to be transcribed while excluding transcription from the other. Another possibility is that some step in the processing that is the same for both of the preprohormones involves the neurophysin moiety.

Fig. 4. Comparison of the human (H), rat (R), mouse (M) and cow (C) exon B nucleotide sequences over the first 177 bases

OT vs OT \ VP vs VP	H	R	M	C
H	**6**	15	11	6
R	16	**5**	10	14
M	15	9	**3**	9
C	10	13	11	**0**

The diagonal (bold numbers) is the number of mismatches between the OT and VP exon B sequence within a species. The lower left comparisons are OT exon B mismatches between specie and the upper right comparisons are for the corresponding VP mismatches.

Examination of the promoter regions of the two genes shows relatively little homology between them, but significant homology across species is present for either gene, especially within the 200 bp upstream of the initiation codon (Figs. 5 and 6). 'CAT' boxes are absent but modified 'TATA' (Goldberg-Hogness) boxes are present in both. The rat 5′ untranslated region of the VP gene contains elements reminiscent of cyclic AMP (cAMP) and glucocorticoid response elements at −227 and −622, respectively, and AP-2 binding sites at −345, −146, −95 and −84 (Mohr and

Richter, 1990; Fig. 5). The relationships of these sites to the physiological response of the VP gene remains to be determined as only in the bovine promoter has a cAMP-responsive element been characterized (Pardy *et al.*, 1992). However, increased VP expression caused by hyperosmolality in the rat is accompanied by increased cAMP levels in the SON and PVN (Young *et al.*, 1987*a*; Carter and Murphy, 1989*a*). cAMP can stimulate VP expression in a small cell lung cancer line and human VP promoter-driven luciferase activity in transfected P19 embryonal cell cultures (Verbeeck *et al.*, 1990). VP expreession is also stimulated by cAMP in primary dissociated cultures from 14 day old fetal rats (Oeding *et al.*, 1990). Glucocorticoids inhibit VP expression in parvocellular neurons of the PVN (see below). cAMP-enhancing agents and the synthetic glucocorticoid, dexamethasone, synergistically elevate, whereas phorbol ester treatment decreases, VP mRNA levels in a small cell lung carcinoma line, but the relevance to VP gene activation *in vivo* is not clear (Verbeeck *et al.*, 1991). Similar concerns

Fig. 5. Alignments of the promoter regions of mammalian VP genes

```
Rat        AAAGCTCA-C CTGGTAT--C GATGCTTCCT CTTATCCTAC ACCACAACTA ACAAACCTG- CCC-CACCTC CT-----GGT CCTGACCCTG CTGCAGACCT
Mouse      AAAGCCCA-C CTGATAT--G GGTGCTTCCT CTCATCCTAT ACCACAACTA ACAA-CCTG- CCCACCCCTT CT-----GGT GCTGACCCTG CTCCATACGT
Human      GACCCTTATC TTGTTGTCAC AGGGCTAACC CTCCCCACAC ACC-TGGCTG TC--CCCTGC CCCACAGCTC CTAGCCAGGG CCTGTCCCTG CCCCTAACAC
Cow        ---------- ---------- ---------- ---------- ---------- ---------- ---------- ---------- ---------- ----------
Consensus   a  c  a c  tg t t        gct  c  ct    c  a  acc     ct    c    cctg   ccc c  ct  ct      gg   ctg ccctg c  c  ac      -396

Rat        GCTAGTCCTT GGTGAAT-GA GACCT----G GGGACCCCTC TAGTCTGTTG AGAGCTGCTG AAATGCTCAA CTATGATTTC CAGGTGACCC TC---AA---
Mouse      GCCAGCCCTT GCTGGATGGG GGCCT----G TGGACCCCTT TAGTCTGCTG AGAGCAGCTG GAATATTCAA CTATGATTTC CAGGTGACCC TC---CA---
Human      GCAGCCCCTC GGTTCCTCTT ACCCTCTTCT TGGACCCTAC --GCCATTTC TGGGTAATCA GAAAGCTCAG AGATGGTCTC CAGGTGACCC CC---CA---
Cow        ---------- ---------- ---------- ---------- ---------- ---------- --------GG AGACAGTTTC CAGGTGACCC CCCCTCATTC
Consensus  gc    cct  g t   t       cct        ggaccc       g c    t    g g        aa    tca    Atg TTTC CAGGTGACCC  C    cA      -292

Rat        ---GTCGGCT CACCTCCCTG ATTGCACAGC AC--CAATCA CTGTGG-CGG TGGCTCCCGT CACACGGTGG CCA------- G----TGACA GCCTG-ATGG
Mouse      ---GTCGGCT CACCTCACTG ATCGCACAGC AC--CAATCA CTGTGGGCAG TGGCTCCTGT CAGACGGTGG CCG------- G----TGACA GCCTGCATGG
Human      ---GTCTGTC CCTCTCCCTG AATCCAGAGC GCTGCAGTCA CAGT-AGAGG CAATTGCTGT CATTCCGCGG CC-CA----T GACCATG--- ----GGC-GG
Cow        CCCGTCTGTC CACCTCCCCA AGTCTAGAGA GCCGCAATCA TAGC-CGCAG CAGCTCCTGT CACACCGCAG CCACACCGCT GCCTATGACA GCCTGGA-GG
Consensus     GTC G    CacCTCcCtg A t cA AGc  C  CAaTCA c GtG gc G   gcTcCtGT CA  C G gG CC          G    TGaca gccTG a GG   -196

Rat        CTGGCTCCCC TCCTCCACCA CCCTCTGCAT TGACAGGCCC ACGTGTGTCC CCAGATGCCT GAATCACTGC TGACAGCTTG GGACCTGTCA GCTGTGGGCT
Mouse      CTGGCTCCCC TCCTCCACCA CCCTCTGCAC TGACACGCCC ACGTGTGTCC CCAGATGCCT GAATCACTGC TGACAGCTTG GGACCTGTCG GCTG-AGGCT
Human      CCCGTACCCC TCCCCC--GA TCCCCTGCAC AGACAGGCCC ACGTGTGTCC CCAGATGCCT GAATCACTGC TGACCGCTGG GGACCTGGCG GCCGTGGGCT
Cow        CCAGCATCCC TCCCCCACCG TCCCCTGC-- --ACAGGTCC ACGTGCGTCC CCAGATGCCT GAATCACTGC TGACGCCTGG GGACCTGGAG GCCACGGGCT
Consensus  C  Gc cCCC TCC CCacca  CC CTGCa   gACAgGcCC ACGTGtGTCC CCAGATGCCT GAATCACTGC TGAC gCT G GGACCTG cg GC     GGCT    -96

Rat        CCTGGGGAGC CACTGGGGAG GGGGTTAGCA GCCACGCTGT CGCCTCCTAG CCAACACCTG CAGACATAAA TAGACAGCCC AGCC-CGCTC AG-GC
Mouse      CCTGGGGAGC CACTGGGGAG GGGGTTAGCA GCCACGTTGT CGCCTCCTAG GGAACACCTG CAGACATAAA TAGGCAGCCC AGCC-CGCCA AG-GC
Human      CCTGGGGAGC CACTGGGGAG GGGGT-GGCG GCCGCG-TCT CGCCTCCACG GGAACACCTG CGGACATAAA TAGGCAGCC- AGCAGAGGCA GCAGC
Cow        CCTGGGGAGC CACTGGGGAG GGGGT-GGCG GCCACG---T CACTTCAGAG GGAACACCTG CAGACATAAA TAGGCAGCC- AGCGAAGACA TC-GC
Consensus  CCTGGGGAGC CACTGGGGAG GGGGT  GC  GCCaCG t T CgCcTCc aG ggAACACCTG CaGACATAAA TAGgCAGCC  AGC   G ca    GC         -1
```

As the exact transcriptional start sites are not established for all species, − 1 is somewhat arbitrarily chosen. The consensus row indicates 100% (capital letters) or 80% (small letters) identities. The modified TATA box is underlined in the consensus row − 24 to − 30. Potential AP-2 binding sites are also underlined. Alignments were performed by the GeneWorks program (IntelliGenetics, Mountain View, CA, U.S.A.). References: rat (Schmitz et al.*, 1991); mouse (Hara* et al.*, 1990); human (Sausville* et al.*, 1985; unpublished data kindly supplied by Dr. J. Battey); and cow (Land* et al.*, 1982).*

Fig. 6. Alignments of the promoter regions of mammalian OT genes

```
Rat        ACG--G-G-- CTAGTTAGAA GAAGGG-CTG AGCTGCACTC AGAGATTGAA CAAGA----- TCCTTT---T CCCCATCTAA GCTTTAGATC CACGT-AAGG
Mouse      ACG--G-G-- CTAGTTAGAA GA---G-CTG AGGTGCATTC AGAGATTGAA CAAGACGCCA TCTTTT---T CCCCATCTAA GTTTCAGGTC CATGT-AAGG
Human      ---------- ---------- ---------- ---------- ---------- ---------- ----GG---A TCCTGCCAGA GCCTC-CTCC CACCTGGAGG
Cow        AGGAAGAGAA CTTGCTTGCA GAGGGGAGCT AGGGAGTGTA TGTGGTTGGC CAAGAGGGCA GATGGGGAAG TTCTGAACGA ATCTC-TTCT CTCCTGGAGG
Sheep      AGGAAGAGAA CTTGCTTGCA GAGGGGAGCT AGGGGGTGTG TGTGGTTGGC CAAGAGGGCA GATGGGGGAG TCCTAAACCA ATCTC-TTCT CGCCTGGA-G
Consensus  a-g--g-g-- ct-g-t-g-a ga---g---- ag------t- -g-g-ttg-- caaga----- ---------- -cC------A ---Tc----- C-c-T--AgG   -401

Rat        GCTTCC-TCA CCCACTTGCG CTATCCAA-- CCTCATTCT- GAGGTA---- ----TTGGAT TTCTATGAAA AAACAGCTCT TGGCTAGGCT GCACCTCTAC
Mouse      GCTCCC-TCA CTCACTGGCC CTATCCTG-- CCTTATTCT- GAGATA---- ----TTGGAT ATCTGTG-AA AAACAGCTCC TGGCTAGGGC GCACCTCCAA
Human      GGTCCCAGCG TCCACCTTCC CTGCCCCAGC CCCCCTCCTC GAGGTACTGG GAGGCTGGAT A--------- --A-AGTCTT CGGCTGGGCC ACACC-CCAC
Cow        GC-CCGAGTG TTCA----TC CTG----ACC CCTCA-CCCC AAGGGATTGA GAG--TGGGC -TCC------ --ATAGTCCT CGGTTGGGCC GCAGC----C
Sheep      GC-CCGAGTG TTCA----TC CTG----ACC CCTCA-CCCC AAAGGAATGA GAG------- ---------- ----AGTCCT CGGTTGGGCC GCAGC----C
Consensus  Gc-cC----- --CA-----c CT-----a-- CCtca--C-- -Agg-A---- -----tgg-- ---------- --a-AG--ct -GG-T-GGcc gCA-C----c   -301

Rat        CCC---CTC- C-AAGTCT-- CTTTATC-CT CTTGTAGCT- --TAGG---C CT--CC---- ---C---CTT CT---AG-GC TGTGTCCCTT TTGAGCT-CA
Mouse      CCC---CTC- CCAAGTCT-- CTCTAGC-CT CTTGTAGCC- --TAGG---C CA--CC---- ---C---CTT CC---AG-GC TGCTTCTCTT TTGAGTTCCA
Human      CCCAAATTCT CCCTGTCCCA CCCTAGTGCC CAGGCCACC- --CCGG---C CTGCTC---- ---C---CTT CCGCAAG-GC ACCTCACCTT CTGTGC-CCA
Cow        CCCATA--CT CCCTGTCCCA CCCTAGGACC TAGGCCACCG CTTAGGCTCC CTGCCCAGAG CTGCTTGCTT CCCCCA-AGC AGCACTCGCT ATCATCTCCC
Sheep      CCCATA--CT CCCTGTCCCA CCGTA-GACC CAGGCCACCG CACAGGCTCC CTGCCCAGAG CTGCTTGCTT CCCCTAGCGG CGCGCTCGCT ATCATCTCCC
Consensus  CCC-----C- Cc--GTC--- C--TA---C- c--G---Cc- ---MGG---C Ct--cC---- ---C---CTT Cc---Ag-Gc -gc---c--T -T-a-ctcC-   -201

Rat        G-GTCATTAG CTG--AGGCG --GTGACCTT GACCCCAGCC CAGACCCTGC AAATGAAG-G GCCTGCTTCT AAACAGTGTG GAACAGTTTG ACCCA-AGAG
Mouse      G-GTCATTAG CAG--AGACG --ATGACCTT GACCCTAGCC CAGACCCTGC AAATGAAG-G GCCTGCCTCT AAACAGCGTG GAACAATTTC ACCCA-AGAG
Human      G-ACCATTAG CCA--ACGCG --GTGACCTT GACCCCGGCC CAGGCCCTGC TAATGAAGAG GAAAGCCCGT ACGCA-CTCG G-----CCTG ACCCACGGCG
Cow        GGACCATTAG CCATTAGCCG ACATAACCTT GACCC-GGCA CAGCCTCTGC AAATGAGGGG GCGCGCC-GG GGGC-GCA-G G----GCCTG ACCCGCAGCG
Sheep      GTAACATTAG CCATTAGCAG ACATAACCTT GACCC-GGCA CAGCCCCTGC AAATGAGGGG GCGCGCC-GG GGGC-GCA-G G----GCCTG ACCGGC-G-G
Consensus  G---CATTAG C----Ag-cG ---T-ACCTT GACCC--GC- CAG-CcCTGC aAATGA-G-G Gc--GCc--- ---C-gc--G G-------Tg ACCc---G-G   -101

Rat        ACCTGGCTGT GACCAGTCAT GCAGTCACCC TCTT-AGACT GGGCCCCACC ATGGCAGTGG ACAAGGCATA AAAAGGTCGG TCTGGGCTGG AGAA------
Mouse      ACCT-TCTGT GACCAGTCAT GCTGTCACCC TCTTTAGACA GTGCTCCACC ATGGCAGTG- -CCAGACATA AAAAGGTCGG TCTGGGCCGG AGA-------
Human      ACCC-TCTGT GACCAATCAT ACTACCAACC TC-TTAAACA GAGCTCCACC GACGCAAT-G CCCAGGCATA AAAAGGCCAG GCCG----AG AGACCGCC--
Cow        GCGCG-CTGT GACCAGTCAT GCGGCTGCCC TCTTA-GACA CCGTTCCGCC CACGCGGCCG -CCGGGC-TT AAAAGGCCAG ACC---CGAG AGACGGCCGC
Sheep      GCGCG-CTGT GACCAGTCAT GCGGCTGCCC TCTTAAGACA CCGTT-CGCC CACGCGG-CG -CCGGGC-TT AAAAGGCCAG ACC---CGAG AGACGGCCGC
Consensus  -C----CTGT GACCARTCAT gC-g---cCC TCtT-agACa --G-tcC-CC ---GC-g--g -Cc-GgC-T- AAAAGG-C-G -C----c--G AGA-------   -1
```

As the exact transcriptional start sites are not established for all species, − 1 is somewhat arbitrarily chosen. The consensus row indicates 100% (capital letters) or 80% (small letters) identities. The modified TATA box is underlined in the consensus row − 27 to − 33. The TGACC half sites involved in oestrogen, retinoic acid and thyroxine responses and discussed in the text are also underlined. Alignments were performed by the GeneWorks program (IntelliGenetics, Mountain View, CA, U.S.A.). References: rat (Schmitz et al., *1991); mouse (Hara* et al., *1990); human (Sausville* et al., *1985; unpublished data kindly supplied by Dr. J. Battey); cow (GenBank Accession # X58474); and sheep (GenBank Accession # X55131).*

arise from the use of cultured diencephalic cultures (from 14 day old rat fetuses) that contained predominantly magnocellular neurons whose levels of VP transcripts are increased by the glucocorticoid antagonist, RU38486 (Schilling *et al.*, 1991). *In vivo*, magnocellular VP expression is not affected by glucocorticoids (see below). In contrast to the interaction between cAMP and glucocorticoids in the lung carcinoma cell line, no such interaction was observed in the diencephalic cultures.

A number of potential regulatory sites have been identified in the 5′ untranslated region of the OT gene (Fig. 6). The most thoroughly studied is the oestrogen responsive element (ORE). Both the rat and human OT promoters contain the imperfect palindrome GGTGACCTTGACC that confers oestrogen-responsiveness in various transfected cell lines (Burbach *et al.*, 1990; Richard and Zingg, 1990). This sequence contains two TGACC motifs and a TCACC (read from the opposite strand). The corresponding element in the bovine OT promoter, differing in three positions, is not responsive to oestrogen (Adan *et al.*, 1991). In addition, in the rat, there are two TGACC motifs present 3′ to the full ORE, the one more 5′ providing significant oestrogen responsiveness alone (Adan *et al.*, 1991; Mohr and Schmitz, 1991). Despite the apparent absence of oestrogen receptors from the majority of magnocellular OT neurons in the PVN and from all of them in the SON of both male and female rats (Burbach *et al.*, 1990; Simerly *et al.*, 1990), expression of the OT gene there increases during puberty (Miller *et al.*, 1989*a*; Chibbar *et al.*, 1990) and during chronic oestrogen treatment of ovariectomized rats (Chung *et al.*, 1991). Women on oral contraceptives and men given diethylstilbestrol have elevated levels of plasma OT neurophysin (Robinson, 1975). The relevant site of oestrogen action remains to be determined. A sufficient level of receptors may be present in the PVN and SON but remain below current levels of detection. Alternatively, this steroid may act at any number of remote locations expressing the oestrogen receptor gene with its effects transmitted to the PVN and SON by synaptic transmission (compare Simerly *et al.*, 1990, and Hatton, 1990) as suggested by a study that blocked oestradiol-induced elevations in plasma OT with deafferentations rostral to the optic chiasm in ovariectomized rats (Yamaguchi *et al.*, 1979).

Two thyroid response elements have also been identified in the rat promoter (Fig. 6), one of which binds the c-*erb*A/α-thyroid hormone receptor and is also found in the human promoter, has two tandem TGACC repeats and is located within the ORE (Adan *et al.*, 1992). The rat's other responsive element is located between -195 and -172 and contains the TGACC motif. The co-expression in the heterologous expression systems of both oestrogen and α-thyroid hormone receptors inhibits their individual stimulation via this element (Adan *et al.*, 1992). The α-thyroid hormone receptor gene is expressed within magnocellular neurons of the PVN and SON (Bradley *et al.*, 1989, 1992) and thyroid hormone administration elevates hypothalamic OT mRNA levels, pituitary OT content, and blood OT levels (Adan *et al.*, 1992). Thus, the OT gene appears to be a physiological target gene for thyroid hormone action. As with the actions of oestrogens, the significance of these observations for human physiology remains to be determined. Curiously, hypothyroidism appears to elevate plasma VP levels in rats (although not neurophysin levels; Seif *et al.*, 1979) and sheep (Skowsky and Fisher, 1977). A thyroid-responsive element has not been defined in this gene; however, the effects of hypothyroidism on VP levels appear to be secondary to peripheral effects.

Recent data also suggests that the human and rat OREs harbour retinoic acid responsive elements. When rat OT promoter bases -170 to -158 are placed adjacent to a thymidine kinase promoter-driven luciferase, there is weak inhibition by retinoic acid when co-transfected with the human

α-retinoic acid receptor (Lipkin *et al.*, 1992) Normal oestradiol stimulation in the presence of the oestrogen receptor is inhibited by the retinoic acid receptor in the presence of retinoic acid. Even the normally strong *trans*-activation domain from the herpes simplex viral protein, VP16, can not overcome the inhibition when fused to the retinoic acid receptor (Lipkin *et al.*, 1992). *In vivo*, however, it may be that retinoic acid is a positive *trans*-activator as the two other TGACC repeats found further towards the 3′ co-operate with the 5′ ones to stimulate strongly transcriptional activity when co-transfected with retinoic acid receptors (α, β or γ) or when transfected into P19 embryonic cells that contain α-retinoic acid receptors (Richard and Zingg, 1991*a*; Adan, 1992). In addition, the retinoic acid receptors act co-operatively with the oestrogen receptor on the rat upstream elements when the downstream elements are present (Adan, 1992). Similarly, only when all four TGACC repeats are present in the human promoter does the oestrogen receptor co-operate with the retinoic acid receptor; otherwise, there is inhibition (Richard and Zingg, 1991*a*). DNase footprinting using recombinant human β-retinoic acid or oestrogen receptors showed overlapping, but different, protected sequences about bases −184 to −148 and −106 to −91 containing TGACC motifs (Adan, 1992). The physiological significance of these findings is unknown, although vitamin A deficiency or supplementation with retinoic acid did not affect OT expression (Adan, 1992).

Members of a new family of *trans*-acting factors, the POU domain-containing proteins that are distantly related to the homeobox proteins, are expressed in the PVN, SON and SCN (He *et al.*, 1989; Le Moine and Young, 1992). One of them, RHS2, is detected in VP but not OT magnocellular neurons (Le Moine and Young, 1992). It will be of interest to learn if these POU proteins regulate OT or VP expression, and if they do so through interactions with the factors described above.

The use of transgenic animal technology is only in its infancy with respect to VP and OT. Transgenic VP expression in the PVN and SON is produced with a metallothionein–VP fusion gene that contains only 35 bp of 5′ flanking VP DNA with 800 bp of the metallothionein promoter and 64 bp of its first exon (Habener *et al.*, 1989). In contrast, 3.55 kb of rat VP gene with 1.4 kbp of promoter did not appropriately express VP in transgenic mice (Young *et al.*, 1990*b*). These data imply that a tissue-specifier was omitted that was compensated for by the ‘creation of combination of *cis*-element (i.e. metallothionein and VP sequence) that can be deciphered as transcriptionally competent by *trans*-acting factors present (Habener *et al.*, 1989)’ in the hypothalamus. A preliminary report of a rat transgene consisting of a 5 kb VP genome with 3 kb of 5′ flanking sequence describes expression of rat transcripts in the hypothalamus and temporal lobes (Graff *et al.*, 1989). Another initial report of a 13.4 kb bovine VP genomic transgene (9 kb 5′ flanking) describes expression within the mouse SON that increased 1.6-fold after drinking 2% saline for 1 week and within the mouse ovary (Ang *et* *al.*, 1991). Tissue-specific expression in the brain has not been produced with chimeric transgenes composed of the VP promoter and heterologous reporter constructs (Murphy *et al.*, 1987; Russo *et al.*, 1988). One of these used 1.25 kb of promoter to drive expression of the large T-antigen and resulted in tumours of pancreas and anterior pituitary (Murphy *et al.*, 1987).

Rat OT transgene expression (with 360 bp of promoter) in mouse occurs in a tissue- and physiological (lactation)-specific way with an OT-VP minilocus (Young *et al.*, 1990*b*). These mice translate and transport to the pituitary the rat OT neurophysin coded for by the transgene, as well. In fact, the rat transgene neurophysin is found in the same granules as the endogenous neurophysin from the cell body to the posterior pituitary (Belenky *et al.*, 1992). A bovine OT transgene (4.2 kb with 600 bp 5′ flanking sequence), although not leading to appropriate brain expression, did recapitulate bovine testicular expression in the mouse (Ang *et al.*, 1991). These types of studies establish minimal sequences needed for expression and subsequent alterations in these sequences will permit functional dissection of the genes. However, these studies also emphasize that regulation of VP and OT expression resides both within and outside the proximal promoter region, preventing a simple, directed expression of a fusion gene to the PVN, SON or SCN, for example (above references and W. S. Young, unpublished work).

Expression of OT and VP in the central nervous system

It is well established that hyperosmolality and hypovolemia elevate levels of VP and that lactation elevates levels of OT in plasma. Once the coding sequences for the preprohormones were determined, the actual levels of the transcripts encoding VP and OT could be studied under different conditions through the use of nuclear run-on, Northern and hybridization histochemical (*in situ* hybridization) analyses. Initial studies *in vitro* confirmed that increased transcript levels accompany increased plasma peptide levels (Majzoub *et al.*, 1983; Burbach *et al.*, 1984, 1986; Sherman *et al.*, 1986; Zingg *et al.*, 1986). Interestingly, in the rat, VP and OT transcripts newly synthesized in response to hyperosmolality, pregnancy and lactation have increased poly(A) tails (Zingg *et al.*, 1988; Carter and Murphy, 1989*c*; Zing and Lefebvre, 1989). These changes are rapid (as early as 1 h after injections of intraperitoneal hypertonic saline) and occur in the absence of increased mRNA levels (Murphy and Carter, 1990; Carter and Murphy, 1991*b*), even in the SCN (Carter and Murphy, 1990*b*). The poly(A) tails of OT and VP transcripts also increase during development in the pig hypothalamus (Rehbein and Richter, 1990). Changes in poly(A) tail length in the SCN follow a circadian rhythm, with shorter transcripts present during the dark phase (Robinson *et al.*, 1988; Carter and Murphy, 1989*b*). The mechanism for the increase in poly(A) tail length is not known, but could result from the presumably more rapid elongation of cytoplasmic mRNA or by *de novo* synthesis of mRNA. In the rat SCN the shorter deadenylated transcripts are abundant in the nucleus during the dark phase (Carter and Murphy, 1991*a*), perhaps indicating that subsequent polyadenylation occurs during the light phase to increase mRNA stability and initiation of translation. Studies in other mammalian and non-mammalian systems have shown that polyadenylation can increase stability of mRNA and increase translational initiation (Jackson and Standart, 1990), but these potential effects need to be addressed directly for VP and OT. Increased initiation of translation would, of course, increase the efficacy of response to a new stimulus or stage of development. Interestingly, starvation has been shown recently to decrease poly(A) tail length in rats (Chooi *et al.*, 1992).

Hybridization histochemistry (Fig. 7) provides cellular resolution to the quantitation of mRNA levels. Rats that are dehydrated or given 2% saline chronically to drink show elevated levels of transcripts for VP and OT (Uhl *et al.*, 1985; McCabe *et al.*, 1986; Sherman *et al.*, 1986; Lightman and Young, 1987; Van Tol *et al.*, 1987) and between 4 and 10% of these responsive neurons project to the spinal cord or medullary vagal complex (Pretel and Piekut, 1989). Increases in heteronuclear RNA are detectable as soon as 30 min after an acute salt load using a VP intronic probe, before any change in VP mRNA is detectable (Herman *et al.*, 1991). This use of an intronic probe as well as the use of nuclear run-on assays (Murphy and Carter, 1990;

Fig. 7. Expression of OT in the PVN and SON of the rat hypothalamus

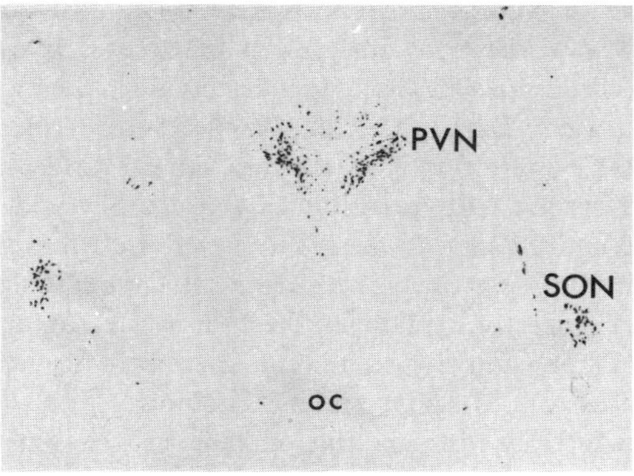

OT transcripts were detected using a digoxigenin-labelled riboprobe. The digoxigenin was detected using an antibody conjugated to alkaline phosphatase. oc, optic chiasm.

Herman *et al.*, 1991) reveal that transcription rates are increased during this hyperosmolar stimulus. The onset of the VP gene's responsiveness to hyperosmolality in the SON appears to develop between gestational days 19 and 21 (E19 and E21), after the onset of its expression (Reppert *et al.*, 1983). Humans that are hypovolemic prior to death also have elevated levels of VP transcripts in the SON (Rivkees *et al.*, 1989).

Of note is that numerous genes for other neuropeptides and proteins are also expressed in these nuclei and are similarly up-regulated in response to chronic hyperosmolality. These include carboxypeptidase H, corticotropin-releasing factor (CRF), cholecystokinin, dynorphin, galanin and tyrosine hydroxylase (Sherman *et al.*, 1986, 1988; Young, 1986, 1987*b*, 1990*a*; Lightman and Young, 1987; Bondy *et al.*, 1989*b*; Meister *et al.*, 1990) as well as one for a ribosomal RNA (Kawata *et al.*, 1988). The co-ordinate control of these genes for neurohormone production undoubtedly leads to finer regulation of axonal release in the posterior pituitary (e.g. Falke, 1988; Bondy and Gainer, 1989; Bondy *et al.*, 1989*a*), and perhaps regulates pituicytes as well. Interestingly, hyperosmolality also results in detectable co-expression of VP and OT in 1–2% of the same magnocellular neurons (Kiyama and Emson, 1990). Levels of transcripts encoding several proteins that may be involved in regulating gene expression are elevated in response to hyperosmolality: protein kinase C-ζ (Young, 1989), and the inhibitory and stimulatory α-subunits of the GTP-binding protein (Young *et al.*, 1987*a*). With regard to the latter, both the levels of cAMP and the stimulatory α-subunit of the GTP-binding protein are also elevated in the SON in response to hyperosmolality (Young *et al.*, 1987*a*; Carter and Murphy, 1989a). Meister *et al.* (1990) and Sherman *et al.* (1988) have examined in detail the relationships between mRNA and peptide levels after salt-loading and support the general rule that elevations in transcript levels lead to increased translation products. The level of c-*fos* is also elevated by chronic hyperosmolality (Sager *et al.*, 1988) and dehydration (Carter and Murphy, 1990*a*), as well as acutely after intraperitoneal hypertonic saline (Carter and Murphy, 1990*a*; Hamamura *et al.*, 1992). Furthermore, c-*fos*'s heterodimeric partner at the AP-1 DNA regulatory element, c-*jun*, is also up-regulated by dehydration and acute hypertonic saline (Carter and Murphy, 1990*a*) and at least part of their up-regulation is due to increased transcription (Carter and Murphy, 1990*a*). Of interest is that chronic ethanol ingestion by rats did not alter plasma VP and reduced levels of VP transcripts despite increased plasma osmolality (Ishizawa *et al.*, 1990).

As noted above, the Brattleboro rat has diabetes insipidus owing to a single base deletion in the second exon of the VP gene (Schmale and Richter, 1984). In this model of chronic hyperosmolality, mRNA levels for OT, cholecystokinin, dynorphin, galanin and tyrosine hydroxylase are increased (Young *et al.*, 1986, 1987*b*; Rökaeus *et al.*, 1988; Sherman *et al.*, 1988). Despite lower levels of VP transcripts (compared with the parent strain, Long-Evans), the Brattleboro rat can elevate them in response to chronic intermittent salt-loading (Sherman *et al.*, 1988). Future studies may indicate a common *cis*-acting element in those genes at which the hyperosmolar signal generated by hyperosmolality acts, but an examination of several genes that respond similarly did not reveal an obvious candidate. It seems more likely that hyperosmolality sets off a cascade of intracellular signals that activate the various responsive genes through different *cis*-acting elements.

The expression of genes within the magnocellular neurons of the PVN and SON during the oestrous cycle, pregnancy and lactation have been studied less extensively. Studies of extracted RNA from the rat SON or whole hypothalamus revealed 2–3-fold increases in OT and VP mRNA within the last 3 days of pregnancy which were maintained during lactation (Van Tol *et al.*, 1987; Zingg and Lefebvre, 1988). OT mRNA levels peak at 1.5–2-fold higher at oestrus than during other parts of the cycle (Van Tol *et al.*, 1987). Hybridization histochemical studies confirmed higher levels of OT transcripts in pregnant rats (McCabe *et al.*, 1986) and higher levels of OT, dynorphin and VP transcripts in lactating rats (Lightman and Young, 1987). Lactating mice also have higher levels of transcripts for OT (Young *et al.*, 1990*b*) and lactating rats show VP expression in about 17% of the OT neurons (Mezey and Kiss, 1991), perhaps to compensate for fluid loss.

Developmental studies have also been rather limited. VP transcripts are detected at E16 and E18 in the rat SON and PVN, respectively. OT transcripts are detected at E17 and E18, respectively (Reppert and Uhl, 1987; Laurent *et al.*, 1989). Development of VP expression in the PVN and SON of the Brattleboro rat lags behind that of the heterozygous littermate, suggesting that VP itself may be involved in a positive feedback loop (Van Tol *et al.*, 1986). In contrast, OT expression is higher in the Brattleboro rat throughout development into the adult (Young *et al.*, 1986; Van Tol *et al.*, 1990).

The effects of other physiological states or perturbations have been studied even more sporadically in the magnocellular neurons of the PVN and SON. As discussed above with regard to the OT promoter's ORE, ovariectomy prevents the increase in OT mRNA levels during sexual maturation of female rats (Miller *et al.*, 1989*a*). Conversely, 2 days or 2 months of oestradiol doubles OT mRNA levels in the SON (and magnocellular neurons of the anterior commissural nucleus) of ovariectomized rats (Chung *et al.*, 1991).

VP is also expressed in the parvocellular corticotropin-releasing factor neurons of the PVN after adrenalectomy (Tramu *et al.*, 1983; Kiss *et al.*, 1984; Sawchenko *et al.*, 1984; Wolfson *et al.*, 1985; Young *et al.*, 1986) and, probably, after stress (Lightman and Young, 1988). Molecular biological techniques, especially hybridization histochemistry, have confirmed the expression of VP in other parvocellular neurons of the brain. SCN transcript levels of VP have been reported to be highest in the morning (Uhl and Reppert, 1986) or late afternoon (Burbach *et al.*, 1988). The rhythm first appears at E21 (Reppert and Uhl, 1987) and is maintained in Brattleboro rats that do not translate their aberrant mRNA (Uhl and Reppert, 1986). Furthermore, rates of transcription are reduced during the dark phase (Carter and Murphy, 1991*a*). VP expression is also seen in the bed nucleus of the stria terminalis (BNST) and the medial amygdala, but not the locus coeruleus, through hybridization histochemistry (Young and Zoeller, 1987; Miller *et al.*, 1988; Urban *et al.*, 1990). VP mRNA levels in the BNST are stimulated by testosterone (Miller *et al.*, 1989*b*) but in the BNST and medial amygdala are unaffected by dexamethasone in the presence of normal testosterone levels (Urban *et al.*, 1991). The effects of testosterone on VP expression confirm the sex differences and steroid effects described earlier with immunohistochemical analyses of the BNST and its projections (van Leeuwen *et al.*, 1985; de Vries *et al.*, 1986).

Expression of OT and VP in the periphery

The synthesis of OT and VP in several peripheral tissues has been confirmed recently, although the role of these peptides in these tissues is not clear. The preponderance of evidence suggests that VP is not expressed in the anterior pituitary (Levy and Lightman, 1989; Murphy *et al.*, 1989; McCabe *et al.*, 1990; Mohr *et al.*, 1990; however, see Loh *et al.*, 1988; Terrier *et al.*, 1991). The presence of VP transcripts in the posterior pituitary of rats, especially after salt-loading, is more secure (Lehmann, 1988; Levy and Lightman, 1989; Murphy *et al.*, 1989; McCabe *et al.*, 1990; Mohr *et al.*, 1990). These transcripts originate in the PVN and SON as revealed by their loss after stalk transection, their presence in the median eminence, and the absence of primary transcripts in the posterior pituitary (Murphy *et al.*, 1989; Levy *et al.*, 1990; McCabe *et al.*, 1990; Mohr *et al.*, 1990, 1991). In addition, OT transcripts of hypothalamic origin have been detected in the posterior pituitary of male rats (Mohr *et al.*, 1991) and in the median eminence and posterior pituitary of lactating rats (Jirikowski *et al.*, 1990). The latter authors confirmed their *in situ* localization with polymerase chain reactions (PCR) and, although they also did not detect primary transcripts in the posterior pituitary, they did not report primary transcripts in the hypothalamus either. The poly(A) tails of the VP and OT transcripts shorten as they are transported to the posterior pituitary (Mohr *et al.*, 1991). Electron micrographic studies revealed the OT transcripts only in neurohypophysial axon terminals and not pituicytes (Jirikowski *et al.*, 1990). It is not known if they are co-localized with VP or OT. That these transcripts may be released and taken up by nerve terminals and/or retrogradely transported to the cell bodies for protein synthesis is a fascinating possibility. Indeed, a recent study demonstrated that when total hypothalamic mRNA or synthetic VP transcripts are injected into Brattleboro rat hypothalamus among the magnocellular axons, VP immunoreactivity appears in the PVN and SON, and later in the posterior pituitary with subsequent increased urine osmolality

(Jirikowski *et al.*, 1992). Previously, it would not have been considered possible to explain the co-expression of VP and OT within the same magnocellular neurons (Kiyama and Emson, 1990; Mezey and Kiss, 1991), especially during lactation, as the result of transport of transcripts to the posterior pituitary with subsequent re-uptake by axons. However, this explanation must now be considered! Also for consideration is why is there an accumulation of transcripts in the posterior pituitary if release from vesicles is postulated?

OT and VP transcripts and immunoreactivity have been detected in the human thymus (Geenen *et al.*, 1987) and rat adrenal gland (Foo *et al.*, 1991), placenta (Lefebvre *et al.*, 1992*a*) and uterus (Lefebvre *et al.*, 1992*b*). In the placenta, mRNA levels decrease from day 14 to day 21 of gestation without affecting peptide levels. OT expression occurs in the trophoblastic epithelium covering septa of the labyrinth and cytotrophoblastic elements and giant cells of the maternally derived basal zone of the placenta. More impressive is the massive increase in OT expression in the rat uterus in the epithelial cell layer of the endometrium. Transcript levels for OT increase more than 150-fold during pregnancy and fall rapidly after delivery (Lefebvre *et al.*, 1992*b*). Perhaps the OT gene's ORE is actually responsive in this tissue.

Expression within gonadal tissues has been studied in more detail. Although VP transcripts are not detected consistently in human, baboon, bull or mouse testes, rat testes do contain VP transcripts (Ivell *et al.*, 1990*a*; Ang *et al.*, 1991). Interestingly, the great majority of the rat testicular transcripts exclude the first exon present in the hypothalamic version, and contain sequence located further upstream (Foo *et al.*, 1991; Lefebvre and Zingg, 1991). These transcripts appear by the 40th day postnatally and efforts to alter the levels of these transcripts or detect any translation have failed (Foo *et al.*, 1991). The function of these aberrant transcripts is unknown but may be provided through further study of the 5′ sequence. Vasopressin transcripts have been detected in the cow ovary at very low levels (Ivell and Richter, 1984*a*).

OT transcripts are detected in rat and human testes (by PCR) and in bovine Sertoli cells of the testis, but not in the mouse testis (Ivell *et al.*, 1990*a*; Ang *et al.*, 1991; Foo *et al.*, 1991). The structures of these transcripts appear to be the same as the hypothalamic ones. Within ovarian tissues, hypothalamic-type OT transcripts are present in the corpus luteum of the baboon, human, sheep and cow (Ivell and Richter, 1984*a*; Jones and Flint, 1988; Ivell *et al.*, 1990*a*, *b*). OT expression is highly up-regulated after ovulation (Ivell *et al.*, 1985, 1990*b*; Jones and Flint, 1988). Cells from bovine follicles that had contained high levels of oestradiol, insulin and insulin-like growth factor-I are able to stimulate OT biosynthesis (Holtorf *et al.*, 1989), although oestradiol does not directly affect OT gene expression in isolated cell preparations (Furuya *et al.*, 1990). This is probably not unexpected given the divergence of the ORE in the cow (see above). However, evidence has been gathered for potential interactions between a bovine luteal protein, whose concentration varies through the oestrus cycle, and another member of the steroid superfamily, COUP (Walther *et al.*, 1991). OT transcripts are also present in bovine granulosa cells and increase after the luteinizing hormone surge (Voss and Fortune, 1992). Furthermore, isolated granulosa cells *in vitro* demonstrate increased OT mRNA levels after exposure to luteinizing hormone (Voss and Fortune, 1992).

Future directions

The creation of a cell line derived from OT or VP neurons would be a major step forward in our efforts to understand the expression of these genes. The effects of various neurotransmitters and media tonicity could be much more easily studied. Although a neuroblastoma cell line has been found to express OT at low levels and has been used to locate *cis*-acting elements in the OT promoter, it was not reported whether or not the *trans*-acting factors that bind to those elements are cell line- or hypothalamic magnocellular tissue-specific (Richard and Zingg, 1991 *b*). Obviously, a cell line derived from OT or VP neurons would greatly simplify the study of *cis*-acting elements and purification of their DNA-binding proteins. Failing this, greater use of primary hypothalamic cultures, preferably enriched for OT or VP neurons, or slice explants (Wray *et al.*, 1991) should be pursued.

As noted above, transgenic mouse experiments are still in their infancy. After a rat OT transgene sequence that is expressed appropriately was defined (Young *et al.*, 1990*b*), studies were initiated to examine the influence of various mutations (e.g. of the ORE) on the transgene's expression. It will be of interest to learn how the knowledge that has been gained through *in vitro* experiments relates to *in vivo* expression. Do any of the *cis*-acting elements that have been defined so far also regulate cell specificity? Similar experiments with VP transgenes await elucidation of the necessary sequence for proper CNS expression.

Much knowledge is lacking on how various behaviours and perturbations regulate VP and OT expression and through which neuronal pathways. The measurement of expression of 'third messengers' such as c-*fos* should help define components of these pathways. Some components of the osmoregulatory circuit, including the subfornical organ and organum vasculosum of the lamina terminalis, have been studied in this way (Hamamura *et al.*, 1992). The significances of the identified *cis*-acting elements are still largely unclear. Are they important for regulating OT and VP expression in the CNS in any situations, or are they for regulation of expression peripherally? There is still much to learn about the regulation of peptide biosynthesis (largely neglected in this review). What enzymes are involved in post-translational processing and how are they regulated? What are the precise roles of the co-localized neuropeptides?

Receptors for vasopressin, the V1a (Morel *et al.*, 1992) and V2 (Birnbaumer *et al.*, 1992; Lolait *et al.*, 1992) subtypes, and for oxytocin (Kimura *et al.*, 1992) have been cloned and investigations can proceed to determine their role in disease and normal physiology. For example, a defective V2 receptor is responsible for an inherited form of nephrogenic diabetes insipidus (Lolait *et al.*, 1992; Rosenthal *et al.*, 1992). It will be interesting to learn, for example, how many receptor subtypes actually exist, how and where their syntheses regulated, what is the relationship between transcript and binding levels, and which CNS pathways are activated through these receptors. There appears to be an excellent correlation in the CNS, liver and kidney between VP ligand subtype (V1a and V2) binding and hybridization histochemical localization of the receptor transcripts (Ostrowski *et al.*, 1992). It will also be fascinating to learn how the receptors have evolved in relation to their cognate peptides, and how the hormones interact with their receptors.

The next several years will be extremely enlightening as these and other questions are studied. The VP and OT genes have served as model systems for the study of preprohormone processing, transport and release. This neuroendocrine system has exemplified the close correlation between mRNA levels and peptide release. It is likely that, despite the relative paucity of CNS VP or OT cells and current lack of a satisfactory tissue culture system which has hindered our understanding of the regulation of the genes at the molecular level, continued efforts will demonstrate these genes' importance for understanding CNS gene regulation in general.

I gratefully appreciate the critical reading and unpublished data by Dr. J. P. H. Burbach. I am also grateful to Dr. J. Battey for supplying unpublished human OT and VP promoter sequences.

References

Adan, R. (1992) Oxytocin and vasopressin genes: transcriptional regulation and signal transduction. Thesis, University of Utrecht, The Netherlands

Adan, R. A. H., Walther, N., Cox, J. J., Ivell, R. and Burbach, J. P. H. (1991) Comparison of the estrogen responsiveness of the rat and bovine oxytocin gene promoters. Biochem. Biophys. Res. Commun. **175**, 117–122

Adan, R. A. H., Cox, J. J., van Kats, J. P. and Burbach, J. P. H. (1992) Thyroid hormone regulates the oxytocin gene. J. Biol. Chem. **267**, 3771–3777

Ang, H.-L., Ungefroren, H., De Bree, F., Foo, N.-C., Carter, D., Burbach, J. P., Ivell, R. and Murphy, D. (1991) Testicular oxytocin gene expression in seminiferous tubules of cattle and transgenic mice. Endocrinology **128**, 2110–2117

Bahnsen, U., Oosting, P., Swaab, D. F., Nahke, P., Richter, D. and Schmale, H. (1992) A missense mutation in the vasopressin-neurophysin precursor gene cosegregates with human autosomal dominant neurohypophyseal diabetes insipidus. EMBO J. **11**, 19–23

Bargmann, W. and Scharrer, E. (1951) The origin of the posterior pituitary hormones. Am. Sci. **39**, 255–259

Belenky, M., Castell, M., Young, W. S. I., Gainer, H. and Cohen, S. (1992) Ultrastructural immunolocalization of rat oxytocin-neurophysin in transgenic mice expressing the rat oxytocin gene. Mol. Brain Res. **583**, 279–286

Berecek, K. H. and Swords, B. H. (1990) Central role for vasopressin in cardiovascular regulation and the pathogenesis of hypertension. Hypertension **16**, 213–224

Birnbaumer, M., Seibold, A., Gilbert, S., Ishido, M., Barberis, C., Antaramian, A., Brabet, P. and Rosenthal, W. (1992) Molecular cloning of the receptor for human antidiuretic hormone. Nature (London) **357**, 333–335

Bondy, C. A. and Gainer, H. (1989) Corticotropin-releasing hormone stimulates neurohypophysial hormone release through an interaction with the intermediate lobe of the pituitary. J. Neuroendocrinol. **1**, 5–8

Bondy, C. A., Jensen, R. T., Brady, L. S. and Gainer, H. (1989*a*). Cholecystokinin evokes secretion of oxytocin and vasopressin from rat neural lobe independent of external calcium. Proc. Natl. Acad. Sci. U.S.A. **86**, 5198–5201

Bondy, C. A., Whitnall, M. H. and Brady, L. S. (1989*b*) Regulation of carboxypeptidase H gene expression in magnocellular neurons: response to osmotic stimulation. Mol. Endocrinol. **3**, 2086–2092

Bradley, D. J., Young, W. S. I. and Weinberger, C. (1989) Differential expression of alpha and beta thyroid hormone receptor genes in rat brain and pituitary. Proc. Natl. Acad. Sci. U.S.A. **86**, 7250–7254

Bradley, D. J., Towle, H. C. and Young, W. S. I. (1992) Spatial and temporal expression of α and β thyroid hormone receptor mRNAs, including the β2 subtype, in the developing mammalian nervous system. J. Neurosci. **12**, 2288–2302

Braverman, L. E., Mancini, J. P. and McGoldrick, D. M. (1965) Hereditary diabetes insipidus: a system disease. Ann. Intern. Med. **63**, 503–508

Brownstein, M. J., Russell, J. T. and Gainer, H. (1980) Synthesis, transport and release of posterior pituitary hormones. Science **207**, 373–387

Burbach, J. P. H., De Hoop, M. J., Schmale, H., Richter, D., De Kloet, E. R., Ten Haaf, J. A. and De Wied, D. (1984) Differential responses to osmotic stress of vasopressin-neurophysin mRNA in hypothalamic nuclei. Neuroendocrinology **39**, 582–584

Burbach, J. P. H., Van Tol, H. H. M., Bakkus, M. H. C., Schmale, H. and Ivell, R. (1986) Quantitations of vasopressin mRNA and oxytocin mRNA in hypothalamic nuclei by Southern hybridization assays. J. Neurochem. **47**, 1814–1821

Burbach, J. P. H., Liu, B., Voorhuis, T. A. M. and Van Tol, H. H. M. (1988) Diurnal variation in vasopressin and oxytocin messenger RNAs in hypothalamic nuclei of the rat. Mol. Brain Res. **4**, 157–160

Burbach, J. P. H., Adnan, R. A. H., van Tol, H. H. M., Verbeeck, M. A. E., Axelson, J. F., van Leeuwen, F. W., Beekman, J. M. and Geert, A. B. (1990) Regulation of the rat oxytocin gene by estradiol. Examination of promoter activity in transfected cells and of mRNA and peptide levels in the hypothalamo-neurohypophyseal system. J. Neuroendocrinol. **2**, 633–639

Carter, D. A. and Murphy, D. (1989*a*). Cyclic nucleotide dynamics in the rat hypothalamus during osmotic stimulation: *in vivo* and *in vitro* studies. Brain Res. **487**, 350–356

Carter, D. A. and Murphy, D. (1989*b*) Diurnal rhythm of vasopressin mRNA species in the rat suprachiasmatic nucleus: independence of neuroendocrine modulation and maintenance in explant culture. Mol. Brain Res. **6**, 233–239

Carter, D. A. and Murphy, D. (1989*c*) Independent regulation of neuropeptide mRNA level and poly(A) tail length. J. Biol. Chem. **264**, 6601–6603

Carter, D. A. and Murphy, D. (1990*a*). Regulation of c-*fos* and c-*jun* expression in the rat supraoptic nucleus. Cell. Mol. Neurobiol. **10**, 435–444

Carter, D. A. and Murphy, D. (1990*b*) Vasopressin mRNA in parvocellular neurons of the rat suprachiasmatic nucleus exhibits increased poly(A) tail length following water deprivation. Neurosci. Lett. **109**, 180–185

Carter, D. A. and Murphy, D. (1991*a*) Nuclear mechanisms mediate rhythmic changes in vasopressin mRNA expression in the rat suprachiasmatic nucleus. Mol. Brain Res. **12**, 315–321

Carter, D. A. and Murphy, D. (1991*b*) Rapid changes in poly(A) tail length of vasopressin and oxytocin mRNAs form a common early component of neurohypophyseal peptide gene activation following physiological stimulation. Neuroendocrinology **53**, 1–6

Chen, L., Rose, J. P., Breslow, E., Yang, D., Chang, W.-R., Furey, W. F. J., Sax, M. and Wang, B. C. (1991) Crystal structure of a bovine neurophysin II dipeptide complex at 2.8 Å determined from the single-wavelength anomalous scattering signal of an incorporated iodine atom. Proc. Natl. Acad. Sci. U.S.A. **88**, 4240–4244

Chibbar, R., Toma, J. G., Mitchell, B. F. and Miller, F. D. (1990) Regulation of neural oxytocin gene expression by gonadal steroids in pubertal rats. Mol. Endocrinol. **4**, 2030–2038

Chooi, K. F., Carter, D. A. and Murphy, D. (1992) Decrease in hypothalamic vasopressin mRNA poly(A) tail length following physiological stimulation. Cell. Mol. Neurobiol. **12**, 557–567

Chung, S. K., McCabe, J. T. and Pfaff, D. W. (1991) Estrogen influences on oxytocin mRNA expression in preoptic and anterior hypothalamic regions studied by *in situ* hybridization. J. Comp. Neurol. **307**, 281–295

Dale, H. H. (1909) The actions of extracts of the pituitary body. Biochem. J. 4, 427–447

de Kloet, E. R., Voorhuis, D. A. M., Boschma, Y. and Elands, J. (1986) Estradiol modulate density of putative 'oxytocin receptors' in discrete rat brain regions. Neuroendocrinology 44, 415–421

de Vries, G. J., Duetz, W., Buijs, R. M., Heerkhuize, J. and Vreeburg, J. T. M. (1986) Effects of androgens and estrogens on the vasopressin and oxytocin innervation of the adult brain. Brain Res. 399, 296–302

de Wied, D. (1971) Long term effect of vasopressin on the maintenance of a conditioned avoidance response in rats. Nature (London) 232, 58–60

Dombrowski, B. A., Mathias, S. L., Nanthakumar, E., Scott, A. F. and Kazazian, H. H. J. (1991) Isolation of an active human transposable element. Science 254, 1805–1808

Du Vigneaud, V. (1954) Hormones of the posterior pituitary gland: oxytocin and vasopressin. Harvey Lect. 50, 1–26

Evans, A. P., Ivell, R., Van Leeuwen, F. W., Corner, M. and Burbach, J. P. H. (1991) Sequence analysis of vasopressin mRNAs in the homozygous Brattleboro rat which codes for glycopeptide immunoreactivity. Soc. Neurosci. Abstr. 17, 380

Falke, N. (1988) Dynorphin (1–8) inhibits stimulated release of oxytocin but not vasopressin from isolated neurosecretory endings of the rat neurohypophysis. Neuropeptides 11, 163–167

Figueroa, J., Morley, S. D., Heierhorst, J., Krentler, C., Lederis, K. and Richter, D. (1989) Two isotocin genes are present in the white sucker *Catostomus commersoni* lacking introns in their protein coding regions. EMBO J. 8, 2873–2877

Foo, N.-C., Carter, D., Murphy, D. and Ivell, R. (1991) Vasopressin and oxytocin gene expression in rat testis. Endocrinology 128, 2118–2128

Furuya, K., McArdle, C. A. and Ivell, R. (1990) The regulation of oxytocin gene expression in early bovine luteal cells. Mol. Cell. Endocrinol. 70, 81–88

Gainer, H., Altstein, M., Whitnall, M. H. and Wray, S. (1988) The biosynthesis of oxytocin and vasopressin. In The Physiology of Reproduction (Knobil, E. and Neill, J., eds.), pp. 2265–2282, Raven Press, New York

Geenen, V., Legros, J. J., Franchimont, P., Defresne, M.-P., Boniver, J., Ivell, R. and Richter, D. (1987) The thymus as a neuroendocrine organ. Synthesis of vasopressin and oxytocin in human thymic epithelium. Ann. N.Y. Acad. Sci. 496, 56–66

Graff, J. M., Stumpo, D. J. and Blackshear, P. J. (1989) Molecular cloning, sequence, and expression of a cDNA encoding the chicken myristoylated alanine-rich C kinase substrate (MARCKS). Mol. Endocrinol. 3, 1903–1906

Habener, J. F., Cwikel, B. J., Hermann, H., Hammer, R. E., Palmiter, R. D. and Brinster, R. L. (1989) Metallothionein-vasopressin fusion gene expression in transgenic mice. J. Biol. Chem. 264, 18844–18852

Hamamura, M., Nunez, D. J. R., Leng, G., Emson, P. C. and Kiyama, H. (1992) c-*fos* may code for a common transcription factor within the hypothalamic neural circuits involved in osmoregulation. Brain Res. 572, 42–51

Hara, Y., Battey, J. and Gainer, H. (1990) Structure of mouse vasopressin and oxytocin genes. Mol. Brain Res. 8, 319–324

Hatton, G. I. (1990) Emerging concepts of structure–function dynamics in adult brain: the hypothalamo-neurohypophysial system. Progr. Neurobiol. 34, 437–504

He, X., Treacy, M., Simmons, D. M., Ingraham, H. A., Swanson, L. W. and Rosenfeld, M. G. (1989) Expression of a large family of POU-domain regulatory genes in mammalian brain development. Nature (London) 340, 35–42

Herman, J. P., Schäfer, M. K.-H., Watson, S. J. and Sherman, T. G. (1991) *In situ* hybridization analysis of arginine vasopressin gene transcription using intron-specific probes. Mol. Endocrinol. 5, 1447–1456

Holtorf, A.-P., Furuya, K., Ivell, R. and McArdle, C. A. (1989) Oxytocin production and oxytocin messenger ribonucleic acid levels in bovine granulosa cells are regulated by insulin and insulin-like growth factor-I: dependence on developmental status of the ovarian follicle. Endocrinology 125, 2612–2620

Hyodo, S., Kato, Y., Ono, M. and Urano, A. (1991) Cloning and sequence analyses of cDNAs encoding vasotocin and isotocin precursors of chum salmon, *Oncorhynchus keta*: evolutionary relationships of neurohypophysial hormone precursors. J. Comp. Physiol. B 160, 601–608

Ishizawa, H., Dave, J. R., Liu, L.-I., Tabakoff, B. and Hoffman, P. L. (1990) Hypothalamic vasopressin mRNA levels in mice are decreased after chronic ethanol ingestion. Eur. J. Pharmacol. 189, 119–127

Ito, M., Mori, Y., Oiso, Y. and Saito, H. (1991) A single base substitution in the coding region for neurophysin II associated with familial central diabetes insipidus. J. Clin. Invest. **87**, 725–728

Ivell, R. and Richter, D. (1984*a*) The gene for the hypothalamic peptide hormone oxytocin is highly expressed in the bovine corpus luteum: biosynthesis, structure and sequence analysis. EMBO J. **3**, 2351–2354

Ivell, R. and Richter, D. (1984*b*) Structure and comparison of the oxytocin and vasopressin genes from rat. Proc. Natl. Acad. Sci. U.S.A. **81**, 2006–2010

Ivell, R., Brackett, K. H., Fields, M. J. and Richter, D. (1985) Ovulation triggers oxytocin gene expression in the bovine ovary. FEBS Lett. **190**, 263–267

Ivell, R., Schmale, H., Krisch, B., Nahke, P. and Richter, D. (1986) Expression of a mutant vasopressin gene: differential polyadenylation and read-through of the mRNA 3′ end in a frameshift. EMBO J. **5**, 971–977

Ivell, R., Furuya, K., Brackmann, B., Dawood, Y. and Khan-Dawood, F. (1990*a*) Expression of the oxytocin and vasopressin genes in human and baboon gonadal tissues. Endocrinology **127**, 2990–2996

Ivell, R., Hunt, N., Abend, N., Brackman, B., Nollmeyer, D., Lamsa, J. C. and McCracken, J. A. (1990*b*) Structure and ovarian expression of the oxytocin gene in sheep. Reprod. Fertil. Res. **2**, 703–711

Jackson, R. J. and Standart, N. (1990) Do the poly(A) tail and 3′ untranslated region control mRNA translation? Cell **62**, 15–24

Jard, S. and Jamison, R. (eds.) (1991) Vasopressin, John Libbey Eurotext, Paris

Jirikowski, G. F., Sanna, P. P. and Bloom, F. E. (1990) mRNA coding for oxytocin is present in axons of the hypothalamo-neurohypophysial tract. Proc. Natl. Acad. Sci. U.S.A. **87**, 7400–7404

Jirikowski, G. F., Caldwell, J. D., Häussler, H. U. and Pedersen, C. A. (1991) Mating alters topography and content of oxytocin immunoreactivity in male mouse brain. Cell Tissue Res. **266**, 399–403

Jirikowski, G. F., Sanna, P. P., Maciejewski-Lenoir, D. and Bloom, F. E. (1992) Reversal of diabetes insipidus in Brattleboro rats: intrahypothalamic injection of vasopressin mRNA. Science **255**, 996–998

Johnson, A. E., Coirini, H., Insel, T. R. and McEwen, B. S. (1991) The regulation of oxytocin receptor binding in the ventromedial hypothalamic nucleus by testosterone and its metabolites. Endocrinology **128**, 891–896

Jones, D. S. C. and Flint, A. P. F. (1988) Concentrations of oxytocin-neurophysin prohormone mRNA in corpora lutea of sheep during the oestrus cycle and in early pregnancy. J. Endocrinol. **117**, 409–414

Kawata, M., McCabe, J. T., Harrington, C., Chikaraishi, D. and Pfaff, D. W. (1988) *In situ* hybridization analysis of osmotic stimulus-induced changes in ribosomal RNA in rat supraoptic nucleus. J. Comp. Neurol. **270**, 528–536

Kimura, T., Tanizawa, O., Mori, K., Brownstein, M. J. and Okayama, H. (1992) Structure and expression of a human oxytocin receptor. Nature (London) **356**, 526–529

Kiss, J. Z., Mezey, É. and Skirboll, L. (1984) Corticotropin-releasing factor-immunoreactive neurons of the paraventricular nucleus become vasopressin positive after adrenalectomy. Proc. Natl. Acad. Sci. U.S.A. **81**, 1854–1858

Kiyama, H. and Emson, P. C. (1990) Evidence for the co-expression of oxytocin and vasopressin messenger ribonucleic acids in magnocellular neurosecretory cells: simultaneous demonstration of two neurophysin messenger ribonucleic acids by hybridization histochemistry. J. Neuroendocrinol. **2**, 257–260

Land, H., Schutz, G., Schmale, H. and Richter, D. (1982) Nucleotide sequence of cloned cDNA encoding bovine arginine vasopressin-neurophysin II precursor. Nature (London) **295**, 299–303

Land, H., Grez, M., Ruppert, S., Schmale, H., Rehbein, M., Richter, D. and Schutz, G. (1983) Deduced amino acid sequence from the bovine oxytocin-neurophysin I precursor of cDNA. Nature (London) **302**, 342–344

Laurent, F. M., Hindelang, C., Klein, M. J., Stoekel, M. E. and Felix, J. M. (1989) Expression of the oxytocin and vasopressin genes in the rat hypothalamus during development: an *in situ* hybridization study. Dev. Brain Res. **46**, 145–154

Le Moine, C. and Young, W. S., III (1992) *RHS2*, a POU domain-containing gene, and its expression in developing and adult rat. Proc. Natl. Acad. Sci. U.S.A. **89**, 3285–3289

Lefebvre, D. L. and Zingg, H. H. (1991) Novel vasopressin gene-related transcripts in rat testis. Mol. Endocrinol. **5**, 645–652

Lefebvre, D. L., Giaid, A. and Zingg, H. H. (1992*a*) Expression of the oxytocin gene in the rat placenta. Endocrinology **130**, 1185–1192

Lefebvre, D. L., Giaid, A., Bennett, H., Lariviere, R. and Zingg, H. H. (1992*b*) Oxytocin gene expression in rat uterus. Science **256**, 1553–1555

Lehmann, E. (1988) Untersuchungen zur expression des vasopressingens in der ratte. Thesis, University of Heidelberg

Leng, G. and Bicknell, R. J. (1986) The neurohypophysis. In Neuroendocrinology (Lightman, S. L. and Everitt, B. J., eds.), pp. 177–206, Blackwell, Boston

Levy, A. and Lightman, S. L. (1989) Posterior pituitary vasopressin mRNA in slat-loaded rats: effects of pituitary stalk transection. J. Physiol. **418**, 140P

Levy, A., Lightman, S. L., Carter, D. A. and Murphy, D. (1990) The origin and regulation of posterior pituitary vasopressin ribonucleic acid in osmotically stimulated rates. J. Neuroendocrinol. **2**, 329–334

Lewin, B. (ed.) (1990) Genes IV, pp. 636–638, Oxford University Press

Lightman, S. L. and Young, W. S., III (1987) Vasopressin, oxytocin, dynorphin, enkaphalin, and corticotrophin releasing factor mRNA stimulation in the rat. J. Physiol. (London) **394**, 23–39

Lightman, S. L. and Young, W. S., III (1988) Corticotrophin-releasing factor, vasopressin and pro-opiomelanocortin mRNA responses to stress and opiates in the rat. J. Physiol. (London) **403**, 511–523

Lipkin, S. M., Nelson, C. A., Glass, C. K. and Rosenfeld, M. G. (1992) A negative retinoic acid response element in the rat oxytocin promoter restricts transcriptional stimulation by heterologous transactivation domains. Proc. Natl. Acad. Sci. U.S.A. **89**, 1209–1213

Loh, Y. P., Castro, M. G., Zeng, F.-J. and Patel-Vaidya, U. (1988) Presence of pro-vasopressin mRNA, neurophysin and arginine vasopressin in mouse anterior pituitary cells and AtT-20 corticotrophic tumour cell line. J. Mol. Endocrinol. **1**, 39–48

Lolait, S. J., O'Carroll, A.-M., McBride, O. W., Konig, M., Morel, A. and Brownstein, M. J. (1992) Cloning and characterization of a vasopressin V2 receptor: chromosomal localization of gene suggests link to hereditary nephrogenic diabetes insipidus. Nature (London) **357**, 336–339

Majzoub, J. A., Rich, A., van Boom, J. and Habener, J. F. (1983) Vasopressin and oxytocin mRNA regulation in the rat assessed by hybridization with synthetic oligonucleotides. J. Biol. Chem. **258**, 14061–14064

McCabe, J. T., Morrell, J. I. and Pfaff, D. W. (1986) Measurement of expression of the vasopressin and oxytocin genes in single neurons by *in situ* hybridization. In Neuroendocrine Molecular Biology (Fink, G., Harmar, A. J. and McKerns, K. W., eds.), pp. 219–229, Plenum, New York

McCabe, J. T., Lehmann, E., Chastrette, N., Hänze, J., Lang, R. E., Ganten, D. and Pfaff, D. W. (1990) Detection of vasopressin mRNA in the neurointermediate lobe of the rat pituitary. Mol. Brain Res. **8**, 325–329

Meister, B., Cortés, R., Villar, M. J., Schalling, M. and Hökfelt, T. (1990) Peptides and transmitter enzymes in hypothalamic magnocellular neurons after administration of hyperosmotic stimuli: comparison between messenger RNA and petide/protein levels. Cell Tissue Res. **260**, 279–297

Mezey, É. and Kiss, J. Z. (1991) Coexpression of vasopressin and oxytocin in hypothalamic supraoptic neurons of lactating rats. Endocrinology **129**, 1814–1820

Miller, M. A., Zoeller, R. T. and Dorsa, D. M. (1988) Detection of vasopressin messenger RNA in cells within the bed nucleus of the stria terminalis by *in situ* hybridization histochemistry. Neurosci. Lett. **94**, 264–268

Miller, F. D., Ozimek, G., Milner, R. J. and Bloom, F. E. (1989*a*) Regulation of neuronal oxytocin mRNA by ovarian steroids in the mature and developing hypothalamus. Proc. Natl. Acad. Sci. U.S.A. **86**, 2468–2472

Miller, M. A., Urban, J. H. and Dorsa, D. M. (1989*b*) Steroid dependency of vasopressin neurons in the bed nucleus of the stria terminalis by *in situ* hybridization. Endocrinology **125**, 2335–2340

Mohr, E. and Richter, D. (1990) Sequence analysis of the promoter region of the rat vasopressin gene. FEBS Lett. **260**, 305–308

Mohr, E. and Schmitz, E. (1991) Functional characterization of estrogen and glucocorticoid responsive elements in the rat oxytocin gene. Mol. Brain Res. **9**, 293–298

Mohr, E., Schmitz, E. and Richter, D. (1988) A single rat genomic DNA fragment encodes both the oxytocin and vasopressin genes separated by 11 kilobases and oriented in opposite transcriptional directions. Biochimie **70**, 70–74

Mohr, E., Zhou, A., Thorn, N. A. and Richter, D. (1990) Rats with physically disconnected hypothalamo-pituitary tracts no longer contain vasopressin-oxytocin gene transcripts in the posterior pituitary lobe. FEBS Lett. **263**, 332–336

Mohr, E., Fehr, S. and Richter, D. (1991) Axonal transport of neuropeptide encoding mRNAs within the hypothalamo-hypophyseal tract of rats. Eur. Mol. Biochem. Org. J. **10**, 2419–2424

Morel, A., O'Carroll, A.-M., Brownstein, M. J. and Lolait, S. J. (1992) Molecular cloning and expression of a rat V1a arginine vasopressin receptor. Nature (London) **356**, 523–526

Morley, S. D., Schönrock, C., Heierhorst, J., Figueroa, J., Lederis, K. and Richter, D. (1990) Vasotocin genes of the teleost fish *Catostomus commersoni*: Gene structure exon–intron boundary, and hormone precursor organization. Biochemistry **29**, 2506–2511

Murphy, D. and Carter, D. (1990) Vasopressin gene expression in the rodent hypothalmus: transcriptional and posttranslational responses to physiological stimulation. Mol. Endocrinol. **4**, 1051–1059

Murphy, D., Bishop, A., Rindi, G., Murphy, M. N., Stamp, G. W. H., Hanson, J., Polak, J. M. and Hogan, B. (1987) Mice transgenic for a vasopressin–SV40 hybrid oncogene develops tumors of the endocrine pancreas and the anterior pituitary. A possible model for human multiple endocrine neoplasia type 1. Am. J. Pathol. **129**, 552–565

Murphy, D., Levy, A., Lightman, S. and Carter, D. (1989) Vasopressin RNA in the neural lobe of the pituitary: dramatic accumulation in response to salt loading. Proc. Natl. Acad. Sci. U.S.A. **86**, 9002–9005

Nojiri, H., Ishida, I., Miyashita, E., Sato, M., Urano, A. and Deguchi, T. (1987) Cloning and sequence analysis of cDNAs for neurohypophysial hormones vasotocin and mesotocin for the hypothalamus of toad, *Bufo japonicus*. Proc. Natl. Acad. Sci. U.S.A. **84**, 3043–3046

Oeding, P., Schilling, K. and Schmale, H. (1990) Vasopressin expression in cultured neurons is stimulated by cyclic AMP. J. Neuroendocrinol. **2**, 859–865

Oliver, G. and Schäfer, E. A. (1985) On the physiological action of extracts of pituitary body and certain other glandular organs. J. Physiol. (London) **18**, 277–279

Ostrowski, N. L., Lolait, S. J., Bradley, D. J., O'Carroll, A.-M., Brownstein, M. J. and Young, W. S., III (1992) Distribution of V1a and V2 vasopressin receptor messenger ribonucleic acids in rat liver, kidney, pituitary and brain. Endocrinology **131**, 533–536

Ott, I. and Scott, J. C. (1911) The action of the infundibulin upon the mammary secretion. Proc. Soc. Exp. Biol. Med. **8**, 48–49

Pardy, K., Adan, R. A. H., Carter, D. A., Seah, V., Burbach, J. P. H. and Murphy, D. (1992) The identification of a *cis*-acting element involved in cyclic 3′,5′-adenosine monophosphate regulation of bovine vasopressin gene expression. J. Biol. Chem. **267**, 21746–21752

Pretel, S. and Piekut, D. T. (1989) Mediation of changes in paraventricular vasopressin and oxytocin mRNA content to the medullary vagal complex and spinal cord of the rat. J. Chem. Neuroanat. **2**, 327–334

Reeves, W. B. and Andreoli, T. E. (1992) The posterior pituitary and water metabolism. In Williams Textbook of Endocrinology (Wilson, J. D. and Foster, D. W., eds.), pp. 311–356, W. B. Saunders, Philadelphia

Rehbein, M. and Richter, D. (1990) Poly(A) tail length of oxytocin- and vasopressin-encoding mRNAs increases during development in the porcine hypothalamus. J. Mol. Endocrinol. **4**, 151–158

Rehbein, M., Hillers, M., Mohr, E., Ivell, R., Morley, S., Schmale, H. and Richter, D. (1986) The neurohypophyseal hormones vasopressin and oxytocin: precursor structure, synthesis, and regulation. Biol. Chem. Hoppe-Seyler **367**, 695–704

Renaud, L. P. and Bourque, C. W. (1991) Neurophysiology and neuropharmacology of hypothalamic magnocellular neurons secreting vasopressin and oxytocin. Prog. Neurobiol. **36**, 131–169

Repaske, D. R., Phillips, J. A. I., Kirby, L. T., Tze, W. J., D'Ercole, A. J. and Battey, J. (1990) Molecular analysis of autosomal dominant neurohypophyseal diabetes insipidus. J. Clin. Endocrinol. Metab. **70**,

Reppert, S. M. and Uhl, G. R. (1987) Vasopressin messenger ribunucleic acid in supraoptic and suprachiasmatic nuclei: appearance and circadian regulation during development. Endocrinology **120**, 2483–2487

Reppert, S. M., Schwartz, W. J., Artman, H. G. and Fisher, D. A. (1983) Comparison of the temporal profiles of vasopressin and oxytocin in the cerebrospinal fluid of the cat, monkey and rat. Brain Res. **261**, 341–345

Richard, S. and Zingg, H. (1990) The human oxytocin gene promoter is regulated by estrogens. J. Biol. Chem. **265**, 6098–6103

Richard, S. and Zingg, H. H. (1991*a*) Identification of a retinoic acid response element in the human oxytocin gene promoter which overlaps with an estrogen response element. Soc. Neurosci. Abstra. **17**, 1286

Richard, S. and Zingg, H. H. (1991*b*) Identification of *cis*-acting regulatory elements in the human oxytocin gene promoter. Mol. Cell. Neurosci. **2**, 501–510

Rivkees, S. A., Chaar, M. R., Hanley, D. F., Maxwell, M., Reppert, S. M. and Uhl, G. R. (1989) Localization and regulation of vasopressin mRNA in human neurons. Synapse **3**, 246–254

Robinson, A. G. (1975) Isolation, assay, and secretion of individual human neurophysins. J. Clin. Invest. **55**, 360–367

Robinson, B. G., Frim, D. M., Schwartz, W. J. and Majzoub, J. A. (1988) Vasopressin mRNA in the suprachiasmatic nuclei: daily regulation of polyadenylate tail length. Science **241**, 342–344

Rökaeus, Å., Young, W. S., III and Mezey, É. (1988) Galanin coexists with vasopressin in the normal rat hypothalamus and galanin's synthesis is increased in the Brattleboro (diabetes insipidus) rat. Neurosci. Lett. **90**, 45–50

Rosenthal, W., Seibold, A., Antaramian, A., Lonergan, M., Arthus, M.-F., Hendy, G. N., Birnbaumer, M. and Bichet, D. G. (1992) Molecular identification of the gene responsible for congenital nephrogenic diabetes insipidus. Nature (London) **359**, 233–235

Russell, J. T., Brownstein, M. J. and Gainer, H. (1981) Biosynthesis of neurohypophyseal polypeptides: the order of peptide components in pro-pressophysin and pro-oxyphysin. Neuropeptides **2**, 59–65

Russo, A. F., Crenshaw, E. B. I., Lira, S. A., Simmons, D. M., Swanson, L. W. and Rosenfeld, M. G. (1988) Neuronal expression of chimeric genes in transgenic mice. Neuron **1**, 311–320

Sager, S. M., Sharp, F. R. and Curran, T. (1988) Expression of c-*fos* protein in brain: metabolic mapping at the cellular level. Science **240**, 1328–1331

Sausville, E., Carney, D. and Battey, J. (1985) The human vasopressin gene is linked to the oxytocin gene and is selectively expressed in a cultured lung cancer cell line. J. Biol. Chem. **260**, 10236–10241

Sawchenko, P. E., Swanson, L. W. and Vale, W. W. (1984) Co-expression of corticotropin-releasing factor and vasopressin immunoreactivity in parvocellular neurosecretory neurons of the adrenalectomized rat. Proc. Natl. Acad. Sci. U.S.A. **81**, 1883–1887

Schäfer, E. A. and Mackenzie, K. (1911) The action of animal extracts on milk secretion. Proc. R. Soc. London B **84**, 16–22

Schilling, K., Schmale, H., Oeding, P. and Pilgrim, C. (1991) Regulation of vasopressin expression in cultured diencephalic neurons by glucocorticoids. Neuroendocrinology **53**, 528–535

Schmale, H. and Richter, D. (1984) Single base deletion in the vasopressin gene is the cause of diabetes insipidus in Brattleboro rats. Nature (London) **308**, 705–709

Schmale, H., Ivell, R., Briendl, M., Darmer, D. and Richter, D. (1984) The mutant gene from diabetes insipidus (Brattleboro) rat is transcribed but the message is not efficiently translated. EMBO J. **3**, 3289–3293

Schmitz, E., Mohr, E. and Richter, D. (1991) Rat vasopressin and oxytocin genes are linked by a long interspersed repeated DNA element (LINE): sequence and transcriptional analysis of LINE. DNA Cell Biol. **10**, 81–89

Schumacher, M., Coirini, H., Frankfurt, M. and McEwen, B. (1989) Localized actions of progesterone in hypothalamus involve oxytocin. Proc. Natl. Acad. Sci. U.S.A. **86**, 6798–6801

Seif, S. M., Robinson, A. G., Zenser, T. V., Davis, B. B., Huellmantel, A. B. and Haluszczak, C. (1979) Neurohypophyseal peptides in hypothyroid rats: plasma levels and kidney response. Metabolism **28**, 137–143

Sherman, T. G., Civelli, O., Douglass, J., Herbert, E. and Watson, S. J. (1986) Coordinate expression of hypothalamic pro-dynorphin and pro-vasopressin mRNAs with osmotic stimulation. Neuroendocrinology **44**, 222–228

Sherman, T. G., Day, R., Civelli, O., Douglas, J., Herbert, E., Akil, H. and Watson, S. J. (1988) Regulation of hypothalamic magnocellular neuropeptides and their mRNAs in the Brattleboro rat: coordinate responses to further osmotic challenge. J. Neurosci. **8**, 3785–3796

Sherwood, N. M. and Parker, D. B. (1990) Neuropeptide families: an evolutionary perspective. J. Exp. Zool. Suppl. **4**, 63–71

Simerly, R. B., Chang, C., Muramatsu, M. and Swanson, L. W. (1990) Distribution of androgen and estrogen receptor mRNA-containing cells in the rat brain: an *in situ* hybridization study. J. Comp. Neurol. **294**, 76–95

Skowsky, W. R. and Fisher, D. A. (1977) Arginine vasopressin secretion in thyroidectomized sheep. Endocrinology **100**, 1022–1026

Snijdewint, F. G. M., Van Leewen, F. W. and Boer, G. J. (1989) Ontogeny of vasopressin and oxytocin binding sites in the brain of Wistar and Brattleboro rats as demonstrated by light microscopical autoradiography. J. Chem. Neuroanat. **2**, 3–17

Sofroniew, M. V. (1985) Vasopressin, oxytocin and their related neurophysins. In Handbook of Chemical Neuroanatomy (Björklund, A. and Hökfelt, T., eds.), pp. 93–165, Elsevier, New York

Summar, M. L., Phillips, J. A. I., Battey, J., Castiglione, C. M., Kidd, K. K., Maness, K. J., Weiffenbach, B. and Gravius, T. C. (1990) Linkage relationships of human arginine vasopressin-neurophysin-II and oxytocin-neurophysin-I to prodynorphin and other loci on chromosome 20. Mol. Endocrinol. **4**, 947–950

Swaab, D. F., Pool, C. W. and Nijveldt, F. (1975) Immunofluorescence of vasopressin and oxytocin in the rat hypothalamo-neurohypophysial system. J. Neural Transmission **36**, 195–215

Terrier, C., Chabot, J.-G., Pautrat, G., Jeandel, L., Gray, D., Lutz-Bucher, B., Zingg, H. H. and Morel, G. (1991) Arginine-vasopressin in anterior pituitary cells: *in situ* hybridization of mRNA and ultrastructural localization of immunoreactivity. Neuroendocrinology **54**, 303–311

Tramu, G., Croix, C. and Pillez, A. (1983) Ability of CRF immunoreactive neurons of the paraventricular nucleus to produce a vasopressin-like material. Neuroendocrinology **37**, 467–469

Treip, C. S. (1984) The hypothalamus and pituitary gland. In Greenfield's Neuropathology (Adams, J. H., Corsellis, J. A. N. and Duchen, L. W., eds.), pp. 749–778, John Wiley, New York

Tribollet, E., Barberis, C., Jard, S., Dubois-Dauphin, M. and Dreifuss, J. J. (1988) Localization and pharmacological characterization of high affinity binding sites for vasopressin and oxytocin in the rat brain by light microscopic autoradiography. Brain Res. **442**, 105–118

Uhl, G. R. and Reppert, S. M. (1986) Suprachiasmatic nucleus vasopressin messenger RNA: circadian variation in normal and Brattleboro rats. Science **232**, 390–393

Uhl, G. R., Zingg, H. H. and Habener, J. F. (1985) Vasopressin mRNA *in situ* hybridization: localization and regulation studied with oligonucleotide probes in normal and Brattleboro rat hypothalamus. Proc. Natl. Acad. Sci. U.S.A. **82**, 5555–5559

Urano, A., Hyodo, S. and Suzuki, M. (1992) Molecular evolution of neurohypophysial hormone precursors. Prog. Brain Res. **92**, 39–46

Urban, J. H., Miller, M. A., Drake, C. T. and Dorsa, D. M. (1990) Detection of vasopressin mRNA in cells of the medial amygdala but not the locus coeruleus by *in situ* hybridization. J. Chem. Neuroanat. **3**, 277–283

Urban, J. H., Miller, M. A. and Dorsa, D. M. (1991) Dexamethasone-induced suppression of vasopressin gene expression in the bed nucleus of the stria terminalis and medial amygdala is mediated by changes in testosterone. Endocrinology **129**, 109–116

van Leeuwen, F. W., Caffe, A. R. and De Vries, J. (1985) Vasopressin cells in the bed nucleus of the stria terminalis of the rat: sex differences and the influence of androgens. Brain Res. **325**, 391–394

van Leeuwen, F., van der Beek, E., Seger, M., Burbach, P. and Ivell, R. (1989) Age-related development of a heterozygous phenotype in solitary neurons of the homozygous Brattleboro rat. Proc. Natl. Acad. Sci. U.S.A. **86**, 6417–6420

Van Tol, H. H. M., Voorhuis, T. A. M., Snijdewint, F. G. M., Boer, G. J. and Burbach, J. P. H. (1986) Vasopressin gene expression is attenuated in the fetal Brattleboro rat. FEBS Lett. **204**, 101–105

Van Tol, H. H. M., Voorhuis, D. T. A. M. and Burbach, J. P. H. (1987) Oxytocin gene expression in discrete hypothalamic magnocellular cell groups is stimulated by prolonged salt loading. Endocrinology **120**, 71–76

Van Tol, H. H. M., Snijdewint, F. G. M., Boer, G. J. and Burbach, J. P. H. (1990) Hypothalamic oxytocin mRNA during development of the Brattleboro rat: comparison to vasopressin mRNA. Neurosci. Res. Commun. **7**, 25–33

Vandesande, F. and Dierickx, K. (1975) Identification of the vasopressin producing and of the oxytocin producing neurons in the hypothalamic magnocellular neurosecretory system of the rat. Cell Tissue Res. **164**, 153–162

Verbeeck, M. A. E., Adan, R. A. H. and Burbach, J. P. H. (1990) Vasopressin gene expression is stimulated by cyclic AMP in homologous and heterologous expression systems. FEBS Lett. **272**, 89–93

Verbeeck, M. A. E., Sutanto, W. and Burbach, J. P. H. (1991) Regulation of vasopressin messenger RNA levels in the small cell lung carcinoma cell line GLC-8: interactions between glucocorticoids and second messengers. Mol. Endocrinol. **5**, 795–801

Voss, A. K. and Fortune, J. E. (1992) Oxytocin/neurophysin-I messenger ribonucleic acid in bovine granulosa cells increases after the luteinizing hormone (LH) surge and is stimulated by LH *in vitro*. Endocrinology **131**, 2755–2762

Walther, N., Wehrenberg, U., Brackmann, B. and Ivell, R. (1991) Mapping of the bovine oxytocin gene control region: identification of binding sites for luteal nuclear proteins in the 5′ non-coding region of the gene. J. Neuroendocrinol. **3**, 539–549

Wolfson, B., Manning, R. W., David, L. G., Arentzen, R. and Baldino, F., Jr. (1985) Colocalization of corticotropin releasing factor and vasopressin mRNA in neurones after adrenalectomy. Nature (London) **315**, 59–61

Wray, S., Kusano, K. and Gainer, H. (1991) Maintenance of LHRH and oxytocin neurons in slice explants cultured in serum-free media: effects of tetrodotoxin on gene expression. Neuroendocrinology **54**, 327–339

Xiong, Y. and Eickbush, T. H. (1990) Origin and evolution of retroelements based upon their reverse transcriptase sequences. EMBO J. **9**, 3353–3362

Yamaguchi, K., Akaishi, T. and Negoro, H. (1979) Effect of estrogen treatment on plasma oxytocin and vasopressin in ovariectomized rats. Endocrinol. Jpn. **26**, 197–205

Young, W. S., III (1986) Corticotropin-releasing factor mRNA in the hypothalamus is affected differently by drinking saline and by dehydration. FEBS Lett. 208, 158–162

Young, W. S., III (1989) Levels of transcripts encoding a member of the protein kinase C family in the paraventricular and supraoptic nuclei are increased by hyperosmolality. J. Neuroendocrinol. 1, 1–4

Young, W. S., III and Zoeller, R. T. (1987) Neuroendocrine gene expression in the hypothalamus: *in situ* hybridization histochemical studies. Cell. Mol. Neurobiol. 7, 353–366

Young, W. S., III, Mezey, É. and Siegel, R. E. (1986) Vasopressin and oxytocin mRNAs in adrenalectomized and Brattleboro rats: analysis by quantitative *in situ* hybridization histochemistry. Mol. Brain Res. 1, 231–241

Young, W. S., III, Shepard, E. A. and Burch, R. M. (1987*a*) Plasma hyperosmolality increases G-protein and cAMP synthesis in the paraventricular and supraoptic nuclei. Mol. Endocrinol. 1, 884–888

Young, W. S., III, Warden, M. and Mezey, É. (1987*b*) Tyrosine hydroxylase mRNA is increased by hyperosmotic stimuli in the paraventricular and supraoptic nuclei. Neuroendocrinology 46, 439–444

Young, W. S., III, Horvath, S. and Palkovits, M. (1990*a*). The influences of hyperosmolality and synaptic inputs on galanin and vasopressin expression in the hypothalamus. Neuroscience 39, 115–125

Young, W. S., III, Reynolds, K., Shepard, E. A., Gainer, H. and Castel, M. (1990*b*) Cell-specific expression of the rat oxytocin gene in transgenic mice. J. Neuroendocrinol. 2, 917–925

Zingg, H. H. and Lefebvre, D. L. (1988) Oxytocin and vasopressin gene expression during gestation and lactation. Mol. Brain Res. 4, 1–6

Zingg, H. H. and Lefebvre, D. L. (1989) Oxytocin mRNA: increase of polyadenylate tail size during pregnancy and lactation. Mol. Cell. Endocrinol. 65, 59–62

Zingg, H. H., Lefebvre, D. and Almazan, G. (1986) Regulation of vasopressin gene expression in rat hypothalamic neurons. J. Biol. Chem. 261, 12956–12959

Zingg, H. H., Lefebvre, D. L. and Almazan, G. (1988) Regulation of poly(A) tail size of vasopressin mRNA. J. Biol. Chem. 263, 11041–11043

For additional discussion of evolution of vasopressin and oxytocin family members, see also: Acher, R. (1985) Biosynthesis, processing and evolution of neurohypophysial hormone precursors in Neurosecretion and the Biology of Neuropeptides (Kobayashi, H., Bern, H. A. and Urano, A., eds.), pp. 11–25, Springer, Berlin

Transcriptional and post-transcriptional regulation of the human neuropeptide Y gene

C. D. Minth

Wright State University, Department of Biochemistry, School of Medicine, College of Science and Mathematics, 234 Biological Sciences, Dayton, OH 45435-0001, U.S.A.

Introduction

Neuropeptide Y (NPY) is one of the most abundant and widely expressed peptides present in the central and peripheral nervous systems (McDonald, 1988). It functions either as a neurotransmitter or as a peptide hormone to affect a wide variety of physiological processes. When released from nerve terminals commonly in conjunction with norepinephrine, it acts as a potent vasoconstrictor, potentiating the effects of this neurotransmitter (Gu *et al.*, 1983; Tuor *et al.*, 1985). It has also been implicated in the control of circadian rhythm, depression and satiety (Zini *et al.*, 1984; Allen *et al.*, 1985; Wahlestedt *et al.*, 1993). Recently, part of the cascade by which NPY exerts its effects was elucidated by the cloning and characterization of its Y1 receptor (Larhammar *et al.*, 1992). This receptor belongs to the G-protein-associated superfamily and upon hormonal stimulation transduces a signal resulting in decreased cellular production of cyclic AMP or increased intracellular calcium levels (Herzog *et al.*, 1992). Similarly, studies regarding the mechanisms responsible for regulating transcription of the NPY gene are underway. An understanding of these processes will allow us to define better the physiological roles of NPY.

The study of the expression of the human NPY gene has elucidated regulatory mechanisms both general and specific in nature. Since this gene is expressed in a wide variety of neuronal tissues, some regulatory signals are by necessity going to be permissive, at least in the sense that they are present in several different cell types. In contrast, the exclusive expression of this gene in neuronal tissues or tissues of neuronal origin dictates an ultimate requirement for transcriptional specificity (Adrian *et al.*, 1983; Varndell *et al.*, 1984; Gray and Morley, 1986; Stone *et al.*, 1986). This pattern of gene expression is illustrative of the complexity of the nervous system thus making the NPY gene a good model for studying the basis of neuronal diversity.

The 5′-flanking sequence of the NPY gene contains *cis*-acting elements that regulate transcription directed by RNA polymerase II. A characterization of these sequences reveals the interactions of proteins capable of both positively and negatively influencing the activity of the transcription complex (Minth and Dixon, 1990). The availability and activity of these DNA-binding proteins determine the transcriptional activity of this gene ultimately specifying its expression pattern.

A variety of agents including nerve growth factor, 12-*O*-tetradecanoyl phorbol-13-acetate (TPA), and forskolin have been shown to increase the steady-state levels of rat NPY mRNA (Sabol and Higuchi, 1990; Magni and Barnea, 1992). Parallel studies on the human gene indicate that it is similarly regulated by these compounds and that the increase in steady state levels of NPY mRNA is affected by two disparate mechanisms: an increase in the transcriptional activity of the gene and an increase in the stability of the mRNA encoding NPY (C. D. Minth *et al.*, unpublished work). The latter level of regulation is responsible for most of the mRNA increase resulting in the augmentation of the smaller transcriptional increase. Using both control mechanisms, neuronal cells are capable of fine tuning NPY mRNA levels in response to a variety of stimuli.

The study of the regulation of the human NPY gene has been aided by the availability of several cell lines of neuronal origin which express NPY to varying degrees. These cells were cultured from neuroblastomas, childhood tumours commonly arising in the sympathetic nervous system (Seeger *et al.*, 1977; Spinelli and Ishii, 1983). The expression of the NPY gene was studied using LA-N-5 cells, a neuroblastoma cell line originally derived from a bone marrow metastasis. These cells are representative of sympathetic-like neurons expressing NPY and noradrenergic markers. While experiments using cell lines are valuable for identifying *cis*-acting sequences and *trans*-acting factors which influence gene transcription, they have limitations. For example, LA-N-5 cells over-express NPY and have lost important growth controls. Additionally, the promiscuous expression of NPY–chloramphenicol acyltransferase (CAT) fusion constructs in HeLa cells, an epithelial-like cell line which does not express NPY, further complicates the determination of the elements responsible for tissue-specific expression. To circumvent this problem, studies utilizing transgenic mice were initiated to define further the promoter sequences necessary for appropriate tissue-specific expression *in vivo*.

Analysis of the 5′-flanking sequence of the NPY promoter

NPY belongs to a family of peptide hormones including pancreatic polypeptide (PP) and peptide YY (PYY). The genes encoding these peptides have been isolated and characterized (Leiter *et al.*, 1985; Krasinski *et al.*, 1990). Comparisons of the structures of these genes result in a highly conserved genomic organization represented schematically by the diagram of the human NPY gene (Fig. 1A). Furthermore, the chromosomal localizations of the human NPY and PP genes have been determined. The NPY gene is located on chromosome 7 while the PP gene is present on chromosome 17. Several other gene families including type 1 collagen, actin, erb proto-oncogenes, and protein kinases segregate with these genes suggesting that this family arose from a tandem duplication of an original linkage group followed by chromosomal translocation (Takeuchi *et al.*, 1986). Despite the similarities in gene organization, these peptides have a divergent expression pattern. While NPY is mainly expressed in neuronal tissues, PP is expressed mainly in the islet cells of the pancreas and PYY is localized primarily to cells in the ileum and colon. Therefore, promoter elements dictating tissue-specific expression have evolved concomitantly with the specialized functions of these peptides.

The NPY gene has been isolated and characterized from at least three sources: human, rat and *Xenopus* (Minth *et al.*, 1986; Larhammar *et al.*, 1987; Higuchi *et al.*, 1988; W. L. Taylor, personal communication). The alignment of the 5′-flanking sequences is shown in Fig. 1B. The human and rat sequences are 70% identical over the indicated region. Sequences upstream of −176 display significant diversity which continues throughout the remainder of the 5′-flanking sequences. Comparison of the human versus *Xenopus* sequences resulted in only 44% identity with most of the conserved nucleotides occurring in well delineated groups. The sequence of greatest identity among these species started at the TATA box and extended approximately 30 bp upstream. This included the CT-rich sequence shown to be important for basal level transcription of the human NPY gene (Minth and Dixon, 1990). Further upstream, the sequence centred around −82 was also conserved. Base on the hypothesis that conservation of sequence implies functional importance, the contributions of these regions to human NPY gene regulation will be discussed in the following sections.

Computer analysis of the 5′-flanking sequence identified consensus binding sites for several previously characterized DNA-binding proteins (Fig. 2). The sequence between −246 and the TATA box is GC-rich (75%) and by inspection contains several putative Sp1 and AP-2 binding sites. Sp1 is a protein which was originally identified as a transcriptional activator of the SV40 early promoter (Gidoni *et al.*, 1984). Its consensus binding site is GC-rich and consists of the following nucleotides: (G/T)GGGCGGPuPuPy, where Pu represents any purine and Py any pyrimidine. Similarly, AP-2 was found to be a transcriptional activator of the metallothionein IIA gene. Characterization of its DNA-binding site resulted in the following consensus sequence, (T/C)C(C/G)CC(A/C)N(G/C)(C/G)(G/C), N representing any nucleotide (Imagawa *et al.*, 1987). This sequence is also GC-rich and contains multiple ambiguities. The region between −86 and −62 consists of an imperfect palindrome which contains partial NF-1

Fig. 1. Comparison of the human, rat and *Xenopus* NPY genes

(a)

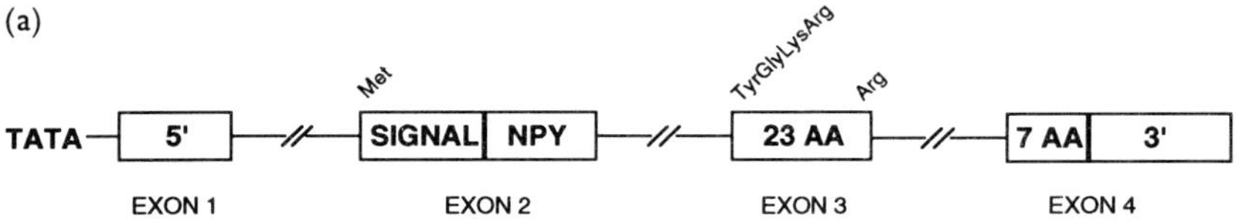

(b)

```
HUMAN           TTGGGGTGTGGGTGGCT-CCCTAAGTC-GAC--ACTCGTG-CGGCTGCGG  -132
RAT             TTGGGGTGTGGGTGGCT-CCAGACGCC-GCC--ACTCGAGACGGCTGTGG  -131
XENOPUS         -------------GTCTGCC--ACATCTAACTTATTAGTGACG--TGTGA  -120
                             * ** **  *   *   *  * * * * **  ** *

HUMAN           TTCCAGCCCCCTCCCCCCGCCACTCAGGGGCGGGAAGTGGC-GGG-TGGG   -84
RAT             CTCCAGCCTCCTCCCCC-GCTGAT-GGGGGCGGGAAGTGGC-TG--TGGG   -86
XENOPUS         A----------------CTCTGTAGAGAAGAGAGGAGGTGCATGGCTTCT   -86
                                   *      *  * * * **  **  *  *

HUMAN           AGTCACCCAAGCGTGAC--TGCCC--GAGGCCCCTCCTGCCGCGGCGAGG   -38
RAT             AGTCACCCGGGCGTGAC--TCCCCCCGAGGCCCCTCCTGCCGCGACAAGG   -38
XENOPUS         AGTCACAGAGGC---ACAATTGACTC-AAGCCCCTCCTGCC-CAA-AAGG   -42
                ******    **   **  *   *   * ************ *    ***

                        TATA
HUMAN           AAGCTCCATAAAAGCCCTGTCGCGACCCGCTCT-CTGCACCCCATCC--G   +10
RAT             GCGCTCCATAAAAGCCCGTTGGCGACCCGCTCTAC-GCATCCCA-CC--G    +9
XENOPUS         AAGAGA-ATAAAAG---------GAACAGCT-GGCTGAGAACCTGCCAGG    -3
                  *    *******         ** * ***   * *    **  **  *

HUMAN           CTGGCTCTCA   +20
RAT             GTGGAGCTCA   +19
XENOPUS         ATTGTGATCA    +8
                 * *   ***
```

(a) Schematic diagram of the human NPY gene. Exons are represented by boxes and intervening sequences by lines interrupted with breaks. The positions of the signal peptide, mature hormone and C-terminal peptides are indicated. The positions of the initiator Met and the C-terminal processing site(s) are shown. (b) Alignment of the 5′-flanking sequences of these genes. The alignment was performed using the CLUSTAL program (Intelligenetics, Mountain View, CA). Numbering of the nucleotides relative to the transcription initiation site (+1) is indicated to the right and the species to the left. Transcription start sites are represented by bold letters and the TATA box is indicated above the sequence. Identical nucleotides are indicated by an asterisk below the sequence. Dashes indicate gaps introduced into a sequence to produce maximal alignment.

Fig. 2. Nucleotide sequence of the human NPY promoter

```
Δ1078
GATCTCAGTCCACTGCATAAAATCTTGATCCTGTAATAATAGTTTCTGTATCTTGCATAT   -1019

TCATTCAACAGGTTTAACGCGATGAGCAAATTAATGTTCATCGTTTTTAACATGTTTCAT   -959

CTTAATCAGAACCCACATTCTCAACGTTAATTGAACGTACATAGGACTATACAAGGGTTA   -899

GTAAATAAGACAGAAACTGTTGCTCATTTAACCACCGTCACTTTGGACCAAAAAAGAAAA   -839
                                            Δ796
AATATATATTTTTAAAATTGAGCTTAAAAGAGTCTCTAGAAGCTGGAAGCGTGGCTCTTT   -779
                                   NRE
TTCAGCAAACTGGGGGAATAGGTTTACCGTGTTCCCCCTCTGGGGAATTTTGAGTCGCCA   -719
                Δ710
CACTCATGTCTCGACCGAGCCTGGCTCGCTGCGTCTGAGCGAGTACTTGAGGAAGGCTGA   -659
           AP-4
TCTAGAAAAACCAGCTGAGAGAAGGGGCAGAAGCCCCTGAAACCACGGGCGGGGGTGGGG   -599

TGGGGAGCGCACGTTTGGGACCCTCTAGCCGGAGACTTCCGAGCTGCCTCCGACTTGTTC   -539
     Δ533                           Δ499
                     HLH                            HLH
TAAGTACAGGAAAAATCTGTGCGCCCAGTTGCCTCACTCCAACAGCGCGCAGTTGTGCCC   -479
                                                    Δ428
GGCGAGGATGCCGCGCTAGTCGTGGAGATGCCCCACCACAAAGAGGATTCAGGTGCTTCC   -419
                                                    HLH

TACTCCGGCACCCAGTGGGCTCGTAGTCCTGTTGGCAGGAGACAAGAATCGTCTGGGCTG   -359

CTCCTATCTCTGGCAGGACTAGACGGGGCGTGAAGGAAAGAAGGAAAGAAGGAAAGCAGG   -299
                                                    Δ246
GATCGGGCACTGCCCGAGGGCAGATACTTGGGCTTTGGTGTTGTCCAGCGCGCTCGGAGT   -239
                                                     GC Box
GCGCTGCCTCGCTCACGCGGTCCCAGGCCCCGCTTCTTCAGGCAGTGCCTGGGGCGGGAG   -179
Δ178                            Δ143
                         AP-4                         CT Box
GGTTGGGGTGTGGGTGGCTCCCTAAGTCGACACTCGTGCGGCTGCGGTTCCAGCCCCCTC   -119
Δ118                              Δ83                    Δ63
              GC Box      GC Box        AP-2     AP-1     AP-2
CCCCCGCCACTCAGGGGCGGGAAGTGGCGGGTGGGAGTCACCCAAGCGTGACTGCCCGAG   -59
          Δ51                Δ30          NF-1
   CT Box        AP-2             TATA                        +1
GCCCCTCCTGCCGCGGCGAGGAAGCTCCATAAAAGCCCTGTCGCGACCCGCTCTCTGCAC   +2

CCCATCCGCTGGCTCTCACCCCTCGGAGACGCTCGCCCGACAGCATAGT        CAT   +51
```

Nucleotides are numbered relative to the initiation of transcription (+ 1). Deletion end points are designated by an arrow and bold print. Sequences homologous to consensus sequences for the TATA box, CT boxes, GC boxes and negative regulatory element (NRE), as well as for the DNA-binding proteins AP-2, AP-1, NF-1 and AP-4 are boxed. The nucleotides comprising the core consensus sequences are italicized.

(TGG[A/C]N_5GCCAA), AP-1 (TGACTCA), and AP-2 consensus sequences (Jones *et al.*, 1987; Lee *et al.*, 1987). In addition, two CT boxes are present at −118 and −51. The CT box at −51 was previously shown to interact with a protein related in binding specificity to Sp1 (Minth and Dixon, 1990). The close proximity of these sites suggests that protein–protein interactions could potentially play an important regulatory role and that competition between factors for binding to mutually exclusive sites may result in an additional level of transcriptional control.

The sequence becomes AT-rich between −1078 and −796 (67%). A computer program designed to calculate regions of DNA secondary structure predicts that the region centred around −780 could potentially form an alternative DNA structure known as bent DNA (Eckdahl and Anderson, 1987). The consensus binding sites for several helix–loop–helix DNA-binding proteins defined as the CANNTG motif are found interspersed throughout this region (Gregor *et al.*, 1990). Finally, the sequence centred around −740 binds a potential negative regulator of NPY transcription (C. D. Minth *et al.*, unpublished work). The following sections discuss in greater detail the characterization of several of these putative regulatory sequences.

Transient expression of NPY–CAT fusion constructs *in vivo* and *in vitro*

NPY–CAT fusion constructions were generated by Exonuclease III digestion of the plasmid pCATNPYΔ1078 that contains the nucleotides spanning −1078 to +51 of the 5′-flanking sequence of the human NPY gene fused to the coding sequence for the bacterial reporter gene CAT (Minth and Dixon, 1990). The plasmid RSV*LacZ* encoding the β-galactosidase gene under the control of the Rous Sarcoma virus long terminal repeat was included an an internal control during transfections. These constructions were introduced into LA-N-5 cells using a modified $CaPO_4$ transfection procedure (Chen and Okayama, 1987). A schematic diagram of the NPY–CAT fusion constructions for the extended deletion series is depicted in Fig. 3 (left) while their corresponding relative CAT conversion represented as CAT activity normalized by division with the β-galactosidase activity, is shown in Fig. 3 (right). When the relative CAT conversion for the Δ246 construct is set to 100%, the assay profile suggests both positive and negative regulation of NPY transcription. Sequences from −428 to −246 as well as from −1078 to −701 decrease transcription as measured by this assay and may contain negative regulatory elements. Conversely, the sequence between −701 and −533 contains information which increases transcription, as deletion of this sequence produces a large decrease in activity. Lastly, the sequence between −246 and −30 functions to influence transcription positively and contains the information necessary for basal level expression.

Fig. 3. Transient expression of NPY–CAT fusion constructions

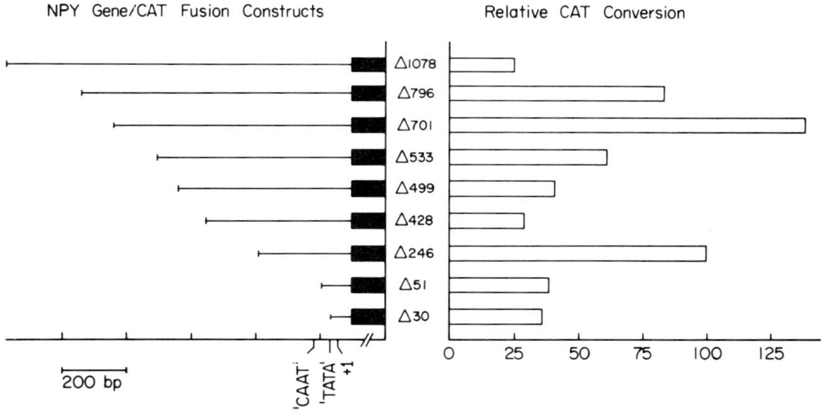

(Left) Schematic diagram of the 5′-flanking sequences for the NPY–CAT deletion constructions. The solid boxes represent 51 bp of the first non-translated exon of the NPY gene, the coding sequences for the CAT gene, the SV40 small T antigen intervening sequence, and the SV40 polyadenylation signal. The solid line represents NPY promoter sequences. (Right) Relative CAT conversion obtained from transient transfection of LA-N-5 cells with the corresponding NPY fusion construct and the control plasmid pRSVLacZ. These values are a ratio of CAT activity to β-galactosidase activity measured at 420 nm. The activities are presented relative to the Δ246 deletion. Data represent an average of two to five transfections for each construct. The variability of CAT activity between experiments was less than 15%.

Determination of the minimal information necessary for transcription of this gene required the construction of an additional deletion series extending from −246 to −30 (Fig. 4). Addition of sequences extending from −63 to −51 resulted in approximately a 3-fold increase in promoter activity. This level of activity is maintained until the addition of the sequence between −143 and −118 which increases the relative CAT activity by an additional 2-fold. The proximal sequence (−63 to −51) contains adjacent binding sites for a Sp1-like molecule and AP-2. The contribution of each of these proteins to NPY gene transcription is currently under investigation. The sequence centred around −118 contains an additional CT box sequence which positively regulates NPY transcription and potentially forms a complex similar to that present at −51. These data suggest that the CT box elements are likely to be important in the expression of this gene.

Surprisingly, measurable CAT activity was generated by the Δ30 construct which only contains the TATA box, 24 bp of non-transcribed sequences, and 51 bp of exon 1, thus raising the possibility that this activity

Fig. 4. Transient expression of NPY–CAT fusion constructions between −246 and −30

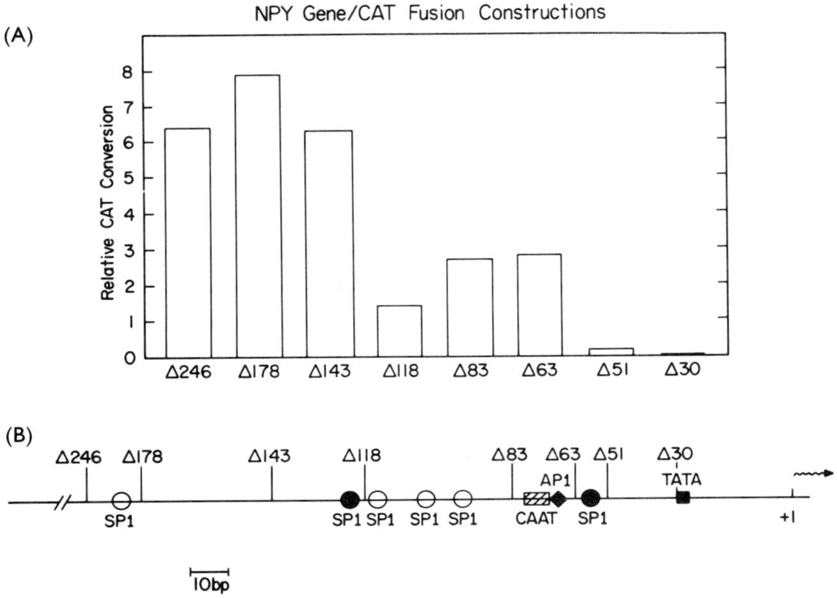

(A) Open bars represent the relative CAT conversion obtained from transient transfections of LA-N-5 cells with the deletions ranging from −246 to −30 and the plasmid pRSVLacZ. The values are a ratio of CAT activity to β-galactosidase activity relative to the Δ246 deletion. Data represent an average of four to 11 transfections for each construct. (B) Schematic diagram of cis*-acting elements between −246 and −30. Positions of the deletions are denoted by vertical lines. The TATA box is represented as a solid box, the CT sequences are depicted by solid circles, and the GC-rich Sp1 consensus sequences are represented by open circles. The putative AP-1 and CAAT consensus sequences are represented by a solid diamond and hatched box, respectively.*

was generated due to spurious transcription initiation directed by vector sequences. Sequences 3′ to the TATA box could, however, also be important for transcription initiation. Resolution of this question required the determination of the transcription start site utilized by these fusion constructions. Due to the labile nature of the NPY–CAT mRNA and the inefficient transfection of neuronal cells, a primer extension or S1 analysis could not readily be performed. To resolve this problem, the activity of the fusion constructs was assayed using *in vitro* transcription competent extracts made from LA-N-5 cell nuclei. A plasmid, MCAT, under the control of the adenovirus major late promoter (Ad5 MLP) sequences (−270 to +32) was included in all transcription reactions to serve as an internal control. The NPY–CAT fusion constructions were identical to those used in the *in vivo* transfection studies. The transcriptional capability of each template was quantified by primer extension analysis using a primer complementary to CAT coding sequences. Transcripts of 131 nucleotides were generated from the correct start site for the NPY gene and transcripts of 110 nucleotides from the MLP's start site (Fig. 5). Using this type of analysis, it was evident that the CAT activity generated by the Δ30 construct represented an improperly initiated transcript. Similarly, minimal activity was generated by the Δ51 construct and addition of the sequences between −63 and −51 permitted efficient initiation of basal level transcription at the appropriate start site. The activity generated by the other constructs correlates well with the activity observed in the transient transfection analysis. Overall, the sequence between −246 and −51 supports transcription of this gene, and harbours the potential for regulation via DNA–protein interactions.

Dissection of proximal promoter elements

To determine specifically which regions of the proximal 5′-flanking DNA were involved in DNA–protein interactions, a footprinting assay was performed (Minth and Dixon, 1990). The fragment spanning −118 to +51 was labelled on the non-coding strand, incubated with LA-N-5 nuclear extracts, and treated with varying amounts of DNAse I (Fig. 6). Several regions were protected from DNAse I digestion indicating multiple DNA–protein interactions. The region spanning −61 to −45, which was necessary for initiating basal level transcription, was protected from digestion. This sequence contained the potential consensus binding sites for AP-2 and the CT box binding protein. An additional well-delineated footprint extended through nucleotides −83 to −63, which contained putative binding sites for several DNA-binding proteins including NF-1, AP-1 and AP-2. Further upstream, the GC-rich region between −97 and −84, that could potentially bind Sp1-related molecules, was also protected. Due to their close proximity, protein–protein interactions could add an extra level of complexity in regulating transcriptional activity.

Analyses of the DNA–protein interactions found by footprinting required more precise determinations of the binding sites of the proteins. The region between −65 and −36 was tested for its ability to form specific DNA–protein interactions utilizing the gel retardation assay (Fried and Crothers, 1981). The nucleotide sequence of this region is shown in Fig. 7A and a typical gel retardation analysis using this end-labelled oligonucleotide in the presence of LA-N-5 nuclear extract is depicted in Fig. 7B. Three DNA–protein complexes were formed in the absence of competitor oligonucleotides (Fig. 7B, lanes 1 and 20). Only the most slowly migrating complex designated N1 was specifically competed by unlabelled oligonucleotide −65 to −36 (Fig. 7B, lanes 2–4).

Fig. 5. ***In vitro* transcription analyses**

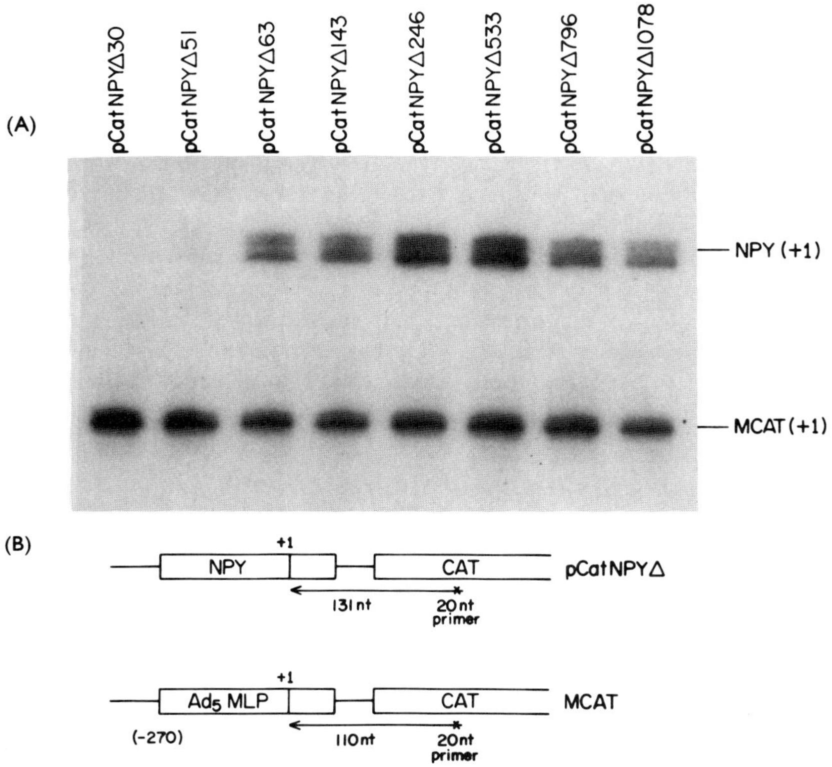

(A) In vitro *transcription products directed from the NPY and MCAT promoters. Each reaction contains LA-N-5 nuclear extract (3 mg/ml), 0.75 µg of the designated NPY/CAT fusion construct and 0.25 µg of MCAT. The positions of the correctly initiated primer extension products for the NPY and MCAT constructs are noted to the right. (B) Schematic diagram of the templates utilized in the* in vitro *transcription analyses. pCatNPYΔ templates contain various amounts of the 5′-flanking sequences of the NPY gene, and MCAT contains the Ad5 major late promoter fused to the bacterial CAT gene. The primer is complementary to the sequences + 15 to + 35 of the CAT transcript. The arrow depicts the predicted primer extension products arising from the* in vitro *synthesized RNA. nt, nucleotide.*

To determine the sequence specificity of the complexes, competition assays were performed with DNA fragments and oligonucleotides containing previously characterized DNA/protein consensus sequences. Both oligonucleotides −65 to −36 and −75 to −46 competed for binding of the NI complex (Fig. 7B, lanes 2–6). The sequences generated from the herpes simplex virus thymidine kinase gene that contains two Sp1 binding sites and

Fig. 6. **DNase I protection analysis of proximal 5′-flanking sequences**

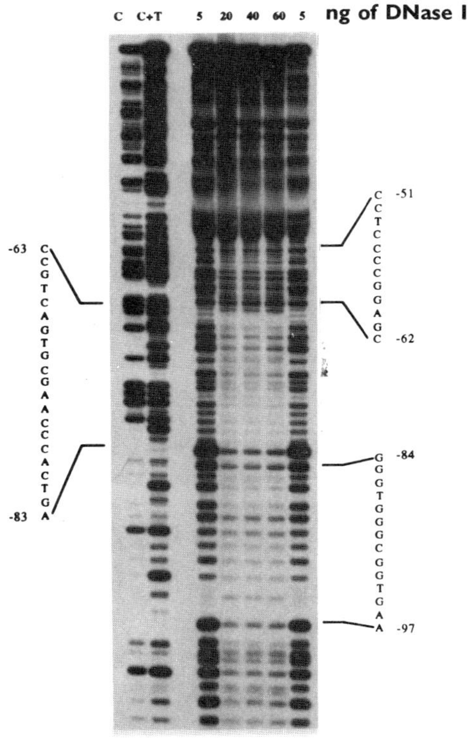

The non-coding strand of the NPY promoter (− 118 to + 51) was labelled and incubated with or without 50 μg of LA-N-5 nuclear extract followed by treatment with the specified amounts of DNase I. The DNA was deproteinized and analysed on a 12% sequencing gel. Maxim–Gilbert sequencing reactions (C and C + T) generated from the same fragment serve as markers. The protected sequences are noted to the left and right of their respective footprints.

the α-actin gene that contains a CA-rich sequence similarly shown to bind Sp1, competed with the labelled NPY oligonucleotide for N1 complex formation (Fig. 7B, lanes 7–10) (Jones *et al.*, 1985; Gustafson and Kedes, 1989). This suggested that the CCCCTC sequence was binding an Sp1-like molecule. The binding site was confirmed by mutating specific residues in this sequence and using the mutated oligonucleotide as a competitor. This sequence was unable to compete for binding (Fig. 7B, lanes 15 and 16).

Further proof of an Sp1-like binding activity was obtained when the oligonucleotide representing the SV40 GC boxes IV and V competed efficiently for the N1 complex (Fig. 7B, lanes 17–19). These GC boxes were

Fig. 7. Gel retardation analysis of the NPY promoter, − 65 to − 36

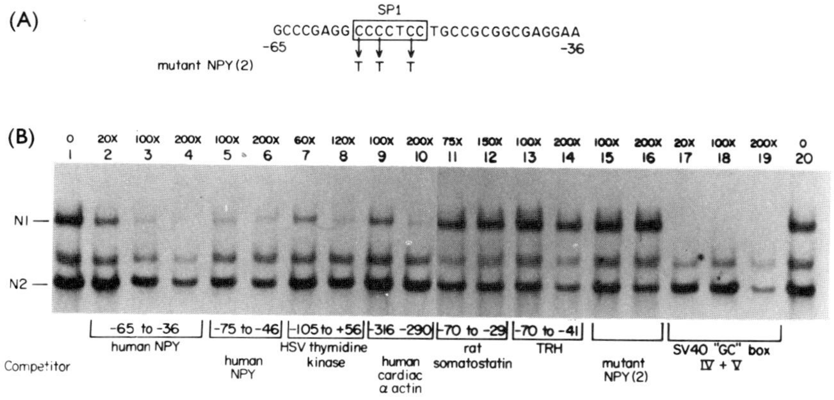

(A) The sequence of the NPY promoter between − 65 and − 36. The Sp1 binding site is boxed, and the C to T nucleotide changes in the mutant NPY (2) oligonucleotide are noted by arrows. (B) Autoradiogram of gel retardation assay. Lanes 1 and 20 were obtained with LA-N-5 nuclear extract and the labelled NPY oligonucleotide (− 65 to − 36) when no competitor DNA is present. Competitor DNAs present in lanes 2–19 are listed below the lanes in which they appear. The fold excess of each competitor is noted above the lane number. Two major complexes formed are designated N1 and N2. Abbreviations used: HSV, herpes simplex virus; TRH, thypotropin-releasing hormone.

previously shown to be high-affinity Sp1-binding sites (Gidoni *et al.*, 1984). Promoter sequences from the rat somatostatin and thyrotropin-releasing hormone genes that contain cAMP-responsive element binding sites did not compete for binding (Fig. 7B, lanes 11–14). Therefore, the NPY sequences either represent a low-affinity Sp1-binding site or a binding site for other members of the Sp1 family of DNA-binding proteins.

The sequence centred around − 61 contains a consensus binding site for the DNA-binding protein AP-2. To determine if both an Sp1-like molecule and AP-2 could bind to these sequences, an oligonucleotide spanning the region between − 75 to − 46 was synthesized and used in a gel retardation assay (Fig. 8C). Competition patterns indicate that only the AP-2 oligonucleotide competes for the more slowly retarded DNA/protein complex, C1 (Fig. 8A). Neither the Sp1 oligonucleotide nor the overlapping oligonucleotide − 65 to − 36 is able to compete with the C1 complex (Fig. 8A, lanes 2 and 6). This suggests that strong binding by AP-2 may preclude the binding of an Sp1-like protein to the neighbouring CT box. Since the CT-box is located near the end of the oligonucleotide − 75 to − 46, it is possible

Fig. 8. Gel retardation analysis of the NPY promoter, −75 to −46

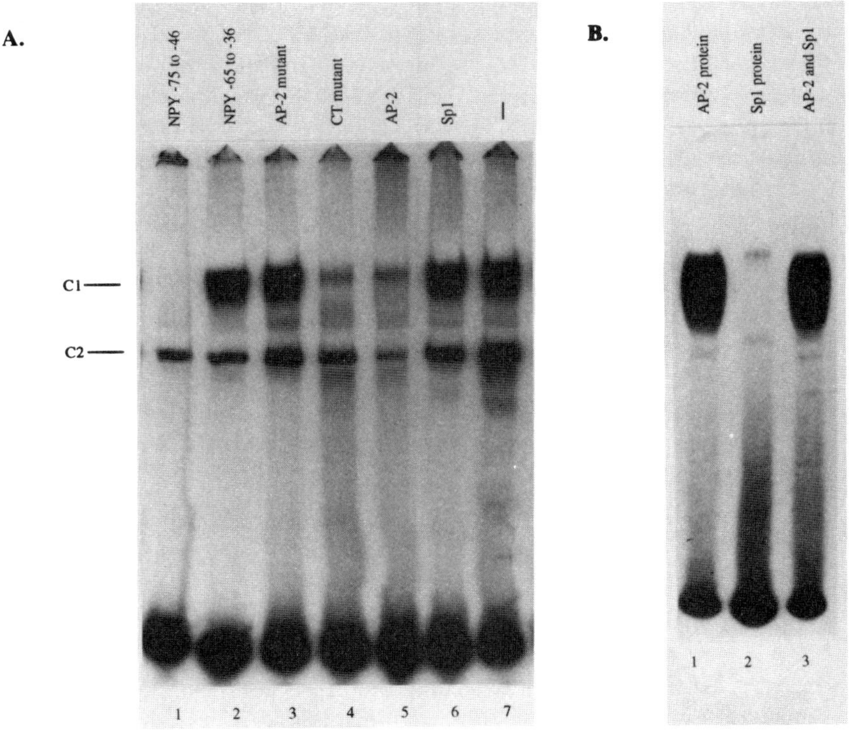

C.

```
                          AP-1       AP-2          CT box
NPY   -75 AAGCG|TGACTG|CCCGAGG|CCCCTCCTG|CCG -46
                 |     |       ||  |  |    |
AP-2 mutant      A     A       AA  |  |    |
                                   |  |    |
CT mutant                          T  T    T
```

(A) Lanes 1–6 were obtained with LA-N-5 nuclear extract and the labelled NPY oligonucleotide (− 75 to − 46) when 0.1 μg of the indicated competitor DNAs are present. Lane 7 contains no competitor DNA. (B) Lanes 1–3 were obtained when the − 75 to − 46 oligonucleotide was incubated with the indicated purified proteins (Promega, Madison, WI, U.S.A.). (C) Sequence of the NPY 5′-flanking sequence between − 75 and − 46. The consensus sequences for the DNA-binding proteins AP-1, AP-2 and the CT-box binding protein are boxed. The consensus sequence for AP-2 is italicized as it overlaps with the other binding sites. The G to A changes in the AP-2 mutant oligonucleotide are indicated by vertical lines as are the C to T changes in the CT box mutant.

that it is no longer able to interact with its respective DNA-binding protein. It is also possible that both proteins could interact with these sequences in a concerted manner or that binding of one protein could preclude the binding of the other. To address this question, purified AP-2 and Sp1 were incubated with the labelled −75 to −46 oligonucleotide and the resulting complexes were resolved by electrophoresis (Fig. 8B). Both AP-2 and Sp1 are capable of interacting with this sequence. However, when added in concert, no super-shifted band is formed suggesting that the complexes are mutually exclusive. The relative amounts of these proteins present in the nucleus in response to various stimuli may modulate the binding pattern present at this site.

The binding site of the factor important for basal level transcription of the NPY gene contained the CT box sequence, CCCCTCC. A South-western analysis was performed using the oligonucleotide −65 to −36 and LA-N-5 cell extract to determine the approximate molecular mass of the protein capable of interacting with this sequence (Minth and Dixon, 1990). A protein of approximately 97 kDa hybridized to the labelled oligonucleotide. This molecular mass was in the same range as that determined for Sp1, 95–105 kDa (Briggs *et al.*, 1986). This analysis, however, does not preclude the binding of AP-2 to this or neighbouring sequences since AP-2 will not bind to its consensus sequence under these conditions.

Regulation by TPA and forskolin

The participation of AP-2 in regulating the expression of the NPY gene is manifested by the ability of TPA and forskolin to increase NPY gene transcription. Forskolin is a diterpene activator of adenylate cyclase while TPA replaces diacylglycerol in activating protein kinase C. In several systems, these activation cascades ultimately result in increased transcription directed in part by interaction of the DNA-binding protein, AP-2, with its consensus-binding site (Imagawa *et al.*, 1987). Experiments performed on PC12 cells treated with TPA and forskolin resulted in an approximately 80-fold increase in steady-state NPY mRNA levels (Higuchi *et al.*, 1988). Further experiments determined that only a small part of this increase, 3–4-fold, was due to a transcriptional increase, the remainder presumably being due to increased mRNA stability (Sabol and Higuchi, 1990). Similar treatment of LA-N-5 cells with these agents for 6–8 h increased NPY mRNA levels by 8–10-fold (Fig. 9). We were unable to demonstrate reliably any increases in transcription in LA-N-5 cells and speculated that small transcriptional increases were being overshadowed by large increases in the stability of the mRNA encoding NPY.

The half-life of NPY mRNA was determined in the LA-N-5 cells that were or were not treated with TPA and forskolin. This experiment utilized the RNA polymerase II inhibitor 5,6-dichloro-1-ribofuranosyl-benzimidazole (DRB) (Helms and Rottman, 1990). The advantages of DRB

Fig. 9. Treatment of LA-N-5 cells with TPA and forskolin

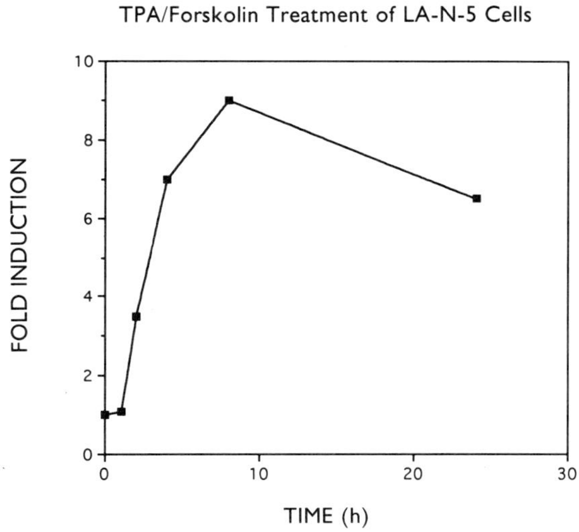

LA-N-5 cells were treated for the indicated times with 0.4 μM-TPA and 10 μM-forskolin. Total RNA was isolated and analysed via Northern analysis. Aliquots from each time point were stained with ethidium bromide to control inequalities in loading. The signal intensities were quantitated by densitometric scanning. The value obtained for the zero time point was set to 1.

include its rapid inhibition of heterogeneous RNA synthesis, the reversible nature of this inhibition, and the drug's inability to alter the rate of cellular protein synthesis. Actinomycin D, another RNA polymerase II inhibitor, prevents attachment of mRNA transcripts to ribosomes thereby affecting protein synthesis and increasing the rate of mRNA decay. The half-life calculations for NPY mRNA are based on first order decay kinetics, i.e. that all the mRNAs encoding NPY have an equal probability of being destroyed with time. The data fit this assumption with the untreated samples displaying a half-life of 7.7 ± 1.7 h while the drug-treated samples resulted in extending the half-life to 41.9 ± 8.8 h. Nevertheless, this dramatic increase cannot account for the entire 8-fold increase in NPY mRNA levels seen upon drug treatment. However, due to the small nature of the transcriptional increase, we were unable to determine the promoter elements responsible using *in vivo* transfection experiments; the variability of the assay precluded consistent results.

It is tempting to speculate that the increase is due to the AP-2 element present at −61 that was previously shown to be a strong AP-2 binding site in the aforementioned gel retardation assay. Lending credence to this theory, the gene encoding PYY, a member of the NPY family, responds to TPA and forskolin in a manner similar to the NPY gene. Experiments performed in HeLa cells suggest that an AP-2 element present in the 5′-flanking sequence of the PYY gene mediates a 6-fold increase in transcription (Krasinski *et al.*, 1990).

The regulation of the steady state NPY mRNA levels both by transcriptional and post-transcriptional mechanisms is economical and allows for additional fine tuning of NPY production. As NPY is often co-localized and released with norepinephrine (NE) in sympathetic nerve terminals (Lundberg *et al.*, 1985), under conditions of extensive neuronal activity stores of NE and NPY may become depleted. NE is synthesized in the nerve terminals via enzymic reactions thereby allowing rapid synthesis and replenishment. The long half-life of NPY mNRA would increase the rate by which the stores of NPY in the nerve terminal could be replaced. That is, the cell would not necessarily have to produce more mRNA to increase the production of this molecule. Control would be at the rate at which the NPY mRNA is translated. Under conditions where even more NPY is needed or in alternative cell types, a transcriptional increase could further boost NPY production. It is, therefore, important to recognize the contributions of both types of regulation in maintaining the appropriate NPY mRNA levels in various cell types.

Transgenic mice

The experiments described above are useful for elucidating the various mechanisms by which the activity of the NPY gene is regulated. However, the *in vivo* transfection studies can be problematic when addressing the question of how tissue specificity is manifested. Other factors such as appropriate chromatin structure, methylation patterns, or chromosomal localization cannot be effectively reproduced in transient transfection studies. This becomes especially apparent when NPY–CAT fusion constructions are introduced into HeLa cells. As mentioned previously, these cells do not produce NPY mRNA but can support the activity of the deletion constructs. In fact, the pattern of chimaeric gene expression obtained in HeLa cells mimics that seen in LA-N-5 cells (data not shown). To identify regions of the NPY gene that dictate tissue-specific transcription, a fusion gene containing 796 bp of 5′-flanking DNA fused to the CAT gene was inserted into the mouse genome (Fig. 10) (Waldbieser *et al.*, 1992). Five founder transgenic mice were produced via pronuclear microinjection into fertilized ova and transgenic lines were established for each founder. Four of the five transgenes were transmitted with an autosomal inheritance pattern while the remaining one appeared to be integrated within the X chromosome.

The pattern of NPY gene expression in the mouse closely parallels the expression in both rats and humans. The tissues expressing the largest

Fig. 10. Schematic diagram of the NPY 5′-flanking sequence used to make transgenic mice

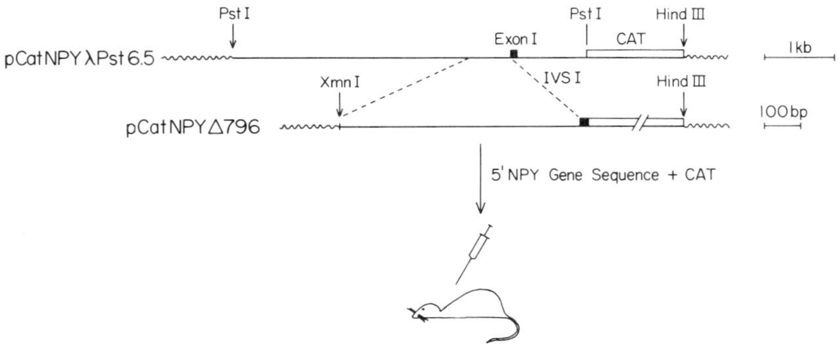

The plasmid pCatNPYλPst 6.5 was digested with XmnI and an approximately 800 bp fragment was inserted upstream of the coding sequences for the bacterial enzyme CAT to create the construct pCatNPYΔ796. The NPY gene's 5′-flanking sequence is indicated by a solid line. Exon I is depicted as a solid box while the CAT coding sequences aer represented by an open box. Vector sequences are indicated by wavy lines. The pCatNPY Δ796 construct was linearized with Hind III prior to injection.

amounts of NPY mRNA include the cerebral cortex, spinal cord, adrenal glands, olfactory bulb and limbic region. The expression of the transgenes was varied in each mouse line presumably due to differing integration sites and differing copy number of the integrated transgene. The X-linked line did not express the transgene and is not included in the discussion. Three of the four autosomally inserted transgenes produced measurable activity. In all cases, the highest activities resided in the olfactory bulb, cerebral cortex, and limbic region. Lower levels of activity were present in the brain stem, spinal cord, eye and adrenal glands of at least one of the transgenic lines (Fig. 11). Therefore, transcriptional activity of the transgene, as determined by the CAT assay, was detected in tissues which normally express NPY. However, the CAT activity observed did not always correlate with the amount of steady-state NPY mRNA levels as measured by slot blots. This was true for the adrenal gland and spinal cord which had large amounts of NPY mRNA but very little CAT activity. Several theories could explain this result. First, the CAT assay measures transcriptional activity while the slot blot measures steady-state mRNA levels. Secondly, the experiments described in the previous section suggest that the stability of the NPY mRNA may be an important control point in regulating the amount of steady-state mRNA present in certain cell types. The instability of the CAT message could exacerbate these differences. Thirdly, the integration site of the transgene

Fig. 11. Representative CAT assays for two transgenic mice lines

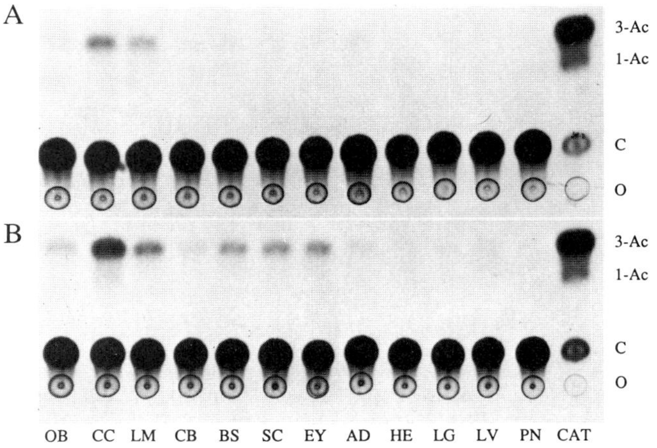

(A) CAT assay for tissues isolated from the 24-9 transgenic mice line. (B) Cat assay performed from the 30-2 transgenic mouse line. In both cases, up to 100 μg of protein from each tissue was incubated overnight with 5 μCi of [^{14}C]chloramphenicol and 4 mM-acetyl CoA at 37 °C. Reaction products were separated by thin-layer chromatography in 95% chloroform:5% methanol (v/v) and exposed to Kodak X-Omat AR film for 2 days. The reaction products identified on the right: C, chloramphenicol; 1-Ac, monoacetylated chloramphenicol; 3-Ac, diacetylated chloramphenicol; O, origin; OB, olfactory bulb; CC, cerebral cortex; LM, limbic region; CB, cerebellum; BS, brain stem; SC, spinal cord; EY, eye; AD, adrenal glands; HE, heart; LG, lung; LV, liver; PN, pancreas.

may influence its expression in certain tissues. Finally, the Δ796 deletion construct may not contain sufficient information for efficient expression in all the NPY-expressing cell types. Additional experiments of this type will allow us to define further the sequences necessary for tissue-specific expression in all the various cell types which express NPY.

Summary

The availability of human neuroblastoma cell lines that express NPY has allowed the dissection of several promoter elements. Further studies will provide additional information about how these various elements either work in concert or as mutually exclusive elements to modulate the expression of this gene. Caution must be used when evaluating these data because, by definition, neuroblastomas represent cells which have lost important growth controls. The NPY gene is over-expressed in these cells suggesting that some factors controlling the expression of this gene have been lost or mutated. Notably, various neuroblastomas, including LA-N-5 cells, have amplified the n-*myc* gene (Schwab *et al.*, 1985). In a survey of neuroblastoma cell lines, we were unable to correlate expression of n-*myc* with either increased or decreased NPY mRNA production (data not shown). Finally, studies using transgenic mice have extended our ability to investigate important *cis*-regulatory regions giving us a better picture of what happens in a more authentic *in vivo* setting.

I thank Robin Lerchen and Drs. Randy Haun and Jack Dixon for critically reading this manuscript. The data communicated prior to publication by Dr. William Taylor was also much appreciated. This work was supported by NS 28496 (C.D.M.), American Heart Association, Ohio Division (C.D.M.), and NIDDKD 18849 (J.E.D.).

References

Adrian, T. E., Allen, J. M., Bloom, S. R., Ghatei, M. A., Rossor, M. N., Roberts, G. W., Crow, T. J., Tatemoto, K. and Polak, J. M. (1983) Neuropeptide Y distribution in human brain. Nature (London) **306**, 584–586

Allen, L. G., Kalra, P. S., Crowley, W. R. and Kalra, S. P. (1985) Comparison of the effects of neuropeptide Y and adrenergic transmitters on LH release and food intake in male rats. Life Sci. **37**, 617–623

Briggs, M. R., Kadonaga, J. T., Bell, S. P. and Tjian, R. (1986) Purification and biochemical characterization of the promoter-specific transcription factor, Sp1. Science **234**, 47–52

Chen, C. and Okayama, H. (1987) High-efficiency transformation of mammalian cells by plasmid DNA. Mol. Cell. Biol. **7**, 2745–2752

Eckdahl, T. T. and Anderson, J. N. (1987) Computer modelling of DNA structures involved in chromosome maintenance. Nucleic Acids Res. **15**, 8531–8545

Fried, M. and Crothers, D. M. (1981) Equilibria and kinetics of *lac* repressor–operator interactions by polyacrylamide gel electrophoresis. Nucleic Acids Res. **9**, 6505–6525

Gidoni, D., Dynan, W. S. and Tjian, R. (1984) Multiple specific contacts between a mammalian transcription factor and its cognate promoters. Nature (London) **312**, 409–413

Gray, T. S. and Morley, J. E. (1986) Neuropeptide Y: anatomical distribution and possible function in mammalian nervous system. Life Sci. **38**, 389–401

Gregor, P. D., Sawadogo, M. and Roeder, R. G. (1990) The adenovirus major late transcription factor USF is a member of the helix–loop–helix group of regulatory proteins and binds to DNA as a dimer. Genes Dev. **4**, 1730–1740

Gu, J., Adrian, T. E., Tatemoto, K., Polak, J. M., Allen, J. M. and Bloom, S. R. (1983) Neuropeptide tyrosine (NPY)—a major cardiac neuropeptide. Lancet **i**, 1008–1010

Gustafson, T. A. and Kedes, L. (1989) Identification of multiple proteins that interact with functional regions of the human cardiac α-actin promoter. Mol. Cell. Biol. **9**, 3269–3283

Helms, S. R. and Rottman, F. M. (1990) Characterization of an inducible promoter system to investigate decay of stable mRNA molecules. Nucleic Acids Res. **18**, 255–259

Herzog, H., Hort, Y. J., Ball, H. J., Hayes, G., Shine, J. and Selbie, L. A. (1992) Cloned human neuropeptide Y receptor couples to two different second messenger systems. Proc. Natl. Acad. Sci. U.S.A. **89**, 5794–5798

Higuchi, H., Yang, H.-Y. T. and Sabol, S. L. (1988) Rat neuropeptide Y precursor gene expression. J. Biol. Chem. **263**, 6288–6295

Imagawa, M., Chiu, R. and Karin, M. (1987) Transcription factor AP-2 mediates induction by two different signal-transduction pathways: protein kinase C and cAMP. Cell **51**, 251–260

Jones, K. A., Yamamoto, K. R. and Tjian, R. (1985) Two distinct transcription factors bind to the HSV thymidine kinase promoter *in vitro*. Cell **42**, 559–572

Jones, K. A., Kadonaga, J. T., Rosenfeld, P. J., Kelly, T. J. and Tjian, R. (1987) A cellular DNA-binding protein that activates eukaryotic transcription and DNA replication. Cell **48**, 79–89

Krasinski, S. D., Wheeler, M. B., Kopin, A. S. and Leiter, A. B. (1990) Pancreatic polypeptide and peptide YY gene expression. In Central and Peripheral Significance of Neuropeptide Y and its Related Peptides, 1st edn., pp. 73–85, The New York Academy of Sciences, New York

Larhammar, D., Blomqvist, A. G., Yee, F., Jazin, E., Yoo, H. and Wahlestedt, C. (1992) Cloning and functional expression of a human neuropeptide Y/peptide YY receptor of the Y1 type. J. Biol. Chem. **267**, 10935–10938

Lee, W., Mitchell, P. and Tjian, R. (1987) Purified transcription factor AP-1 interacts with TPA-inducible enhancer elements. Cell **49**, 741–752

Leiter, A. B., Montminy, M. R., Jamieson, E. and Goodman, R. H. (1985) Exons of the human pancreatic polypeptide gene define functional domains of the precursor. J. Biol. Chem. **260**, 13013–13017

Lundberg, J. M., Saria, A., Franco-Cereceda, A., Hokfelt, T., Terenius, L. and Goldstein, M. (1985) Differential effects of reserpine and 6-hydroxydopamine on neuropeptide Y (NPY) and noradrenaline in peripheral neurons. NS Arch. Pharm. **328**, 331–340

Magni, P. and Barnea, A. (1992) Forskolin and phorbol ester stimulation of neuropeptide Y (NPY) production and secretion by aggregating fetal brain cells in culture: evidence for regulation of NPY biosynthesis at transcriptional and posttranscriptional levels. Endocrinology **130**, 976–984

McDonald, J. K. (1988) NPY and related substances. CRC Crit. Rev. Neurobiol. **4**, 97–135

Minth, C. D. and Dixon, J. E. (1990) Expression of the human neuropeptide Y gene. J. Biol. Chem. **256**, 12933–12939

Minth, C. D., Andrews, P. C. and Dixon, J. E. (1986) Characterization, sequence, and expression of the cloned human neuropeptide Y gene. J. Biol. Chem. **261**, 11974–11979

Sabol, S. L. and Higuchi, H. (1990) Transcriptional regulation of the neuropeptide Y gene by nerve growth factor: antagonism by glucocorticoids and potentiation by adenosine 3′,5′-monophosphate and phorbol ester. Mol. Endocrinol. **4**, 384–392

Schwab, M., Varmus, H. E. and Bishop, J. M. (1985) Human n-*myc* gene contributes to neoplastic transformation of mammalian cells in culture. Nature (London) **316**, 160–162

Seeger, R. C., Rayner, S. A., Banerjee, A., Chung, H., Laug, W. E., Neustein, H. B. and Benedict, W. F. (1977) Morphology, growth, chromosomal pattern, and fibrinolytic activity of two new human neuroblastoma cell lines. Cancer Res. **37**, 1364–1371

Spinelli, W. and Ishii, D. N. (1983) Tumor promoter receptors regulating neurite formation in cultured human neuroblastoma cells. Cancer Res. **43**, 4119–4125

Stone, R. A., Laties, A. M. and Emson, P. C. (1986) Neuropeptide Y and the ocular innervation of rat, guinea pig, cat and monkey. Neuroscience **17**, 1207–1216

Takeuchi, T., Gumucio, D. L., Yamada, T., Meisler, M. H., Minth, C. D., Dixon, J. E., Eddy, R. F. and Shows, T. B. (1986) Genes encoding pancreatic polypeptide and neuropeptide Y are on human chromosomes 17 and 7. J. Clin. Invest. **77**, 1038–1041

Tuor, U. I., Edvinsson, L. and McCulloch, J. (1985) Neuropeptide Y and cerebral blood flow. Lancet **i**, 1271

Varndell, I. M., Polak, J. M., Allen, J. M., Terenghi, G. and Bloom, S. R. (1984) Neuropeptide tyrosine (NPY) immunoreactivity in norepinephrine-containing cells and nerves of the mammalian adrenal gland. Endocrinology **114**, 1460–1462

Wahlestedt, C., Pich, E. M., Koob, G. F., Yee, F. and Heilig, M. (1993) Modulation of anxiety and neuropeptide Y–Y1 receptors by antisense oligodeoxynucleotides. Science **259**, 528–531

Waldbieser, G. C., Minth, C. D., Chrisman, C. L. and Dixon, J. E. (1992) Tissue-specific expression of the human neuropeptide Y gene in transgenic mice. Mol. Brain Res. **14**, 87–93

Zini, I., Pich, M., Fuxe, K., Lenzi, P. L., Agnati, L. F., Harfstrand, A., Mutt, V., Tatemoto, K. and Moscara, M. (1984) Actions of centrally administered neuropeptide Y on EEG activity in different rat strains and in different phases of their circadian cycle. Acta Physiol. Scand. **122**, 71–77

5

Environmental and hormonal control of vasoactive intestinal peptide (VIP) gene expression and changes observed in clinical and experimental pathological conditions

James A. Waschek

The Mental Retardation Research Center and Department of Psychiatry, University of California at Los Angeles, CA 90024-1759, U.S.A.

Introduction

Vasoactive intestinal peptide (VIP), a 28 amino acid peptide originally isolated in the porcine intestine (Said and Mutt, 1970), is one of the most highly utilized regulatory peptides in vertebrates (Fahrenkrug, 1980; Rostene, 1984; Said, 1984; Gozes and Brenneman, 1989; Said, 1991*a*). VIP is expressed in distinct subpopulations of neurons in most major structures in the central nervous system (CNS), and in numerous sites of the peripheral nervous system. The spectrum of its biological activity is broad and includes regulation of blood flow, smooth muscle tone and exocrine secretion, and control of cell phenotype, growth rate, survival and glycogen metabolism. By way of these basic activities, the expression of VIP impacts on cardiovascular, pulmonary, gastrointestinal, neurological, immunological and reproductive function. VIP shares structural similarity in its amino terminal with peptide histidine methionine (PHM), which is encoded on the same gene, secretin, glucagon, growth hormone-releasing hormone, pituitary adenylate cyclase-activating peptide, and the lizard peptide helodermin, which appears to have a mammalian counterpart distinct from known members of the VIP family.

Because of the perceived importance of VIP as a regulatory molecule, numerous investigations have been carried out to examine its expression in a variety of hormonal states and pathological conditions. The purpose of this review is to summarize our current knowledge of the changes that occur during normal homeostasis and in disease, and to summarize studies where mechanisms controlling VIP expression have been investigated at the cellular and molecular level.

Distribution and ontogeny of VIP expression in the nervous system

This article focuses primarily on the regulation of VIP expression, so the reader is referred to other articles for detailed information on the distribution and ontogeny of VIP (Emson *et al.*, 1979; Loren *et al.*, 1979; Fahrenkrug, 1980; Maletti *et al.*, 1980; Sims *et al.*, 1980; Palkovits *et al.*, 1981; Beinfeld *et al.*, 1983; Ottesen, 1983; Rostene, 1984; Said, 1984; Fink *et al.*, 1986; Gozes *et al.*, 1987*b*; Fahrenkrug *et al.*, 1988; Gozes *et al.*, 1988; Gozes and Brenneman, 1989; Goetzl *et al.*, 1990). Particularly large numbers of VIP neurons are found in the hypothalamus [particularly in the suprachiasmatic nuclei (SCN)], the cerebral cortex, amygdala, midbrain periaqueductal grey and spinal cord. VIP is also widely expressed in the peripheral nervous system. VIP-immunoreactive neurons have been found in numerous autonomic ganglia, and within the gut, heart, lung, reproductive system and lymphoid tissue.

Immunohistochemical methods do not always reveal which cells actually produce VIP, because the cell bodies are not always labelled, and because target cells can bind to and internalize the peptide (Muller *et al.*, 1985). To complicate matters, pretreatment of animals with colchicine to reveal neuropeptide-producing cells has been shown to have pharmacological effects on peptide mRNA production (Ceccatelli *et al.*, 1991*a*). Thus, *in situ* hybridization, although not always as sensitive as immunohistochemistry, has been useful in helping determine which cells express a particular gene. Studies employing *in situ* hybridization have generally confirmed the location of VIP-producing cells. For example, in agreement with immunohistochemistry results (Emsom *et al.*, 1979), VIP mRNA was found to be present in non-pyramidal cells in all layers of the cortex but principally in layers II and III (Baldino *et al.*, 1989; Benson *et al.*, 1991). In the rat hypothalamus, VIP mRNA was found to be present in the ventrolateral aspect of the SCN (Card *et al.*, 1988), and in the thalamus, VIP gene transcripts have been localized to the ventrolateral, lateral reticular, ventromedial, and ventrobasal nuclei (Baldino *et al.*, 1989). In the intestine, VIP mRNA has been localized to the mucosal (Meissner's) and Auerbach's plexus (Fink *et al.*, 1986). Certain experimental manipulations cause VIP mRNA to appear in locations where it is not usually detected in normal healthy animals (discussed in subsequent sections). These sites include the pituitary, the parvocellular and magnocellular cells of the hypothalamic paraventricular nucleus (PVN) and supraoptic nucleus (SON), cells of the dorsal root ganglia (DRG), and inflammatory cells.

With respect to ontogeny, VIP immunoreactivity in rodent tissues is very low until birth, except for in the intestine, where VIP expression is appreciable on embryonic day 16 (Emsom *et al.*, 1979; Maletti *et al.*, 1980; Beinfeld *et al.*, 1983). After birth, VIP immunoreactivity in the brain rises rapidly until about day 20, when peptide levels in most structures reach a plateau. *In situ* hybridization and Northern blot analyses (Fink *et al.*, 1986; Gozes *et al.*, 1987*b*, 1988) are in good agreement with ontogeny data obtained by radioimmunoassay and immunohistochemistry.

Hormonal control of VIP expression *in vivo*

A large number of studies document that VIP expression is influenced by multiple hormonal and other environmental stimuli (Table 1). Many observed changes in expression relate to the role of the peptide in reproductive function (reviewed in Ottesen, 1983; Fahrenkrug *et al.*, 1988). During the oestrous cycle VIP immunoreactivity falls appreciably at oestrous in the rabbit oviduct (Menghi *et al.*, 1990), where the peptide may regulate reproductive physiology by inhibiting oviduct motility. During pregnancy, VIP expression is increased up to 10-fold in the spinal cord (Chery-Croze *et al.*, 1986) but is decreased in the myometrium (Ottesen *et al.*, 1982). Increases in

Table 1. Hormonal or environmental conditions in which changes in VIP expression have been documented

Hormonal/environmental condition	Location of change in expression*
Oestrous cycle	Oviduct
Pregnancy	Spinal cord, myometrium
Labour, delivery, immediate post partum	Peripheral blood
Lactation, suckling	PVN, SCN, several thalamic nuclei
Ovariectomy, oestrogen treatment	Hypothalamus, anterior pituitary
Orchectomy, testosterone treatment	Hypothalamus, anterior pituitary
Adrenalectomy, dexamethasone treatment	Anterior pituitary, SCN, PVN, pancreatic islets, cerebral cortex, hippocampus
Hypophysectomy	PVN, SON
Thyroidectomy	Anterior pituitary, adrenal gland, PVN
Constant light/dark, enucleation	SCN, retina

**See text for references and abbreviations.*

plasma VIP levels have been noted during labour, or delivery and early post partum (Ottesen *et al.*, 1982; Holst *et al.*, 1986). Relevant actions of VIP during pregnancy and post partum are relaxation of uterine muscle, enhancement of myometrial blood flow and regulation of prolactin, oxytocin, and vasopressin secretion (Ottesen, 1983; Bardrum *et al.*, 1988; Fahrenkrug *et al.*, 1988). Lactation has been associated with an increase (Mezey and Kiss, 1985) and/or no change (Bartha *et al.*, 1990) in VIP immunoreactivity in the parvocellular neurons of the PVN. *In situ* hybridization and Northern blot analysis indicated that VIP mRNA levels in the rat SCN were elevated approximately two-fold during lactation (Gozes *et al.*, 1989*a*). However, another group (Lamperti *et al.*, 1991) reported no change in VIP mRNA in this structure. Suckling was found to increase plasma VIP levels in dogs (Eriksson *et al.*, 1987); mother rats who had pups removed for 9 days had SCN VIP mRNA levels (measured by dot blot analysis) that were less than 30% of those from mothers who did not have pups removed (Ishihara *et al.*, 1991).

Ovariectomy and oestrogen treatment of rats has given mixed results with respect to VIP expression in the hypothalamus. Three groups observed no changes in VIP immunoreactivity and/or mRNA in the hypothalamus after ovariectomy or after subsequent oestrogen treatment (Lam *et al.*, 1990; O'Halloran *et al.*, 1990; Carrillo *et al.*, 1991). Two other groups, however, observed that oestrogen supplementation of ovariectomized animals caused a 50% reduction in anterior and/or mediobasal hypothalamus VIP levels compared with unsupplemented ovariectomized animals (Maletti *et al.*, 1982; Lasaga *et al.*, 1991). Another group observed decreased VIP mRNA levels in the hypothalamus after ovariectomy and reversal of this change after oestradiol treatment (Gozes *et al.*, 1989*b*).

In contrast to these seemingly disparate findings in the hypothalamus, several groups have shown that long-term oestrogen treatment causes a dramatic increase in VIP immunoreactivity (up to 50-fold) and VIP mRNA (up to 100-fold) in the anterior pituitary in normal and ovariectomized animals (Prysor-Jones *et al.*, 1987; Lam *et al.*, 1990; O'Halloran *et al.*, 1990; Carrillo *et al.*, 1991; Lasaga *et al.*, 1991). Kasper *et al.* (1992) recently examined the time course of VIP mRNA induction in the pituitary and other tissues after oestrogen treatment and noted rapid transient induction of VIP mRNA in the pituitary, peaking 30 to 60 min after a single injection of 17β-oestradiol, and long-term induction in oestrogen-implanted animals. Short-term induction of VIP mRNA was also observed in the hypothalamus, ileum and whole brain (within 24 h), but no long-term changes were observed in these structures in animals with the implant. The effect of ovariectomy alone on VIP expression in the pituitary is less clear, reportedly being unchanged (Lam *et al.*, 1990; Lasaga *et al.*, 1991) or decreased (O'Halloran *et al.*, 1990). Treatment of animals with the anti-oestrogen drug tamoxiphen resulted in decreased pituitary VIP mRNA levels (O'Halloran *et al.*, 1990). The effects of oestrogen on VIP expression in the pituitary were reported to occur specifically in the lactotrophs (Steel *et al.*, 1989); however, other data suggest that this is not the case (Carillo and Phelps, 1992). An autocrine or paracrine action of VIP on prolactin secretion has been suggested (Nagy *et al.*, 1987).

Testosterone may also influence VIP expression. Experimental neonatal androgenization induced by testosterone administration 24 h after birth resulted in anterior pituitary VIP levels that were 60-fold higher than controls 8 weeks later (Watanobe *et al.*, 1991). Testosterone treatment of castrated adult male and female rats resulted in significantly decreased VIP immunoreactivity in the mediobasal hypothalamus and anterior pituitary (Lasaga *et al.*, 1991), the effect being most pronounced in the hypothalamus of castrated males, where exogenous testosterone decreased VIP immunoreactivity to approximately 20% of that in unsupplemented controls. Orchiectomy alone had no effect on VIP levels and/or cell number in the mediobasal hypothalamus or hypothalamic PVN (Bartha *et al.*, 1990), or in other tissues such as the anterior pituitary (Lasaga *et al.*, 1991), hippocampus (Rotsztejn *et al.*, 1980), or adrenal gland (Tsuchiya *et al.*, 1990)

Sexual dimorphic patterns of basal VIP expression have also been reported. In the hypothalamus, random cycling females were found to have 50% higher VIP levels than males, but males were found to have 3–7-fold higher VIP levels in the anterior pituitary as well as a greater number of VIP immunoreactive cells (Riskind *et al.*, 1989; Steel *et al.*, 1989; Carillo and Phelps, 1992). Differences in VIP expression in the pituitary are reflected at the level of the mRNA, whereas no differences of VIP mRNA between the sexes were observed in the hypothalamus and cortex (Lam and Srivastava, 1990).

Adrenalectomy resulted in an increase in VIP immunoreactivity in the parvocellular neurons of the hypothalamic PVN of adult rats, an effect which was reversed by dexamethasone administration (Mezey and Kiss, 1985; Bartha *et al.*, 1990; Watanobe, 1990). Adrenalectomy also resulted in an increase of VIP immunoreactivity in the adult rat anterior pituitary but a decrease in the dorsal and ventral hippocampus, all of which were reversed by dexamethasone (Rotsztejn *et al.*, 1980). The induction of VIP expression in the rat pituitary by adrenalectomy has also been noted at the level of the VIP mRNA (Lam *et al.*, 1992). In this latter work, adrenalectomy resulted in no change in VIP mRNA concentration in the cerebral cortex; however, corticosteroid treatment produced an increase in VIP mRNA in the cortex of both adrenalectomized and sham-operated animals. In neonatal rats, corticosterone implants increased VIP immunoreactivity in the hypothalamus and hippocampus, but not the olfactory bulb, while metyrapone-induced hypocorticism caused VIP immunoreactivity to increase in the hypothalamus and olfactory bulb, but to decrease in the hippocampus (Nobou *et al.*, 1985). In another study, dexamethasone treatment of normal rats resulted in increased VIP immunoreactivity and mRNA levels in pancreatic islets of Langerhans (Jamal *et al.*, 1991).

Hypophysectomy resulted in the appearance of VIP immunoreactivity in the perikarya of the SON and PVN (Koves *et al.*, 1991). This induction may have occurred at the level of the gene because hypophysectomy induced the appearance of VIP mRNA in both the parvocellular and magnocellular neurons of the PVN (Ceccatelli *et al.*, 1991 *b*).

Thyroid status has ben reported to be a significant factor controlling VIP expression in certain tissues. VIP immunoreactivity and mRNA levels in the anterior pituitary became elevated when rats were rendered hypothyroid by surgical or pharmacological means (Jones *et al.*, 1989; Lam *et al.*, 1989; Segerson *et al.*, 1989), effects which were reversed by exogenous thyroid supplementation (Lam *et al.*, 1989; Segerson *et al.*, 1989). Thyroidectomy also caused an increase in VIP immunoreactivity in rat adrenal gland extracts (Tsuchiya *et al.*, 1990). No change in VIP immunoreactivity or mRNA level was observed in the rat hypothalamus after induction of hypothyroidism (Jones *et al.*, 1989; Lam *et al.*, 1989); however, specific changes in VIP mRNA were readily apparent in the hypothalamic PVN measured by *in situ* hybridization (Toni *et al.*, 1992).

Pharmacological manipulation of VIP expression *in vivo*

VIP expression has been shown to be altered *in vivo* by chronic treatment with various pharmacological agents. For example, treatment of rats for 7 days with dopamine agonists and antagonists reduced VIP immunoreactivity in the anterior pituitary to 10–15% of control values (Prysor-Jones *et al.*, 1987). Treatment for 24 h with reserpine, an agent known classically to deplete catecholamine and 5-hydroxytryptamine stores, or with colchicine was found to induce the appearance of VIP mRNA levels in the parvocellular neurons of the PVN (Ceccatelli *et al.*, 1991 *a*). Administration of atropine subcutaneously for 10 days resulted in a significant decrease in VIP content in the rat cerebral cortex (Lapchak and Beaudet, 1990). Treatment with nicotine resulted in the same effect, possibly explained by receptor desensitization. Treatment of rats with atropine also resulted in decreased VIP immunoreactivity in the dorsal raphe, locus coeruleus, and ventrolateral and dorsolateral medulla (Petit *et al.*, 1992). Administration of the histamine-1 antagonist chlorpheniramine resulted in decreased VIP immunoreactivity in the adrenal gland (Tsuchiya *et al.*, 1990).

Regulation of VIP expression and circadian function

VIP peptide and mRNA levels oscillate with 24 h periodicity in the SCN (Gozes *et al.*, 1989*b*; Takahashi *et al.*, 1989; Albers *et al.*, 1990; Okamoto *et al.*, 1991), the structure believed to generate and modulate circadian function in mammals (Meijer and Rietveld, 1989). The VIP-expressing cells in the

SCN are located in the ventrolateral region of this nucleus and receive neural input from two well-defined primary and secondary photic projections, the retinohypothalamic tract and the geniculohypothalamic tract, and from a third pathway originating in the raphe nucleus (Albers *et al.*, 1991). When rats are housed under a normal light/dark cycle, levels of peptide and mRNA in the SCN are highest in the dark (Gozes *et al.*, 1989*c*; Takahashi *et al.*, 1989; Albers *et al.*, 1990; Okamoto *et al.*, 1991), and when housed in constant darkness, VIP immunoreactivity in the SCN is higher than when housed in constant light (Albers *et al.*, 1987). Similarly, VIP peptide and mRNA levels are increased in the SCN of rats after bilateral enucleation (Stopa *et al.*, 1988; Holtzman *et al.*, 1989; Okamoto *et al.*, 1990). Thus, it appears that cells in the SCN might be involved in integrating visual and/or other environmental input into the circadian rhythm (a process termed entrainment). A recent report indicated that after hemisection of the SCN, circadian electrical activity was unperturbed in the ventrolateral portion, but possibly abolished in the dorsomedial portion, suggesting that the ventromedial region contained the cells that drive the rhythm (Tcheng and Gillette, 1990). It is possible that the VIP-producing cells, which are the only cells in the ventrolateral region of the SCN known to exhibit 24 h rhythmicity (Albers *et al.*, 1991), are the actual cells which drive the circadian rhythm.

VIP expression may also change with the light/dark cycle or with an endogenous rhythm in other structures of the brain. For example, in extracts of the rat pineal gland, VIP immunoreactivity was highest during the dark phase (Kakau *et al.*, 1986). Similarly, VIP immunoreactivity in the PVN was highest during the dark cycle (Morin *et al.*, 1991), and was increased in the retina after eyelid fusion (Stone *et al.*, 1988). In contrast VIP mRNA in the cerebral cortex was decreased after enucleation (Holtzman *et al.*, 1989).

Changes in VIP expression in pathological conditions

Changes in patterns of VIP expression have been observed in a number of diseases in man (Table 2), and it has been suggested that these changes contribute to the pathology of the disease or compensate for the pathological state. Some disease states are characterized by a loss in more than one type of peptidergic neuron and may represent a response to degeneration of a target tissue or the absence of a factor that controls the expression of more than one peptide. An example of the latter case is a mouse dwarf strain that contains a mutation in the gene encoding the growth hormone/prolactin gene transcription factor Pit-1 (Li *et al.*, 1990). In these mice, pituitary lactotrophs, somatotrophs and thyrotrophs fail to develop, and animals do not express growth hormone, thyroid-stimulating hormone or prolactin.

Nerve injury

It has been proposed that VIP may play a role in neuronal regeneration following injury probably by way of the ability of the peptide to stimulate

Table 2. Pathological conditions in humans and in animal models in which changes in VIP expression have been reported

Pathological condition	Tissue where change is measured
Human diseases	
Diabetes	Skin, penis
Asthma, cystic fibrosis	Lung
Neuroendocrine tumours	Tumour cells, peripheral blood
Barrette's oesophagus	Peripheral blood
Achalasia, hiatal hernia	Oesophageal sphincter
Infantile pyloric stenosis	Pyloric muscle
Diverticular disease, idiopathic constipation	Colon
Hirschsprungs disease, Crohn's disease, ulcerative colitis	Colon
Chagas' disease	Rectum
Pre-eclampsia	Peripheral blood
Parkinson's disease	Adrenal medulla
Atopic dermatitis, eczema, psoriasis	Skin
Animal models	
Nerve injury	Dorsal root ganglia, spinal cord
Spontaneous hypertension	SCN, thalamic nuclei, cerebral cortex
Murine schistosomiasis	Granulomas (eosinophils)
Experimental and spontaneous diabetes	Skin, eye, urinary bladder, enteric nerves, seminal vesicles, penis, sciatic nerve

cell growth, release neurotrophic factors or stimulate glucose metabolism (Magistretti *et al.*, 1981; Brenneman *et al.*, 1985, 1990; Pincus *et al.*, 1990). After crushing or resection of the sciatic nerve, VIP immunoreactivity rises dramatically in the terminal area of this nerve in the upper spinal dorsal horn (McGregor *et al.*, 1984) and in the corresponding DRG (Shehab and Atkinson, 1986). The appearance of VIP expression in the DRG occurs primarily, if not exclusively, in small type-B sensory neurons which innervate visceral, cutaneous and muscular sensory targets (Kashiba *et al.*, 1992*a*). Local application of either the microtubule inhibitor vinblastine or antibodies to nerve growth factor (NGF) to a section of the rat sciatic nerve mimicked the rise in VIP expression in the dorsal horn after sciatic resection, suggesting that blockade of retrograde axoplasmic transport of NGF or a related factor might up-regulate VIP expression in the DRG cells (Knyihar-Csillik *et al.*, 1991; Kashiba *et al.*, 1992*b*). The increase in VIP immunoreactivity in the DRG after resection of the sciatic nerve has been associated with a large increase in VIP mRNA concentration as measured by Northern blot analysis (Nielsch and Keen, 1989) and *in situ* hybridization (Noguchi *et al.*, 1989). Others have shown that VIP immunoreactivity is increased in the proximal stump after sciatic resection, and that part of this increase might be associated with neurons originating in the sympathetic ganglia (Anand *et al.*, 1990).

Lung diseases

VIP is present in the lung in nerve fibres innervating the smooth muscle, blood vessels and submucosal glands where the peptide may regulate blood flow, airway tone and mucus secretion (reviewed in Said, 1991*b*,*c*). A complete lack of VIP, but not substance P innervation was noted in asthmatic patients (Ollereshaw *et al.*, 1989) and this has stimulated efforts to design stable VIP receptor analogues for the treatment of this disease. In cystic fibrosis, VIP immunoreactivity in the innervation of the acini and ducts of eccrine sweat glands was sharply reduced (Heinz-Erian *et al.*, 1985), possibly relevant to the pathology of this disease.

Neuroendocrine tumours

Several neuroendocrine tumours have been found to secrete large amounts of VIP and other neuropeptides *in vivo* and are associated with a watery diarrhoea syndrome possibly due to the action of VIP on receptors on intestinal endothelial cells that control water and ion transport. VIP-producing tumours include neuroblastomas, pancreatic carcinomas, bronchogenic carcinomas, ganglioneuromas, ganglioneuroblastomas and pheochromocytomas (reviewed in Said and Faloona, 1975; Allen *et al.*, 1985; Scholar and Paul, 1991). We found that a human neuroblastoma cell line expressed VIP and VIP mRNA, but also possessed high affinity VIP receptors coupled to cyclic AMP (cAMP) production, suggesting a possible autocrine action of VIP on tumour growth of these cells (Muller *et al.*, 1989). More recently, several human neuroblastoma tumours were found to express both VIP immunoreactivity and functional VIP receptors (O'Dorisio *et al.*, 1992), and the growth rate of a neuroblastoma cell line co-expressing VIP and VIP receptors was shown to be increased by VIP receptor antagonists (Pence and Shorter, 1992). Non-small cell lung cancer (NSCLC) tumours were also shown to express VIP and functional VIP receptors, and nude mice with NSCLC explants exhibited a 40–60% reduction of tumour volume compared with controls after daily subcutaneous injection of VIP receptor antagonists (Moody *et al.*, 1993). As seen below, cell lines derived from VIP-producing tumours have been useful in studying transcription and post-transcription mechanisms that control VIP expression.

Diabetes

Changes in VIP expression in several tissues have been noted in various diabetic complications. Because the penis is richly innervated by VIP neurons and because VIP appears to be involved in mediating the erectile response, VIP innervation of the penis was investigated as a possible cause of diabetic impotence. VIP immunoreactivity was found to be reduced in penile tissue of several adult diabetic men (Crowe *et al.*, 1983; Karanth *et al.*, 1990) and streptozocin-diabetic rats (Crowe *et al.*, 1983). A reduction of VIP and other neuropeptide immunoreactivity has also been observed in the cutaneous nerves of patients with advanced diabetes (Levy *et al.*, 1989). However, in skin of streptozocin-diabetic rats, increased numbers of VIP-immunoreactive fibres were present around sweat glands and blood vessels 12 weeks after induction of diabetes (Karanth *et al.*, 1990), raising the possibility that VIP expression in some tissues may increase in the early stages of the disease. In this regard, a selective increase in expression of VIP was observed in the enteric nerve of the proximal colon of streptozocin-diabetic rats 8 weeks after induction of diabetes, but thereafter VIP immunoreactivity decreased over the next 13 weeks (Belai *et al.*, 1988). VIP expression is reportedly increased in several other tissues in animal diabetes models, for example in the seminal vesicles (Moss *et al.*, 1987) and the iris (Crowe and Burnstock, 1988) of streptozocin-diabetic rats, and in the eye and urinary bladder of diabetic chinese hamsters (Diani *et al.*, 1985). In contrast, VIP immunoreactivity was found to be decreased in the sciatic nerve but unchanged in the spinal cord in streptozocin-rats (Noda *et al.*, 1990).

Gastrointestinal diseases

Changes in degree of VIP innervation have been reported in a number of gastrointestinal diseases. VIP immunoreactivity in the colonic wall was reported to be selectively increased in patients with idiopathic constipation (Milner *et al.*, 1990), and increased (along with other neuropeptides) in the oesophageal sphincter of patients with hiatal hernia (Aggewstrup *et al.*, 1985). Elevations of VIP in the plasma were reported in patients with severe Barrette's oesophagus (Rossiter *et al.*, 1991). In contrast to these increases, a complete absence of VIP and several other neuropeptides has been observed in aganglionic bowel segments in patients with Hirschsprung's disease (congenital megacolon) (Bishop *et al.*, 1981; Larsson and Sundler, 1983; Taguchi *et al.*, 1983). Similar non-specific decreases in VIP expression have been noted in the pyloric muscle in infants with hypertrophic pyloric stenosis (Wattchow *et al.*, 1987), in the lower oesophagus of patients with achlasia (Aggewstrup *et al.*, 1983), and in rectal biopsies of patients with Chagas' disease (Long *et al.*, 1980). A selective decrease in VIP expression has been reported in the colonic wall of patients with Crohn's disease and ulcerative colitis (Koch *et al.*, 1987), and in patients with diverticular disease (Milner *et al.*, 1990).

Hypertension, Parkinson's disease and skin diseases

In a strain of spontaneously hypertensive rat (SHR), VIP mRNA in the brain was reported to be increased compared with that of non-hypertensive Wistar

Kyoto (WKY) rats (Avidor *et al.*, 1989). The effect was most apparent in the SCN, where a 2-fold elevation was reported. Adult WKY rats grafted with embryonic hypothalamic tissue from SHR became hypertensive and showed a 40% elevation of VIP mRNA in the SCN as measured by *in situ* hybridization (Eilam *et al.*, 1991). In pregnant women, plasma VIP levels were more than 3-times higher in those with hypertensive pre-eclampsia versus normal pregnancies (Holst *et al.*, 1991). Parkinson's disease has been associated with decreased immunoreactive levels of VIP and other neuropeptides in the adrenal medulla (Stoddard *et al.*, 1991). VIP levels in the skin have been reported to be selectively increased in eczema and psoriasis (Anand *et al.*, 1991) and increased in atopic dermatitis (Pincelli *et al.*, 1991).

Granuloma diseases

VIP immunoreactivity and VIP mRNA have been observed in murine granulomas induced by Schistosomiasis mansoni (Weinstock and Blum, 1990). Both peptide and mRNA were localized to the eosinophils which comprised approximately 50% of the inflammatory cells. In addition to the potential significance of VIP in the pathology of granulomas, the detection of VIP mRNA in eosinophils provides evidence that VIP immunoreactivity detected in immune cells represents authentic VIP gene products. Previous evidence that VIP is produced by immune cells is that low levels of VIP immunoreactivity have been detected primarily in granulocytes (reviewed in Goetzl *et al.*, 1990).

Mechanisms controlling the expression of VIP

Establishment of the VIP phenotype

It is generally accepted that neuronal phenotype in vertebrates is determined at least in part by environmental factors (Patterson, 1990). In a relevant example, catecholaminergic sympathetic neurons in the rat destined to innervate sweat glands normally switch to a cholinergic and VIP-expressing phenotype when they reach their targets. In transplant studies, when these neurons are caused to innervate another target that does not normally induce the phenotypic switch, the neurons do not come to express VIP or cholinergic markers, indicating the switch is not preprogrammed (Schotzinger and Landis, 1990). By some interaction, therefore, the sweat gland or its environment must be involved in establishing the expression of VIP and cholinergic markers in these neurons. Other studies demonstrate the potential for secretory factors to effect such a conversion. Fractionation of heart cell-conditioned medium revealed a substance called cholinergic differentiation factor (CDF), now known to be identical to leukaemia inhibitory factor (LIF), that suppresses catecholamine synthesis in cultured rat superior cervical ganglia cells, while stimulating production of VIP and cholinergic markers (Nawa and Patterson, 1990). Similarly, ciliary neurotrophic factor (CNTF) induces cultured cells of the chick sympathetic ganglia to express VIP (Ernsberger *et al.*, 1989). In other examples, NGF caused an increase and dexamethasone a decrease in VIP immunoreactivity in both cultured pheochromocytoma (Tischler *et al.*, 1984) and in human adrenal chromaffin cells (Tischler *et al.*, 1985). These examples suggest how VIP expression might be stimulated by their peripheral targets or local environments.

Tissue culture studies suggest some other ways that VIP expression might be regulated during development. That interleukin-1α and tumour necrosis factor-α selectively increased VIP but not other neuropeptide immunoreactivity in cultured bovine chromaffin cells provides evidence that cytokines in addition to CDF/LIF and CNF regulate VIP and neuronal phenotype (Eskay *et al.*, 1992). In other work, blockade of spontaneous electrical activity in fetal spinal cord/DRG cultures by tetrodotoxin caused a down-regulation of VIP mRNA levels (Agoston *et al.*, 1991), suggesting that VIP gene expression is controlled by neuronal activity. Conversely, application of the excitatory amino acid glutamate receptor agonist *N*-methyl-D-aspartate to these cultures resulted in a 2- to 3-fold increase in VIP mRNA. The morphogen retinoic acid may regulate VIP expression during development, because treatment of a human neuroblastoma cell line with this agent resulted in increased VIP peptide (Waschek *et al.*, 1989) and mRNA levels (Agoston *et al.*, 1992).

Intracellular messengers

It is likely that classical second messenger pathways are utilized to control VIP expression in some of the situations described above and possibly in other processes such as replenishment of peptide after peptide secretion or in adaptation. Cultured adult bovine adrenomedullary chromaffin cells have been used as a model to study transynaptic and second messenger regulation of VIP expression. Treatment of chromaffin cells with nicotine, which presumably mimics splanchnic nerve stimulation, resulted in calcium-dependent secretion of VIP (Eiden *et al.*, 1983). We found that this same treatment also increased VIP peptide biosynthesis, and that this could be blocked by lowering calcium in the medium or by the addition of calcium channel blockers (Waschek *et al.*, 1987; Waschek and Eiden, 1988). Mechanistically, stimulation of nicotinic receptors is thought to cause sodium entry, depolarization, opening of voltage-dependent calcium channels, and entry of calcium into the cell. Calcium appears to act at different targets within the cell to stimulate peptide secretion and biosynthesis. When chromaffin cells were treated with barium ions (which can act as calcium agonists and also depolarize the cell), VIP and met-enkephalin secretion and biosynthesis were greatly enhanced (Waschek and Eiden, 1988). However, in the absence of calcium in the medium, biosynthesis was inhibited but secretion was further enhanced (Waschek and Eiden, 1988). The data suggested that the calcium target used for secretion could be activated by calcium or barium, while the target for biosynthesis exclusively required calcium.

We and others have found that VIP production in bovine chromaffin cells can also be increased by treatment with forskolin, which increases intracellular cAMP and by the phorbol ester, phorbol 12-myristate 13-acetate (PMA), which is thought to mimic diacylglycerol and activate protein kinase C (Pruss *et al.*, 1985; Waschek *et al.*, 1987). Thus, it appears that VIP expression can be stimulated by second messenger pathways activated by calcium, cAMP, and diacylglycerol. The involvement of second messengers in regulating VIP expression has also been studied in electrically-active or tetrodotoxin-blocked mouse fetal spinal cord/DRG cultures. In electrically-active cultures, VIP mRNA levels became increased after treatment with 8-bromo-cAMP or calcium ionophores (Agoston *et al.*, 1991). In contrast, when electrical activity in cultures was blocked by tetrodotoxin, VIP mRNA levels were increased by 8-bromo-cAMP and PMA, but not calcium ionophores. Subnanomolar concentrations of VIP itself were found to induce expression of VIP mRNA in these cultures, supporting a role for autoregulation of the VIP gene. Other culture systems of primary tissue have been used or are amenable to use in studying mechanisms controlling VIP expression. For example, VIP expression has been detected in long-term cultures of dissociated adult rat dorsal root ganglia cells (Schoenen *et al.*, 1989; Mulderry and Lindsay, 1990), fetal rat cerebrocortical and hypothalamic cells (Lorenzo *et al.*, 1989), and in organotypic cultures of neonatal cortex (Gotz and Bolz, 1989) and hypothalamus (Bos and Mirmiran, 1990). Thyroid hormone was recently shown to down-regulate levels of VIP immunoreactivity in cultured fetal cerebrocortical cells (Lorenzo *et al.*, 1992).

Cloned populations of VIP-producing tumour cell lines hold some advantages over primary tissue culture models because they represent a relatively uniform population of VIP-producing cells and thus have been useful in studying mechanisms controlling VIP expression. Similar to cultures of chromaffin cells, cultured neuroblastoma cells respond to cAMP and phorbol esters with increased VIP peptide, and, further, the increase in VIP peptide levels occurs in association with an increase in VIP mRNA (Ohsawa *et al.*, 1985; Waschek *et al.*, 1988). Data from one neuroblastoma indicated that cAMP acts to increase VIP mRNA by increasing VIP gene transcription (Hayakawa *et al.*, 1984). Interestingly, in another neuroblastoma cell line, we found that dibutyryl-cAMP (apparently not acting at a cAMP site) caused a greater than 6-fold induction of VIP immunoreactivity, but no detectable change in mRNA levels (Agoston *et al.*, 1992). Carbachol, which increases phosphatidylinositol turnover in these latter cells (Yu and Sadee, 1986), also increased VIP immunoreactivity but did not change VIP mRNA levels, providing additional evidence that regulation can occur at translational or post-translational steps (discussed below).

Gene organization and transcriptional control

We and others have cloned and characterized the human, rat and mouse VIP genes (Bodner *et al.*, 1985; Tsukada *et al.*, 1985; Linder *et al.*, 1987; Waschek *et al.*, 1988; Yamagami *et al.*, 1988; Giladi *et al.*, 1990; Lamperti *et al.*, 1991; Waschek *et al.*, 1992). The human gene spans about 8.8 kb and includes seven exons (Fig. 1). The rat and mouse genes have a similar organization with seven exons approximately equivalent in size to corresponding exons of the human gene. Several peptides in addition to VIP are encoded on the precursor molecule, some of which have been isolated and shown to have biological activity. Of these, PHM [peptide histidine isoleucine (PHI), in rodents and some other species] has been best characterized and mapped. The distribution of PHI is approximately the same as VIP in most areas of the brain examined although some structures may show significantly higher levels of PHI (Beinfeld *et al.*, 1984). Probes directed to the VIP and PHI portion of the mRNA both recognize a single band at a size corresponding to the full-length mature message on Northern blots (Gozes *et al.*, 1987*b*; Lamperti *et al.*, 1991). This suggests that the preferential appearance of one peptide over the other does not occur because of differential processing of the precursor mRNA, but rather because of post-translational processing of the peptide precursor. Other data do suggest that mRNA processing could be a form of regulation of biosynthesis. A higher percentage of VIP mRNA

Fig. 1. Structure of the human VIP gene

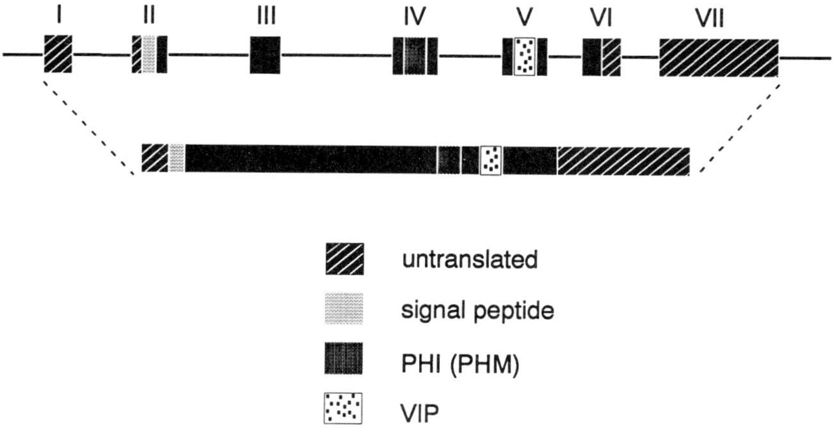

The seven exons of the gene are indicated with the mature mRNA shown below. Various portions of the encoded peptide precursor are denoted. Black portions of the bar indicate additional peptide fragments which are generated by precursor cleavage.

detected in the brain early in development is reportedly in the form of a high molecular mass precursor containing intronic sequences (Gozes *et al.*, 1987*a*). Thus, it is possible that a specific regulatory step in development involves the processing of this presumed heteronuclear precursor.

The full-length mature VIP mRNA in most adult tissues appears as a single band on a Northern blot at a size of 1800 to 2000 nucleotides. However, in the pituitary, another equally prominent band about 600 nucleotides shorter appears (Jones *et al.*, 1989; Segerson *et al.*, 1989; Lam and Srivastava, 1990; Lam *et al.*, 1990). On long exposure of Northern blots, a similar size band also appears as a minor species in RNA from the cortex and other tissues (Lam and Srivastava, 1990). Data suggest that an alternative polyadenylation site in the untranslated region of the VIP gene is used to generate this message (Lamperti *et al.*, 1991). It will be of interest to know if this alternative use of polyadenylation signals results in a quantitative or qualitative change in expression of gene-derived peptides.

We and others have analysed the VIP gene for transcription control sequences in human neuroblastoma, rat pheochromocytoma and other tumour cell lines. Tsukada *et al.*, (1987) studied cAMP induction of the VIP gene by transfecting PC12 cells with a series of plasmids containing up to 2.5 kb of 5′ VIP gene flanking sequences fused to a reporter. Induction by cAMP was found to be conferred by a 17 bp element within the first 100 bp upstream from the transcription start site (Fig. 2). This element contained two copies of a consensus cAMP element found in other cAMP-regulated genes, both of which are required for maximum cAMP induction, the downstream element being most active (Tsukada *et al.*, 1987; Fink *et al.*, 1988). This same 17 bp sequence was found to confer phorbol ester induction in HeLa cells (Fink *et al.*, 1991). The cAMP, but not the phorbol ester response, could be blocked by an inhibitor of protein kinase A, suggesting that the two

Fig. 2. Alignment of the proximal promoter portions of the human, rat and mouse VIP genes

```
TCAAGCCCTATTCATCCCATGGCCGTCATACTGTGACGTCTTTCAGAGCACTTTGTGATTGCTCAGCCCTAAGTATAAGCCCTATAAAATGATGGGCTTTGAAATGCTG  - human
     xx  x            xxxxxx xx             x x xx x  x              x   x x                xx x      x
TCAAGAACTGTTCATCCCATGGGATCACTCATGTGACGTCTTTCGGGGAGCCTTCTGATTGCTCAGCCCGAAGCACAAGCCCTATAAAATGAGAGACTTTGAGATGCTG  - mouse
      x                     xxx              x                       x                       xx
TCAAGACCTGTTCATCCCATGGGATCACCACTGTGACGTCTTTCGAGGAGCCTTCTGATTGCTCAGCCCTAAGCACAAGCCCTATAAAATGAGGAACTTTGAGATGCTG  - rat
```

Sequence mismatches between rodent and human genes are indicated by an X. A heavy bar overlies the cAMP- and phorbol ester-responsive element defined in rat PC12 and human HeLa cells (Tsukada et al.*, 1987; Fink* et al.*, 1988, 1991). The two 5 bp motifs on this element necessary for full cAMP responsiveness are underlined. The lighter and more extended bar overlying the human gene sequences is a region on the gene found to confer combinational cAMP/phorbol ester responsiveness in human neuroblastoma cells (Ohsawa* et al.*, 1987).*

pathways converge beyond protein kinase A activation (Fink *et al.*, 1991). Gel retardation assays indicated that the element could interact with either the cAMP responsive element binding protein or a c-*fos*/c-*jun* complex. As pointed out by Lamperti *et al.* (1991), although the rat and mouse genes are highly homologous to the human gene in the proximal promoter, the rodent sequences diverge from human sequences in the upstream portion of the cAMP/phorbol ester-responsive sequences identified in the above studies. Interestingly, another group found that full combinational phorbol ester/cAMP responsiveness of the human VIP gene in a human neuroblastoma cell line was maintained in the absence of the upstream copy of the element (Ohsawa *et al.*, 1987).

In our work, using subclones of the VIP-expressing human neuroblastoma cell line SK-N-SH, we showed that multiple transcription control sequences span the human VIP gene 5′ flanking sequence (Waschek *et al.*, 1988; Agoston *et al.*, 1990 and Fig. 3). Individual subclones of this cell line displayed large differences in basal VIP expression and phorbol ester responsiveness, so we were able to transfect subclones with various VIP/reporter plasmids to identify VIP gene regulatory elements responsible for these differences. We found that some cAMP and phorbol ester responsiveness resided in an area of the gene containing the cAMP and phorbol ester-

Fig. 3. Segments on the 5′ flanking sequence of human VIP gene found to confer a high level of basal and inducible expression in subclones of the human neuroblastoma SK-N-SH

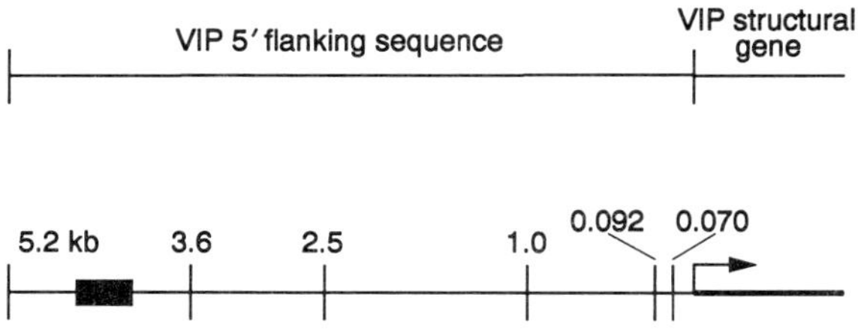

Basal VIP expression - high	++++	+	0	V	+
Basal VIP expression - low	0	0	0	V	+
Phorbol ester PMA, SH-SY-5Y	++++	0	0	0	+
Phorbol ester PMA, SH-IN	0	0	0	VV	0
Forskolin (cAMP)	0	0	0	0	+

(+) indicates that the segment contains a positive response element, (0) indicates no measured effect, and (V) indicates that the segment contains a negative response element. Data from Waschek et al. *(1988) and Agoston* et al. *(1990).*

responsive sequences defined by the studies described above. However, additional regulatory elements resided further upstream, including a sequence located 4.0 to 4.6 kb upstream from the VIP gene transcription start site that conferred a high degree of basal expression and phorbol ester responsiveness. This element exhibited a property of a typical enhancer because its activity increased when it was moved closer to the promoter (Agoston *et al.*, 1990). The sequence of this fragment (Waschek *et al.*, 1992) revealed no nucleotide stretches with significant homology to consensus sequences that bind known transcription factors. However, the fragment contained a closely spaced set of 9 bp sequences repeated approximately 80 bp downstream on the same fragment. We are currently analysing the binding of human neuroblastoma nuclear proteins to this fragment and carrying out further deletion/mutation analyses to determine how these or other sequences on this fragment might control transcriptional activity.

Translation and post-translation regulation

In addition to regulation at the level of transcription, there is evidence that expression of VIP and other VIP gene products is regulated at other steps in the biosynthesis pathway. VIP immunoreactivity in the rat cortex is low at birth, but rises approximately in concert with VIP mRNA levels until peaking at about day 14 to 16 (Emson *et al.*, 1979; Maletti *et al.*, 1980; Beinfeld *et al.*, 1983). However, in the hypothalamus, as pointed out (Gozes and Brenneman, 1989; Gozes *et al.*, 1987*a*), VIP mRNA levels at birth are relatively high, when VIP immunoreactivity is quite low, suggesting that VIP mRNA translation might be negatively controlled in the hypothalamus at early stages of development. As discussed above, in cultured human neuroblastoma cells, we found that VIP peptide levels could be greatly increased without a detectable change in VIP mRNA levels (Agoston *et al.*, 1992). A change in translation efficiency could occur by a subtle change in the mRNA by use of alternative, but closely spaced transcription start sites, splice sites or polyadenylation signals. Alternatively, translation efficiency could be specifically enhanced by the appearance or disappearance of a regulatory factor which binds to VIP mRNA sequences.

Much evidence already exists for post-translational regulation of VIP expression at the level of peptide precursor processing. At least five different VIP precursor products can be detected in the brain in non-equimolar ratios (Fahrenkrug and Emson, 1989). As already discussed, some reports indicate that the immunoreactivity ratio of VIP to PHI varies from tissue to tissue (Beinfeld *et al.*, 1984). Several additional alternatively processed VIP gene products with differing pharmacological profiles have been shown to be produced in some but not other tissues (Yiangou *et al.*, 1987; Gafvelin *et al.*, 1988; Goetzl *et al.*, 1988; Cauvin *et al.*, 1989; Fahrenkrug, 1991). Thus translational and post-translational regulation could affect the ultimate activity of the VIP gene by changing the amount and character of the gene-encoded peptides.

Transgenic mouse models

Further investigation of regulation of gene expression in development and in the adult may be facilitated by the ability to study altered forms of the gene in the intact animal. For example, the transgenic model provides an *in vivo* system to confirm hypotheses generated from transfected cultured cells, and to identify transcription control elements that cannot be assessed in cell culture systems. The usefulness of the model in studying gene regulation, however, goes beyond the study of transcription control sequences. For example, tissue-specific alternative splicing of the calcitonin/calcitonin gene-related peptide gene mRNA has been examined in transgenic mice (Crenshaw *et al.*, 1987), and glucagon prohormone processing was studied in a tumour cell line made from transgenic mice carrying a glucagon promoter/T antigen fusion gene (Powers *et al.*, 1990).

We have begun to apply the transgenic mouse model to study VIP gene transcription control sequences. Transfection studies in human neuroblastoma cells (discussed in preceding section) indicated that the regulatory behaviour and cell-type specific expression of the endogenous VIP gene was fully mimicked by a chimaeric gene construction, which consisted of 5.2 kb of human VIP gene 5′ flanking sequence, the first VIP gene exon, and part of the first intron fused to the reporter gene chloramphenicol acetyltransferase (CAT). This suggested to us that the same construction might also mimic the endogenous gene in the transgenic mouse model. Four sets of mice were produced containing up to 20 copies of this chimaeric gene in their genome, and were analysed for CAT expression in various tissues (Agoston *et al.*, 1990). CAT activity was detected in the intestine of all of these mice or their offspring but not in the intestine of wild-type mice. Thus the VIP gene fragment in this construction contained the sequences necessary to target expression to the intestine. However, overall expression of the transgene was relatively low and CAT activity was not detected in the cerebral cortex, a primary site where VIP is expressed.

We tried to resolve this problem by fusing VIP gene 5′ flanking sequences to the *lacZ* reporter, which would allow histochemical detection of expression. (Antibodies capable of detecting CAT by immunohistochemistry in transgenic mice were not available.) This transgene contained the same 5.2 kb of 5′ flanking sequences that were in the CAT construction, but part of the first untranslated VIP exon and intron sequences were removed. Two founder mice were produced, each containing multiple copies of the transgene. However, β-galactosidase expression could not be detected in the offspring of these mice in either the CNS or the intestine (Agoston, D. and Waschek, J. A., unpublished work). The lack and low level of transgene expression observed with the VIP/*lacZ* and VIP/CAT constructions suggested that the human VIP gene sequences used in the constructions functioned inefficiently in mice or that one or more important enhancers required for CNS expression are located elsewhere on the gene. Future studies utilizing additional human VIP gene or rodent VIP gene sequences may lead to the identification of such sequences.

References

Aggewstrup, S., Uddman, R., Sundler, F., Fahrenkrug, J., Hakanson, R., Sorensen, H. R. and Hambraeus, G. (1983) Lack of vasoactive intestinal polypeptide nerves in esophageal achalasia. Gastroenterology **84**, 924–927

Aggewstrup, S., Uddman, R., Jensen, S. L., Sundler, F., de Muckadell, O. S., Holst, J. J., Hakanson, R., Ekman, R. and Sorensen, H. R. (1985) Regulatory peptides in the lower esophageal sphincter of man. Regul. Peptides **10**, 167–178

Agoston, D. A., Bravo, D. T. and Waschek, J. A. (1990) Expression of a chimeric VIP gene is targeted to the intestine in transgenic mice. J. Neurosci. Res. **27**, 479–486

Agoston, D. V., Eiden, L. E., Brenneman, D. E. and Gozes, I. (1991) Spontaneous electrical activity regulates vasoactive intestinal polypeptide expression in dissociated spinal cord cultures. Mol. Brain Res. **10**, 235–240

Agoston, D., Colburn, S., Kranjiak, K., Fahrenkrug, J. and Waschek, J. (1992) Distinct regulation of vasoactive intestinal peptide (VIP) expression at mRNA and peptide levels in human neuroblastoma cells. Neurosci. Lett. **139**, 213–216

Albers, H. E., Minamitani, N., Stopa, E. and Ferris, C. F. (1987) Light selectively alters vasoactive intestinal peptide and peptide histidine isoleucine immunoreactivity within the rat suprachiasmatic nucleus. Brain Res. **437**, 189–192

Albers, H. E., Stopa, E. G., Zoeller, R. T., Kauer, J. S., King, J. C., Fink, J. S., Mobtaker, H. and Wolfe, H. (1990) Day-night variation in prepro VIP/PHI mRNA within the rat suprachiasmatic nuclei. Mol. Brain Res. **7**, 85–89

Albers, E. H., Liou, S.-Y., Ferris, C. F., Stopa, E. G. and Zoeller, R. T. (1991) The neurochemistry of circadian timing. In The Suprachiasmatic Nucleus: The Minds Clock (Klein, D. C., Moore, R. Y. and Reppert, S. M., eds.), Oxford, New York

Allen, J. M., Hoyle, N. R., Yeats, J. C., Ghatei, M. A., Thomas, D. G. and Bloom, S. R. (1985) Neuropeptides in neurological tumours. J. Neuro-Oncol. **3**, 197–202

Anand, P., Gibson, S. J., Scaravilli, F., Blank, M. A., McGregor, G. P., Appenzeller, O., Dhital, K., Polak, J. M. and Bloom, S. R. (1990) Studies of vasoactive intestinal polypeptide expression in injured peripheral neurons using capsaicin, sympathectomy and mg mutant rats. Neurosci. Lett. **118**, 61–66

Anand, P., Springall, D. R., Blank, M. A., Sellu, D., Polak, J. M. and Bloom, S. R. (1991) Neuropeptides in skin disease: increased VIP in eczema and psoriasis but not axillary hyperhidrosis. Br. J. Dermatol. **124**, 547–549

Avidor, R., Eilam, R., Malach, R. and Gozes, I. (1989) VIP-mRNA is increased in hypertensive rats. Brain Res. **503**, 304–307

Baldino, F., Fitzpatrick-McElligot, S., Gozes, I. and Card, J. P. (1989) Localization of VIP and PHI-27 messenger RNA in rat thalamic and cortical neurons. J. Mol. Neurosci. **1**, 199–207

Bardrum, B., Ottesen, B., Fahrenkrug, J. and Fuchs, A.-R. (1988) Release of oxytocin and vasopressin by intracerebroventricular vasoactive intestinal polypeptide. Endocrinology **123**, 2249–2254

Bartha, L., Kiem, D. T. and Makara, G. B. (1990) The effect of systemically and locally administered steroids on VIP-like immunoreactive cells in the paraventricular nucleus of adrenalectomized rats. Brain Res. **506**, 323–326

Beinfeld, M., Korchak, D. M., Nilaver, G. and O'Dorisio, T. M. (1983) The development of motilin, cholecystokinin and VIP in the forebrain and hindbrain of the rat as determined by RIA. Dev. Brain Res. **10**, 146–150

Beinfeld, M. C., Korchak, D. M., Roth, B. L. and O'Donohue, T. L. (1984) The distribution and chromatographic characterization of PHI (peptide histidine isoleucine amide)-27-like peptides in rat and porcine brain. J. Neurosci. **4**, 2681–2688

Belai, A., Lincoln, J., Milner, P. and Burnstock, G. (1988) Progressive changes in adrenergic, serotonergic, and peptidergic nerves in proximal colon of stretozotocin-diabetic rats. Gastroenterology **95**, 1234–1241

Benson, D. L., Isackson, P. J. and Jones, E. G. (1991) *In situ* hybridization reveals VIP precursor mRNA-containing neurons in monkey and rat neocortex. Mol. Brain Res. **9**, 169–174

Bishop, A. E., Polak, J. M., Lake, B. D., Bryant, M. G. and Bloom, S. R. (1981) Abnormalities of the colonic regulatory peptides in Hirchsprung's disease. Histopathology **5**, 679–688

Bodner, M., Fridkin, M. and Gozes, I. (1985) VIP and PHM-27 sequences are located on two adjacent exons in the human genome. Proc. Natl. Acad. Sci. U.S.A. **82**, 3548

Bos, N. P. A. and Mirmiran, M. (1990) Circadian rhythms in spontaneous neuronal discharges of the cultured suprachiasmatic nucleus. Brain Res. **511**, 158–162

Brenneman, D., Eiden, L. E. and Siegel, R. E. (1985) Neurotrophic action of VIP on spinal cord cultures. Peptides **6**, Suppl. 2, 35–39

Brenneman, D. E., Nicol, T., Warren, D. and Bowers, L. M. (1990) Vasoactive intestinal peptide: a neurotrophic releasing agent and an astroglial mitogen. J. Neurosci. Res. **25**, 386–394

Card, J. P., Fitzpatrick-McElligot, S., Gozes, I. and Baldino, F., Jr. (1988) Localization of vasopressin-, vasoactive intestinal polypeptide-, peptide histidine isoleucine- and somatostatin-mRNA in rat suprachiasmatic nucleus. Cell Tissue Res. **252**, 307–315

Carrillo, A. J. and Phelps, C. J. (1992) Quantification of vasoactive intestinal peptide immunoreactivity in the anterior pituitary glands of intact male and female, ovariectomized, and estradiol benzoate-treated rats. Endocrinology **131**, 964–969

Carrillo, A. J., Doherty, P. C., Guan, X., Sturtevant, J. R. and Walro, D. G. (1991) Preferential increase in pituitary prolactin versus vasoactive intestinal peptide as a function of estradiol benzoate dose in the ovariectomized rat. Endocrinology **128**, 131–138

Cauvin, A., Vandermeers, A., Vandermeers-Piret, M.-C., Rathe, J., Roberecht, P. and Christophe, J. (1989) Peptide histidine isoleucinamide PHI-(1-27)-Gly as a new major form of PHI in the rat small intestine. Endocrinology **125**, 1296–1302

Ceccatelli, S., Cortes, R. and Hokfelt, T. (1991 *a*) Effect of reserpine and colchicine on neuropeptide mRNA levels in the rat hypothalamic paraventricular nucleus. Mol. Brain Res. **9**, 57–69

Ceccatelli, S., Fahrenkrug, J., Villar, M. J. and Hokfelt, T. (1991 *b*) Vasoactive intestinal polypeptide/peptide histidine isoleucine immunoreactive neuron systems in the basal hypothalamus of the rat with special reference to the portal vasculature: an immunohistochemical and *in situ* hybridization study. Neuroscience **43**, 483–502

Chery-Croze, S., Bosshard, A. and Chayvialle, J. A. (1986) Pregnancy-induced increase of spinal cord peptide concentrations in rats. Peptides **7**, 197–200

Crenshaw, E. B., III, Russo, A. F., Swanson, L. W. and Rosenfeld, M. G. (1987) Neuron-specific alternative RNA processing in transgenic mice expressing a metallothionein–calcitonin fusion gene. Cell **49**, 389–398

Crowe, R. and Burnstock, G. (1988) An increase of vasoactive intestinal polypeptide-, but not neuropeptide Y-, substance P- or catecholamine-containing nerves in the iris of the streptozotocin-induced diabetic. Exp. Eye Res. **47**, 751–759

Crowe, R., Lincoln, J., Blacklay, P. F., Pryor, J. P., Lumley, J. S. P. and Burnstock, G. (1983) Vasoactive intestinal polypeptide-like immunoreactive nerves in diabetic penis. Diabetes **32**, 1075–1077

Diani, A. R., Peterson, T., Sawada, G. A., Wyse, B. M., Blanks, M. C., Gerritsen, G. C., Terenghi, G., Varndell, I. M., Polak, J. M., Blank, M. A. and Bloom, S. R. (1985) Elevated levels of vasoactive intestinal polypeptide in the eye and urinary bladder of diabetes and prediabetic Chinese hamsters. Diabetologia **28**, 302–307

Eiden, L. E., Eskay, R. L., Scott, J., Pollard, H. and Hotchkiss, A. J. (1983) Primary cultures of bovine chromaffin cells synthesize and secrete vasoactive intestinal polypeptide (VIP). Life Sci. **33**, 687–693

Eilam, R., Malach, R., Bergmann, F. and Segal, M. (1991) Hypertension induced by hypothalamic transplantation from genetically hypertensive to normotensive rats. J. Neurosci. **11**, 401–411

Emson, P. C., Gilbert, R. F. T., Loren, I., Fahrenkrug, J., Sundler, F. and Schaffalitzky de Muckadell, O. B. (1979) Development of VIP containing neurones in rat brains. Brain Res. **177**, 437–444

Eriksson, M., Linden, A., Stock, S. and Uvnas-Moberg, K. (1987) Increased levels of vasoactive intestinal peptide (VIP) and oxytocin during suckling in lactating dogs. Peptides **8**, 411–413

Ernsberger, U., Sendtner, M. and Rohrer, H. (1989) Proliferation and differentiation of embryonic chick sympathetic neurons: effects of ciliary neurotrophic factor. Neuron **2**, 1275–1284

Eskay, R. L. and Eiden, L. E. (1992) Interleukin-1 α and tumor necrosis factor-α differentially regulate enkephalin, vasoactive intestinal polypeptide, neurotensin, and substance P biosynthesis in chromaffin cells. Endocrinology **130**, 2252–2258

Fahrenkrug, J. (1980) Vasoactive intestinal polypeptide. Trends Neurosci. **3**, 1–2

Fahrenkrug, J. (1991) Glycine-extended processing intermediate of proVIP: a new form of VIP in the rat. Biochem. Biophys. Res. Commun. **178**, 173–177

Fahrenkrug, J. and Emson, P. C. (1989) Characterization and regional distribution of peptides derived from the vasoactive intestinal peptide precursor in the normal human brain. J. Neurochem. **53**, 1142–1148

Fahrenkrug, J., Ottesen, B. and Palle, C. (1988) Vasoactive intestinal polypeptide and the reproductive system. Ann. N.Y. Acad. Sci. **527**, 393–404

Fink, J., Montminy, M. R., Tsukada, T., Hoefler, H., Specht, L. A., Lechan, R. M., Wolfe, H., Mandel, G. and Goodman, R. H. (1986) *In situ* hybridization of somatostatin and VIP mRNA in the rat nervous system. In *In Situ* Hybridization in Brain (Uhl, G., ed.), Plenum, New York

Fink, J. S., Verhave, M., Kasper, S., Tsukada, T., Mandel, G. and Goodman, R. (1988) The CGTCA sequence motif is essential for biological activity of the vasoactive intestinal peptide gene cAMP-regulated enhancer. Proc. Natl. Acad. Sci. U.S.A. **85**, 6662–6666

Fink, J. S., Verhave, M., Walton, K., Mandel, G. and Goodman, R. H. (1991) Cyclic AMP- and phorbol ester-induced transcriptional activation are mediated by the same enhancer element in the human vasoactive intestinal peptide gene. J. Biol. Chem. **266**, 3882–3887

Gafvelin, G., Andersson, M., Dimaline, R., Jornvall, H. and Mutt, V. (1988) Isolation and characterization of a variant form of vasoactive intestinal polypeptide. Peptides **9**, 469–474

Giladi, E., Shani, Y. and Gozes, I. (1990) The complete structure of the rat VIP gene. Mol. Brain Res. **7**, 261–267

Goetzl, E. J., Sreedharan, S. P. and Turck, C. W. (1988) Structurally-distinctive vasoactive intestinal peptides from rat basophilic leukemia cells. J. Biol. Chem. **263**, 9083–9086

Goetzl, E. J., Grotmol, T., Van Dyke, R. W., Turck, C. W., Weshil, B., Galli, S. J. and Sreedharan, S. P. (1990) Generation and recognition of vasoactive intestinal polypeptide by cells of the immune system. Ann. N.Y. Acad. Sci. **594**, 34–44

Gotz, M. and Bolz, J. (1989) Development of vasoactive intestinal polypeptide (VIP)-containing neurons in organotype slice cultures from rat visual cortex. Neurosci. Lett. **107**, 6–11

Gozes, I. and Brenneman, D. E. (1989) VIP: molecular biology and neurobiological function. Mol. Neurobiol. **3**, 201–236

Gozes, I., Giladi, E. and Shani, Y. (1987*a*) Vasoactive intestinal peptide gene: putative mechanism of information storage at the RNA level. J. Neurochem. **48**, 1136–1141

Gozes, I., Shani, Y. and Rostene, W. H. (1987*b*) Developmental expression of the VIP gene in brain and intestine. Mol. Brain Res. **2**, 137–148

Gozes, I., Schachter, P., Shani, Y. and Giladi, E. (1988) VIP gene expression from embryos to aging rats. Neuroendocrinology **47**, 27–31

Gozes, I., Avidor, R., Biegon, A. and Baldino, F., Jr. (1989*a*) Lactation elevates vasoactive intestinal peptide messenger ribonucleic acid in rat suprachiasmatic nucleus. Endocrinology **124**, 181–186

Gozes, I., Shani, Y., Liu, B. and Burbach, J. P. H. (1989*b*) Diurnal variation in vasoactive intestinal polypeptide messenger RNA in the suprachiasmatic nucleus of the rat. Neurosci. Res. Commun. **5**, 83–86

Gozes, I., Werner, H., Fawzi, M., Abdelatty, A., Shani, Y., Fridkin, M. and Koch, Y. (1989*c*) Estrogen regulation of vasoactive intestinal peptide mRNA in rat hypothalamus. J. Mol. Neurosci. **1**, 55–61

Gu, J., Lazarides, M., Pryor, J., Blank, M., Polak, J., Morgan, R., Marangos, P. and Bloom, S. (1984) Decreased vasoactive intestinal polypeptide (VIP) in the penises from impotent men. Lancet **ii**, 315–318

Hayakawa, Y., Obata, K.-I., Itoh, N., Yanaihara, N. and Okamoto, H. (1984) Cyclic AMP regulation of pro-vasoactive intestinal polypeptide/PHM-27 synthesis in human neuroblastoma cells. J. Biol. Chem. **259**, 9207–9211

Heinz-Erian, P., Dey, R. D., Flux, M. and Said, S. I. (1985) Deficient vasoactive intestinal polypeptide innervation in the sweat glands of cystic fibrosis patients. Science **229**, 1407–1408

Holst, N., Jenssen, T. G., Burhol, P. G., Jorde, R. and Maltau, J. M. (1986) Plasma vasoactive intestinal polypeptide, insulin, gastric inhibitory polypeptide, and blood glucose in late pregnancy and during and after delivery. Am. J. Obstet. Gynecol. **155**, 126–131

Holst, N., Qian, P., Aune, B., Jenssen, T. G. and Burhol, P. G. (1991) Increased plasma levels of vasoactive intestinal polypeptide in pre-eclampsia. Br. J. Obstet. Gynecol. **98**, 803–806

Holtzman, R. L., Malach, R. and Gozes, I. (1989) Disruption of the optic pathway during development affects vasoactive intestinal peptide mRNA expression. New Biologist **1**, 215–221

Ishihara, Y., Seo, H., Suganuma, N., Oguri, H., Chihara, K., Matsui, N. and Tomoda, Y. (1991) Suckling stimulates the expression of vasoactive intestinal polypeptide gene in rats. Endocrinol. Jpn. **38**, 159–166

Jamal, H., Jones, P. M., Byrne, J., Suda, K., Ghatei, M. A., Kanse, S. M. and Bloom, S. R. (1991) Peptide contents of neuropeptide Y, vasoactive intestinal polypeptide, and β-calcitonin gene-related peptide and their messenger ribonucleic acids after dexamethasone treatment in the isolated rat islets of Langerhans. Endocrinology **129**, 3372–3380

Jones, P. M., Ghatei, M. A., Steel, J., O'Halloran, D., Gon, G., Legon, S., Burrin, J. M., Leonhardt, U., Polak, J. M. and Bloom, S. R. (1989) Evidence for neuropeptide Y synthesis in the rat anterior pituitary and the influence of thyroid hormone status: comparison with vasoactive intestinal peptide, substance P, and neurotensin. Endocrinology **125**, 334–341

Kakau, K., Tsuchiya, M., Tanizawa, Y., Okuya, S., Inoue, Y., Kaneko, T. and Yanaihara, N. (1986) Circadian cycles in VIP content and VIP stimulation of cAMP accumulation in the rat pineal. Peptides (Suppl. 1), 193–195

Karanth, S. S., Springall, D. R., Francavilla, S., Mirrlees, D. J. and Polak, J. M. (1990) Early increase in CGRP- and VIP-immunoreactive nerves in the skin of streptozotocin-induced diabetic rats. Histochemistry **94**, 659–666

Kashiba, H., Senba, E., Ueda, Y. and Tohyama, M. (1992*a*) Co-localized but target-unrelated expression of vasoactive intestinal polypeptide and galanin in rat dorsal root ganglion neurons after peripheral nerve crush injury. Brain Res. **582**, 47–57

Kashiba, H., Senba, E., Kawai, Y., Ueda, Y. and Tohyama, M. (1992*b*) Axonal blockade induces the expression of vasoactive intestinal polypeptide and galanin in rat dorsal root ganglion neurons. Brain Res. **577**, 19–28

Kasper, S., Popescu, R. A., Torsello, A., Vrontakis, M. E., Ikejiani, C. and Friesen, H. G. (1992) Tissue-specific regulation of vasoactive intestinal peptide messenger ribonucleic acid levels by estrogen in the rat. Endocrinology **130**, 1796–1801

Knyihar-Csillik, E., Kreutzberg, G. W., Raivich, G. and Csillik, B. (1991) A case for transmitter plasticity at the molecular level: axotomy-induced VIP increase in the upper spinal dorsal horn is related to blockade of retrograde axoplasmic transport of nerve growth factor in the peripheral nerve. Acta Histochem. **91**, 77–83

Koch, T. R., Carney, J. A. and Go, V. L. W. (1987) Distribution and quantitation of gut neuropeptides in normal intestine and inflammatory bowel diseases. Digest. Dis. Sci. **32**, 369–376

Koves, K., Arimura, A., Gorcs, T. G. and Somogyvari-Vigh, A. (1991) Comparative distribution of immunoreactive pituitary adenylate cyclase activating polypeptide and vasoactive intestinal polypeptide in rat forebrain. Neuroendocrinology **54**, 159–169

Lam, K. S. L. and Srivastava, G. (1990) Sex-related differences and thyroid hormone regulation of vasoactive intestinal peptide gene expression in the rat brain and pituitary. Brain Res. **526**, 135–137

Lam, K. S., Lechan, R. M., Minamitani, N., Segerson, T. P. and Reichlin, S. (1989) Vasoactive intestinal peptide in the anterior pituitary is increased in hypothyroidism. Endocrinology **124**, 1077–1084

Lam, K. S. L., Srivastava, G., Lechan, R. M., Lee, T. and Reichlin, S. (1990) Estrogen regulates the gene expression of vasoactive intestinal peptide in the anterior pituitary. Neuroendocrinology **52**, 417–421

Lam, K., Srivastava, G. and Tam, P. (1992) Divergent effects of glucocorticoid on the gene expression of vasoactive intestinal peptide in the rat cerebral cortex and pituitary. Neuroendocrinology **56**, 32–37

Lamperti, E. D., Rosen, K. M. and Villa-Komaroff, L. (1991) Characterization of the gene and messages for vasoactive intestinal polypeptide (VIP) in rat and mouse. Mol. Brain Res. **9**, 217–231

Lapchak, P. A. and Beaudet, A. (1990) Cholinergic regulation of vasoactive intestinal polypeptide content and release in rat frontal cortex and hippocampus. J. Neurochem. **55**, 1340–1345

Larsson, L. T. and Sundler, G. M. F. (1983) Peptidergic innervation in hirchsprung's disease. Z. Kinderchir **38**, 301–304

Lasaga, M., Duvilanski, B. H., Seilicovich, A., Afione, S., del Carmen Diaz, M., Pisera, D., Traktenberg, R. and Debeljuk, L. (1991) The effect of gonadal steroids on vasoactive intestinal peptide concentration and release from mediobasal hypothalamus and the anterior pituitary gland. J. Neuroendocrinology **3**, 75–78

Levy, D. M., Karanth, S. S., Springall, D. R. and Polak, J. M. (1989) Depletion of cutaneous nerves and neuropeptides in diabetes mellitus: an immunocytochemical study. Diabetologia **32**, 427–433

Li, S., Crenshaw, E., Rawson, E., Simmons, S., Swanson, L. and Rosenfeld, M. G. (1990) Dwarf locus mutants lacking three pituitary cell types result from mutations in the POU-domain gene pit-1. Nature (London) **347**, 528–533

Linder, S., Barkhem, T., Norberg, A., Persson, H., Schalling, M., Hokfelt, T. and Magnusson, G. (1987) Structure and expression of the gene encoding the vasoactive intestinal peptide precursor. Proc. Natl. Acad. Sci. U.S.A. **84**, 605–609

Long, R. B., Bishop, A. E., Barnes, A. J., Albuquerque, R. M., O'Shaugnessy, D. J., McGregor, G. P., Bannister, R., Polak, J. M. and Bloom, S. R.. (1980) Neural and hormonal peptides in rectal biopsy specimens from patients with Chagas' disease and chronic autooimic failure. Lancet **1**, 559–562

Loren, I., Emson, P. C., Fahrenkrug, J., Bjorklund, A., Alumets, J., Hakanson, R. and Sundler, F. (1979) Distribution of VIP in rat and mouse brain. Neuroscience **4**, 1953–1976

Lorenzo, M. J., Sanches-Franco, F., de los Frailes, M. T., Reichlin, S., Fernandez, G. and Cacicedo,

L. (1989) Synthesis and secretion of vasoactive intestinal polypeptide by rat fetal cerebral cortical and hypothalamic cells in culture. Endocrinology **125**, 1983–1990

Lorenzo, M. J., Sanchez-Franco, F., de los Frailes, M. T., Tolon, R. M., Fernandez, G. and Cacicedo, L. (1992) Thyroid hormones regulate release and content of vasoactive intestinal peptide in cultured fetal cerebral cortical cells. Neuroendocrinology **55**, 59–65

Magistretti, P. J., Morrison, J. H., Shoemaker, W. J., Sapin, V. and Bloom, F. E. (1981) Vasoactive intestinal polypeptide induced glycogenolysis in mouse cortical slices: a possible mechanism for local control of energy metabolism. Proc. Natl. Acad. Sci. U.S.A. **78**, 6535–6539

Maletti, M., Bataille, D., Laburthe, M. and Rosselin, G. (1980) Ontogeny and immunoreactive forms of VIP in rat brain. Acta Endocrinology **93**, 479–487

Maletti, M., Rostene, W. H., Carr, L., Scherrer, H., Rotten, D., Kordon, C. and Rosselin, G. (1982) Interaction between estradiol and prolactin on vasoactive intestinal peptide concentrations in the hyperthalamus and in the anterior pituitary of the female rat. Neurosci. Lett. **32**, 307–313

McGregor, G. P., Gibson, S. J., Sabate, I. M., Blank, M. A., Christofides, N. D., Wall, P. D., Polask, L. M. and Bloom, S. R. (1984) Effect of peripheral nerve section and nerve crush on spinal cord neuropeptides in the rat; increased VIP and PHI in the dorsal horn. Neuroscience **13**, 207–216

Meijer, J. H. and Rietveld, W. J. (1989) Neurophysiology of the suprachiasmatic circadian pacemaker in rodents. Physiol. Rev. **69**, 671–707

Menghi, G., Scocco, P. and Materazzi, G. (1990) Changes in vasoactive intestinal polypeptide (VIP) levels in the rabbit oviduct at different hormonal stages. An immunohistochemical study. J. Anat. **171**, 139–146

Mezey, E. and Kiss, J. (1985) VIP-containing neurons in the paraventricular nucleus may participate in regulating prolactin secretion. Proc. Natl. Acad. Sci. U.S.A. **82**, 245–247

Milner, P., Crowe, R., Kamm, M. A., Lennard-Jones, J. E. and Burnstock, G. (1990) Vasoactive intestinal polypeptide levels in sigmoid colon in idiopathic constipation and diverticular disease. Gastroenterology **99**, 666–675

Moody, T. W., Zia, F., Goldstein, A., Sarin, E., Naylor, P. H., Brenneman, D., Koros, A. M. C., Reubi, J. C., Korman, L. Y., Fridkin, M. and Gozes, I. (1992) VIP analogs inhibit small cell lung cancer growth. Biomed. Res., **13**, Suppl. 2, 131–135

Morin, A., Denoroy, L. and Jouvet, M. (1991) Daily variations in concentration of vasoactive intestinal polypeptide immunoreactivity in discrete brain areas of the rat. Brain Res. **538**, 136–140

Moss, H. E., Crowe, R. and Burnstock, G. (1987) The seminal vesicle in eight and 16 week streptozocin-induced diabetic rats: adrenergic, cholinergic and peptidergic innervation. J. Urol. **138**, 1273–1278

Mulderry, P. K. and Lindsay, R. M. (1990) Rat dorsal root ganglion neurons in culture express vasoactive intestinal polypeptide (VIP) independently of nerve growth factor. Neurosci. Lett. **8**, 314–320

Muller, J.-M., El Battari, A., All-Kye, E., Luis, J., Ducret, F., Pichon, J. and Marvaldi, J. (1985) Internalization of the vasoactive intestinal peptide (VIP) in a human adenocarcinoma cell line (HT2). Eur. J. Biochem. **152**, 107–114

Muller, J.-M., Lolait, S. J., Yu, V. C., Sadee, W. and Waschek, J. A. (1989) Functional VIP receptors in human neuroblastoma subclones that contain VIP precursor mRNA and release VIP-like substances. J. Biol. Chem. **264**, 3647–3650

Nagy, G., Mulchaney, J. J. and Neill, J. D. (1987) Autocrine control of prolactin secretion by vasoactive intestinal peptide. Endocrinology **122**, 364–366

Nawa, H. and Patterson, P. (1990) Separation and partial characterization of neuropeptide-inducing factors in heart cell conditioned medium. Neuron **4**, 269–277

Nielsch, U. and Keen, P. (1989) Reciprocal regulation of tachykinin- and vasoactive intestinal peptide-gene expression in rat sensory neurones following cut and crush injury. Brain Res. **481**, 25–30

Nobou, F., Besson, B., Rostene, W. and Rosselin, G. (1985) Ontogeny of vasoactive intestinal peptide and somatostatin in different structures of the rat brain: effects of hypo- and hypercorticism. Dev. Brain Res. 20, 296–301

Noda, K., Umeda, F., Ono, H., Hisatomi, A., Chijiiwa, Y., Nawata, H. and Ibayashi, H. (1990) Decreased VIP content in peripheral nerve from streptozocin-induced diabetic rats. Diabetes **39**, 608–612

Noguchi, K., Senba, E., Morita, Y., Sato, M. and Tohyama, M. (1989) Prepro-VIP and preprotachykinin mRNAs in the rat dorsal root ganglion cells following peripheral axotomy. Mol. Brain Res. **6**, 327–330

O'Dorisio, M. S., Fleshman, D. J., Qualman, S. J. and O'Dorisio, T. M. (1992) Vasoactive intestinal peptide: autocrine growth factor in neuroblastoma. Regul. Peptides **37**, 213–226

O'Halloran, D. J., Jones, P. M., Ghatei, M. A., Domin, J. and Bloom, S. R. (1990) The regulation of neuropeptide expression in rat anterior pituitary following chronic manipulation of estrogen status: a comparison between substance P, neuropeptide Y, neurotensin, and vasoactive intestinal peptide. Endocrinology **127**, 1463–1469

Ohsawa, K., Hayakawa, Y., Nishizawa, M., Yamagami, T., Yamamoto, H., Yanaihara, N. and Okamoto, H. (1985) Synergistic stimulation of VIP/PHM-27 gene expression by cyclic AMP and phorbol esters in human neuroblastoma cells. Biochem. Biophys. Res. Commun. **132**, 885–891

Ohsawa, K., Yamagami, M., Nishizawa, M., Inque, E., Gotoh, E., Yamamoto, H. and Okamoto, H. (1987) Structure and expression of human vasoactive intestinal peptide/PHM-27 gene. In Hormonal Regulation of Gene Expression, Gunna Symposia on Endocrinology, Vol. 24, pp. 169–181, VNU Sci Press, Utrecht

Okamoto, S., Okamura, H., Takahashi, Y., Akagi, Y., Yanaihara, N. and Ibata, Y. (1990) Contrary effect of eye enucleation on VIP-immunoreactive neurons in the suprachiasmatic nucleus and the superior colliculus of the rat. Neurosci Lett. **112**, 137–142

Okamoto, S., Okamura, H., Miyake, M., Takahashi, Y., Takagi, S., Akagi, Y., Fukui, K., Okamoto, H. and Ibata, Y. (1991) A diurnal variation of vasoactive intestinal peptide (VIP) mRNA under a daily light–dark cycle in the rat suprachiasmatic nucleus. Histochemistry **95**, 525–528

Ollereshaw, S., Jarvis, D., Woolcock, A., Sullivan, C. and Scheibner, T. (1989) Absence of immunoreactive vasoactive intestinal polypeptide in tissue from the lungs of patients with asthma. N. Engl. J. Med. **320**, 1244–1248

Ottesen, B. (1983) Vasoactive intestinal polypeptide as a neurotransmitter in the female genital tract. Am. J. Obstet. Gynecol. **147**, 208

Ottesen, B., Ulrichsen, H., Fahrenkrug, J., Larsen, J. J., Wagner, G., Shierup, L. and Sondergaard, F. (1982) Vasoactive intestinal polypeptide and the female genital tract: relationship to reproductive phase and delivery. Am. J. Obstet. Gynecol. **143**, 414

Palkovits, M., Besson, J. and Rotsztein, W. (1981) Distribution of VIP in intact, stria terminalis transected and cerebral cortex isolated rats. Brain Res. **213**, 455–459

Patterson, P. (1990) Control of cell fate in a vertebrate neurogenic lineage. Cell **62**, 1035–1038

Pence, J. C. and Shorter, N. A. (1992) Autoregulation of neuroblastoma growth by vasoactive intestinal peptide. J. Pediatr. Surg. **27**, 935–944

Petit, J.-M., Denoroy, L. and Jouvet, M. (1992) Effects of chronic atropine administration on regional vasoactive intestinal polypeptide concentrations in rat brain. Eur. J. Pharmacol. **212**, 79–85

Pincelli, C., Fantini, F., Romualdi, P., Lesa, G. and Giannetti, A. (1991) Skin levels of vasoactive intestinal polypeptide in atopic dermatitis. Arch. Dermatol. Res. **283**, 230–232

Pincus, D. W., DiCicco-Bloom, E. M. and Black, I. B. (1990) Vasoactive intestinal peptide regulates mitosis, differentiation and survival of cultured sympathetic neuroblasts. Nature (London) **343**, 564–567

Powers, A., Efrat, S., Mojsov, S., Spector, D., Haebner, J. and Hanahan, D. (1990) Proglucagon processing similar to normal islets in pancreatic α-like cell line derived from transgenic mouse tumour. Diabetes **39**, 406–414

Pruss, R., Moskal, J. R., Eiden, L. E. and Beinfeld, M. C. (1985) Specific regulation of vasoactive intestinal polypeptide biosynthesis by phorbol ester in bovine chromaffin cells. Endocrinology **117**, 1020–1026

Prysor-Jones, R. A., Silverlight, J. J. and Jenkins, J. S. (1987) Hyperprolactinemia reduces vasoactive intestinal peptide in the anterior pituitary glands of rats. Neurosci. Lett. **80**, 333–338

Riskind, P. N., Allen, J. M., Gabriel, S. M., Koenig, J. I. and Audet-Arnold, J. (1989) Sex differences in vasoactive intestinal peptide (VIP) concentrations in the anterior pituitary and hypothalamus of rats. Neurosci. Lett. **105**, 215–220

Rossiter, A., Guelrud, M., Souney, P. F., Mendoza, S., Rossiter, G. and Gelrud, D. (1991) High vasoactive intestinal peptide plasma levels in patients with Barrett's esophagus. Scand. J. Gastroenterol. **26**, 572–576

Rostene, W. (1984) Neurobiological and neuroendocrine functions of the vasoactive intestinal peptide. Prog. Neurobiol. (Oxford) **22**, 103–129

Rotsztejn, W., Besson, J., Briaud, B., Gagnani, L., Rosselin, G. and Kordon, C. (1980) Effect of steroids on VIP in discrete brain regions and peripheral tissues. Neuroendocrinology **31**, 287–291

Said, S. (1984) Vasoactive intestinal polypeptide: current status. Peptides **5**, 143–150

Said, S. I. (1991 *a*) Vasoactive intestinal polypeptide. Biological role in health and disease. Trends Endocrinol. Metab. **2**, 107–112

Said, S. I. (1991 *b*) Vasoactive intestinal polypeptide (VIP) in asthma. Ann. N.Y. Acad. Sci. **629**, 305–318

Said, S. I. (1991 *c*) VIP as a modulator of lung inflammation and airway constriction. Am. Rev. Respir. Dis. **143**, S22–S24

Said, S. I. and Faloona, G. R. (1975) Elevated plasma and tissue levels of vasoactive intestinal polypeptide in the watery-diarrhea syndrome due to pancreatic, bronchogenic, and other tumors. N. Engl. J. Med. **293**, 155–160

Said, S. I. and Mutt, V. (1970) Polypeptide with broad biological activity: isolation from small intestine. Science **169**, 1217–1218

Schoenen, J., Delree, P., Leprince, P. and Moonen, G. (1989) Neurotransmitter phenotype plasticity in cultured dissociated adult rat dorsal root ganglia: an immunocytochemical study. J. Neurosci. Res. **22**, 473–487

Scholar, E. M. and Paul, S. (1991) Stimulation of tumor cell growth by vasoactive intestinal peptide. Cancer **67**, 1561–1564

Schotzinger, R. J. and Landis, S. C. (1990) Acquisition of cholinergic and peptidergic properties by sympathetic innervation of rat sweat glands requires interaction with normal target. Neuron **5**, 91–100

Segerson, T. P., Lam, K. S. L., Cacicedo, L., Minamitani, N., Fink, J. S., Lechan, R. M. and Reichlin, S. (1989) Thyroid hormone regulates vasoactive intestinal peptide (VIP) mRNA levels in the rat anterior pituitary gland. Endocrinology **125**, 2221–2223

Shehab, S. A. S. and Atkinson M. E. (1986) Vasoactive intestinal polypeptide (VIP) increases in the spinal cord after peripheral axotomy of the sciatic nerve originate from primary afferent neurons. Brain Res. **372**, 37–44

Sims, K., Hoffman, D. L., Said, S. I. and Zimmerman, E. A. (1980) VIP in mouse and rat brain: an immunocytochemical study. Brain Res. **186**, 165–183

Steel, J. H., Gon, G., O'Halloran, D. J., Jones, P. M., Yanaihara, N., Bloom, S. R. and Polak, J. M. (1989) Galanin and vasoactive intestinal polypeptide are colocalized with classical pituitary hormones and show plasticity of expression. Histochemistry **93**, 183–189

Stoddard, S. L., Tyce, G. M., Ahlskog, J. E., Zinsmeister, A. R., Nelson, D. K. and Carmichael, S. W. (1991) Decreased levels of [Met]enkephalin, neuropeptide Y, substance P, and vasoactive intestinal polypeptide in Parkinson adrenal medulla. Exp. Neurol. **114**, 23–27

Stone, R. A., Laties, A. M., Raviola, E. and Wiesel, T. N. (1988) Increase in retinal vasoactive intestinal polypeptide after eyelid fusion in primates. Proc. Natl. Acad. Sci. U.S.A. **85**, 257–260

Stopa, E. G., Minamitani, N., Jonassen, J. A., King, J. C., Wolfe, H., Mobtaker, H. and Albers, H. E. (1988) Localization of vasoactive intestinal peptide and peptide histidine isoleucine immunoreactivity and mRNA within the rat suprachiasmatic nucleus. Mol. Brain Res. **4**, 319–325

Taguchi, T., Tanaka, K., Ikeda, K., Matsubayashi, S. and Yanaihara, N. (1983) Peptidergic innervation irregularities in hirchsprung's disease. Virchows Arch. A **401**, 223–235

Takahashi, Y., Okamura, H., Yanaihara, N., Hamada, S., Fujita, S. and Ibata, Y. (1989) Vasoactive intestinal peptide immunoreactive neurons in the rat suprachiasmatic nucleus demonstrate diurnal variation. Brain Res. **497**, 374–377

Tcheng, T. K. and Gillette, M. U. (1990) Electrical characterization of ventrolateral and dorsomedial regions of the suprachiasmatic nucleus. Soc. Neurosci. Abstr. **16**

Tischler, A. S., Lee, Y. C., Periman, R. L., Costopoulos, D., Slayton, V. W. and Bloom, S. R. (1984) Production of 'ectopic' vasoactive intestinal polypeptide-like and neurotensin-like immunoreactivity in human pheochromocytoma cell cultures. J. Neurosci. **4**, 1398–1404

Tischler, A. S., Lee, Y. C., Periman, R. L., Costopoulous, D., Slayton, V. W. and Bloom, S. R. (1985) production of 'ectopic' vasoactive intestinal polypeptide-like immunoreactivity in normal human chromaffin cell cultures. Life Sci. **37**, 1881–1886

Toni, R., Kakucska, I., Mosca, S., Marrama, P. and Lechan, R. M. (1992) Hypothyroidism increases vasoactive intestinal polypeptide (VIP) immunoreactivity and gene expression in the rat hypothalamic paraventricular nucleus. Endocrinology **131**, 976–978

Tsuchiya, T., Suzuki, Y., Suzuki, H., Ohtake, R. and Shimoda, S.-I. (1990) Changes in adrenal neuropeptides content [peptide 7B2, neuropeptide Y (NPY) and vasoactive intestinal polypeptide (VIP)] induced by pharmacological and hormonal manipulations. J. Endocrinol. Invest. **13**, 381–389

Tsukada, T., Horovitch, S. J., Montminy, M. R., Mandel, G. and Goodman, R. H. (1985) Structure of human vasoactive intestinal polypeptide gene. DNA **4**, 292–300

Tsukada, T., Fink, J. S., Mandel, G. and Goodman, R. H. (1987) Identification of a region in the human vasoactive intestinal polypeptide gene responsible for regulation by cyclic AMP. J. Biol. Chem. **262**, 8743–8747

Waschek, J. and Eiden, L. (1988) Calcium requirements for barium stimulation of enkephalin and vasoactive intestinal peptide biosynthesis in adrenomedullary chromaffin cells. Neuropeptides **11**, 39–45

Waschek, J., Pruss, R., Siegel, R., Eiden, L., Bader, M.-F. and Aunis, D. (1987) Regulation of enkephalin, VIP, and chromogrannin biosynthesis in actively-secreting chromaffin cells. Ann. N.Y. Acad. Sci. **493**, 308–323

Waschek, J., Hsu, C.-M. and Eiden, L. (1988) Lineage-specific regulation of the vasoactive intestinal peptide gene in neuroblastoma cells is conferred by 5.2. kilobases of 5′ flanking sequence. Proc. Natl. Acad. Sci. U.S.A. **85**, 9547–9551

Waschek, J. A., Muller, J.-M., Duan, D.-S. and Sadee, W. (1989) Retinoic acid enhances VIP receptor expression and responsiveness in human neuroblastoma cells, SH-SY5Y. FEBS Lett. **250**, 611–614

Waschek, J. A., Krajniak, K. G., Bravo, D. T. and Agoston, D. A. (1992) Regulation of human and mouse vasoactive intestinal peptide genes. Biomed. Res. **13**, Suppl. 2, 1–6

Watanobe, H. (1990) The immunostaining for the hypothalamic vasoactive intestinal peptide, but not for β-endorphin, dynorphin-A or methionine-enkephalin, is affected by the glucorcorticoid milieu in the rat: correlation with the prolactin secretion. Regul. Peptides **28**, 301–311

Watanobe, H., Sasaki, S. and Takebe, K. (1991) Neonatal androgenization increases vasoactive intestinal peptide levels in rat anterior pituitary: possible involvement of vasoactive intestinal peptide in the neonatal androgenization-induced hyperprolactinemia. J. Endocrinol. Invest. **14**, 875–879

Wattchow, D. A., Cass, D. T., Furness, J. B., Costa, M., O'Brien, P. E., Little, K. E. and Pitkin, J. (1987) Abnormalities of peptide-containing nerve fibers in infantile hypertrophic pyloric stenosis. Gastroenterology **92**, 443–448

Weinstock, J. V. and Blum, A. M. (1990) Detection of vasoactive intestinal polypeptide and localization of its mRNA with granulomas of murine schistosomiasis. Cell. Immunol. **125**, 291–300

Yamagami, T., Ohsawa, K., Nishizawa, M., Inque, C., Gotoh, E., Yanaihara, N., Yamamoto, H. and Okamoto, H. (1988) Complete nucleotide sequence of human vasoactive intestinal peptide/PHM-27 gene and its inducible promoter. Ann. N.Y. Acad. Sci. **527**, 87–102

Yiangou, Y., Di Marzo, V., Spokes, R. A., Panico, M., Morris, H. R. and Bloom, S. R. (1987) Isolation, characterization, and pharmacological actions of peptide histidine valine 42, a novel preprovasoactive intestinal peptide-derived peptide. J. Biol. Chem. **29**, 14010–14013

Yu, V. C. and Sadee, W. (1986) Phosphatidylinositol turnover in neuroblastoma cells: regulation by bradykin, acetylcholine, but not μ- and gamma-opioid receptors. Neurosci. Lett. **71**, 219–223

The prepro-neurotensin/ neuromedin N gene: structural organization, regulation of expression and post-transcriptional events

Frédéric Checler

Institut de Pharmacologie Moléculaire et Cellulaire, C.N.R.S. UPR 411,
Université de Nice Sophia Antipolis, 660 Route des Lucioles,
Sophia Antipolis 06560, Valbonne, France

Introduction

Neurotensin (NT) is a tridecapeptide first isolated from bovine hypothalami in 1973 (Carraway and Leeman, 1973). As is the case for most neuropeptides, a peripheral counterpart of NT was later discovered in the gastrointestinal tract of mammals (Kitabgi *et al.*, 1976). In the central nervous system, NT induces hypothermia and a naloxone-insensitive analgesia, and displays the ability to modulate the firing rate and electrophysiological properties of neurons located in various brain regions such as the nucleus accumbens, the locus coeruleus, the cortex, the hippocampus, the lateral thalamus and the striatum (for reviews see Hirsch Fernstrom *et al.*, 1980; Nemeroff *et al.*, 1980, 1983; Saint-Pierre *et al.*, 1984). Furthermore, NT was depicted as an endogenous neuroleptic-like substance with respect to a series of pharmacological observations which included the potentiation of the sedation elicited by barbiturates and ethanol, the decrease of spontaneous or amphetamine-induced locomotor activity and muscle relaxation. The neuroleptic-like pharmacological spectrum of NT which mimics that observed with dopamine antagonists implied NT–dopamine inter-relation in the brain. Such an hypothesis triggered a series of studies which clearly established, on the basis of topographical, neurochemical and functional evidence, that NT closely interacts with dopaminergic systems in the brain (for reviews see Kitabgi, 1989; Kasckow and Nemeroff, 1991).

In the periphery, NT is mainly located in the jejuno–ileal part of the small intestine (Carraway and Leeman, 1976; Holzer *et al.*, 1982), and particularly in discrete cells of the ileal mucosa (Orci *et al.*, 1976; Sundler *et al.*, 1977). NT-like immunoreactivity can be released upon luminal stimulation (Mashford *et al.*, 1978; Rosell and Rökaeus, 1979; Hammer *et al.*, 1982). These observations could probably explain the endocrine-type function of NT on gastric acid and exocrine pancreas secretion. In addition, the paracrine modulation of gut motility by NT has been extensively documented. NT exerts various and complex effects depending on the intestinal layers, organs and species that were examined. Indeed, this peptide can elicit both inhibitory and excitatory responses mediated by myogenic or neurogenic mechanisms, *in vitro* and *in vivo* (for review see Kitabgi, 1982).

The central and peripheral pharmacological effects of NT are mediated via specific receptors that were characterized by means of several radiolabelled ligands (Kitabgi *et al.*, 1977; Lazarus *et al.*, 1977; Mazella *et al.*, 1983; Sadoul *et al.*, 1984) and whose molecular properties were first established by covalent labelling (Mazella *et al.*, 1985*a*, *b*). Two NT binding sites have been described in almost all membrane preparations from central or peripheral origin which are of high affinity (0.1 nM) or of lower affinity (about 2–4 mM) (for review see Vincent *et al.*, 1990). By means of levocabastine, an antihistaminic drug (Schotte *et al.*, 1986) which totally abolishes the interaction of NT with the low-affinity binding sites, it was shown that the high-affinity, NT receptor was the only binding site that was coupled to a modulation of intracellular events, i.e. cyclic AMP, cyclic GMP and phosphatidylinositol turnover (for review see Vincent *et al.*, 1990). This high-affinity levocabastine-insensitive NT receptor has been purified (Mazella *et al.*, 1989; Miyamoto-Lee *et al.*, 1991; Mills *et al.*, 1988) and cloned (Tanaka *et al.*, 1990).

The structure–function relationships of the NT receptors have been extensively documented (for review see Kitabgi *et al.*, 1985). It appears clear that most of the biological activity of the peptide is borne by its *C*-terminal hexapeptide. Table 1 indicates that this region is particularly well conserved in chicken, frog, guinea-pig and alligator NT while several differences exist in the congeners isolated from marsupial possum, and turkey skin or stomach (Table 1).

As expected, there also exists a family of NT-like sequences. Indeed, several studies have led to the identification and purification of NT-related peptides. Xenopsin is an octapeptide isolated from skin extracts of *Xenopus laevis* (Araki *et al.*, 1975). Since this peptide shares some structural features with NT (Table 2), it was thought to correspond to the frog NT (Araki *et al.*, 1979). However, a peptide related to amphibian xenopsin was purified and sequenced from various rat and canine organs (Carraway and Mitra, 1990*b*; Carraway *et al.*, 1990). Therefore, xenopsin is not likely to correspond to the amphibian counterpart of NT. In addition to xenopsin, another peptide isolated from the frog *Rana margaratae* was shown to display the same *C*-terminal pentapeptide as NT but with an additional *C*-terminal dipeptide extension (Table 2; Tang *et al.*, 1990).

Table 1. Comparison of the sequence of NT from various vertebrates

Sequence	Species	Reference
pGlu-Leu-Tyr-Glu-Asn-Lys-Pro-Arg-Arg-Pro-Tyr- Ile -Leu	Bovine	Carraway *et al.* (1973)
pGlu-Leu-His-Val-Asn-Lys-Ala-Arg-Arg-Pro-Tyr- Ile -Leu	Chicken	Carraway and Bhatnagar (1980)
pGlu-Ser-His- Ile - Ser-Lys-Ala-Arg-Arg-Pro-Tyr- Ile -Leu	Frog	Shaw *et al.* (1992)
pGlu-Leu-Tyr-Glu-Asn-Lys-Ser-Arg-Arg-Pro-Tyr- Ile -Leu	Guinea-pig	Shaw *et al.* (1986)
pGlu-Leu-His-Val-Asn-Lys-Ala-Arg-Arg-Val-Tyr- Ile -Leu	Marsupial possum	Shaw *et al.* (1991)
Phe-Glu-Arg -Phe-Gln-Gly-Met-Arg-Thr-Arg-Gly-Pro-Tyr-Phe-Leu	Turkey	Carraway *et al.* (1987*b*)
pGlu-Leu-His-Val-Asn-Lys-Ala-Arg-Arg-Pro-Tyr- Ile -Leu	Alligator	Rodriguez-Bello (1993)

Amino acids distinct from those of the bovine neurotensin sequence are underlined.

Table 2. Sequence of peptides of the NT family

Name	Sequence	Source of purification	Reference
NN	Lys-Ile-Pro-Tyr-Ile-Leu	Porcine spinal cord	Minamino *et al.* (1984)
LANT-6	Lys-Asn-Pro-Tyr-Ile-Leu	Chicken intestine	Carraway and Ferris (1983)
Kinetensin	Ile-Ala-Arg-Arg-His-Pro-Tyr-Phe-Leu	Bovine, canine, human plasma	Carraway *et al.* (1987*a*)
		Human plasma	Mogard *et al.* (1986)
	Val-Ala-Arg-Arg-His-Pro-Tyr-Phe-Leu	Rat plasma	Carraway *et al.* (1987*a*)
Margaratensin	Asp-Lys-Arg-Pro-Tyr-Ile-Leu-His-Glu	Amphibian skin (*Margaratae*)	Tang *et al.* (1990)
Xenopsin	pGlu-Gly-Lys-Arg-Pro-Trp-Ile-Leu	Amphibian skin (*Xenopus*)	Araki *et al.* (1979)
Xenopsin-RP	Phe-His-Pro-Lys-Arg-Pro-Trp-Ile-Leu	Turkey skin and stomach	Carraway and Ferris (1988)
		Rat stomach, liver and brain	Carraway *et al.* (1990)
		Canine stomach	Carraway and Mitra (1990*b*)

Abbreviations used: RP, related peptide.

Neuromedin N (NN) and kinetensin were isolated from spinal cord and plasma of different mammals (Minamino *et al.*, 1984; Mogard *et al.*, 1986; Carraway *et al.*, 1987*a*). Finally, LANT-6 which probably corresponds to the avian counterpart of NN was isolated from chicken intestine (Carraway and Ferris, 1983).

Recent advances concerning the knowledge of NT physiology have been brought about by molecular biology approaches. The present review will deal with the recent data concerning the cloning and structural organization of the prepro-neurotensin precursor (prepro-NT) and its gene. Furthermore, we will examine the regulation of the gene expression together with the distribution and ontogenic profiles of prepro-NT mRNA. Finally, we also report on some of the post-transcriptional events which include the processing and maturation of the precursor together with the proteolytic mechanisms responsible for the inactivation of the neuropeptide.

Xenopsin and NT/NN precursors

Xenopsin precursor

The first precursor of NT-related peptides that was cloned was the prepro-xenopsin, in 1984 (Sures and Grippa, 1984). The precursor was isolated from a cDNA library constructed with *X. laevis* skin poly(A^+) RNA. The full-length cDNA was 490 nucleotides long and encoded a precursor of 80 amino acids (Sures and Grippa, 1984). The organization of the isolated cDNA indicated an *N*-terminal putative signal sequence while xenopsin appeared to be located at the extreme *C*-terminal part of the precursor (Fig. 1). It was interesting to note that xenopsin was not flanked by the typical basic residues generally targeted by proteolytic activities involved in precursor processing. However, there exists a Lys-Arg doublet at position 62–63 of the precursor (Fig. 1). The maturation at this site would result in the liberation of a 17 amino acid *N*-terminally extended form of xenopsin. Alternatively, the possibility of a cleavage at the monobasic arginyl site located at position 68 cannot be excluded (Fig. 1).

Striking similarities were found between the prepro-xenopsin precursor and the precursor of the PYL peptide encoded by *Xenopus* skin mRNA (Hoffman *et al.*, 1983). Thus, there existed a 70% similarity between both mRNAs located at the 1–170 nucleotide sequence of the xenopsin precursor and reflected by a 54% identity in the corresponding amino acid composition.

NT/NN precursor

The cDNA encoding the NT molecule was first cloned from a canine enteric mucosal cell library in 1987 (Dobner *et al.*, 1987) and later from a bovine hypothalamus cDNA library (Kislauskis *et al.*, 1988). The general organization of the bovine prepro-NT precursor is presented in Fig. 1. The most

Fig. 1. Structural organization of the xenopsin and NT/NN precursors

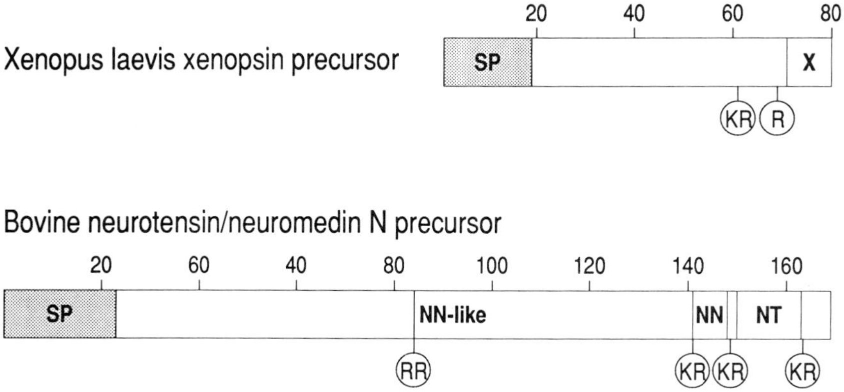

Numbers correspond to the amino acid position from the N-terminal end of the precursors. Circles indicate the position of monobasic or dibasic residues. Abbreviations used: SP, signal peptide; X, xenopsin.

interesting feature concerns the fact that the 170 amino acid precursor encodes NT and also its parent peptide NN (Fig. 1). Single copies of each of the two peptides are located adjacently to each other at the *C*-terminus of the precursor. NT and NN are flanked and separated by identical doublets of the basic amino acids, Lys-Arg. Furthermore, an NN-like sequence occurring in the middle of the precursor and bordered by paired arginyl residues at its *N*-terminus was reported (Dobner *et al.*, 1987). However, it is not yet clear whether such a sequence could be processed since its *C*-terminal part does not reveal nearby basic residues (Fig. 1).

The bovine and canine prepro-NT/NN precursor varied at only nine of the 170 amino acid residues (Kislauskis *et al.*, 1988). RNA blot analysis revealed several mRNA species ranging in size from 500 to 980 nucleotides in the canine mucosa while a single 1.5 kb species was found in bovine hypothalamus poly(A^+) RNA (Kislauskis *et al.*, 1988). The rat sequence of the prepro-NT/NN precursor was more divergent than the canine and bovine sequences (Kislauskis *et al.*, 1988) (77 and 78% identity, respectively) although most of the substitutions between species appeared to be conservative. The rat precursor consisted of 169 amino acids since one of the two methionines that initiated bovine and canine precursors was lacking (Fig. 2). Furthermore, the basic doublet located at the *N*-terminus of the NN-like sequence was Lys-Arg instead of an arginyl doublet (Fig. 2). There was a striking variation between the mRNA species observed in rat intestine

Fig. 2. Comparison of the NT/NN precursors from various species

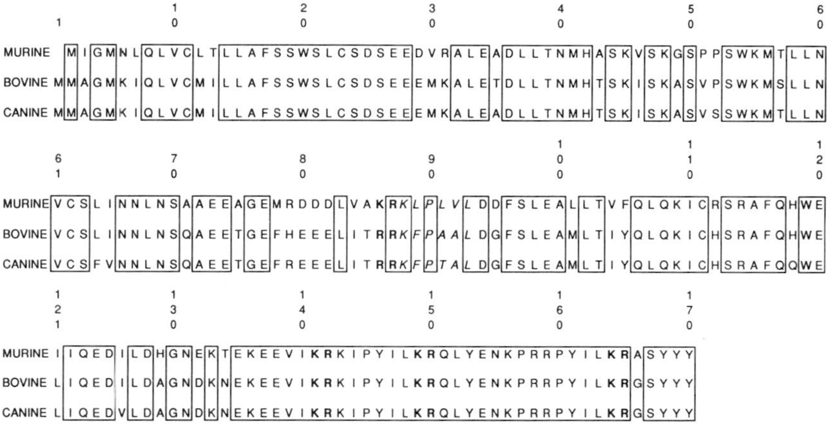

Analogous amino acid residues are boxed.

and brain. A small 1 kb mRNA form appeared predominantly in the intestine while an additional 1.5 kb mRNA was recovered in the brain. These two forms differed in the extent of their 3′ untranslated regions and therefore are probably derived from the rise of distinct polyadenylation signals (Kislauskis *et al.*, 1988). There are very few homologies between the xenopsin and the prepro-NT/NN precursors.

The NT/NN gene

Organization

The rat gene encoding NT and NN indicated that it spanned approximately 10.2 kb (Kislauskis *et al.*, 1988) organized in four exons and three introns (Fig. 3). Exon 1 encoded the signal peptide, while exons 2 and 3 encoded amino acids 24–44 and 45–119 of the rat precursor, respectively (Kislauskis *et al.*, 1988). Exon 4 encoded residues 120–169 and exhibited two poly(A+) addition signals probably responsible for the 1 and 1.5 kb mRNA species encountered in the rat intestine and brain (Kislauskis *et al.*, 1988).

Regulatory elements

Several regulatory elements present in a 200 bp region of the 5′ flanking sequence of the rat NT/NN gene were suspected after sequence analysis (Kislauskis *et al.*, 1988). Particularly interesting was the presence of an AP1 site, two cyclic AMP response elements (CREs) and a glucocorticoid

Fig. 3. Structural organization of the rat NT/NN gene

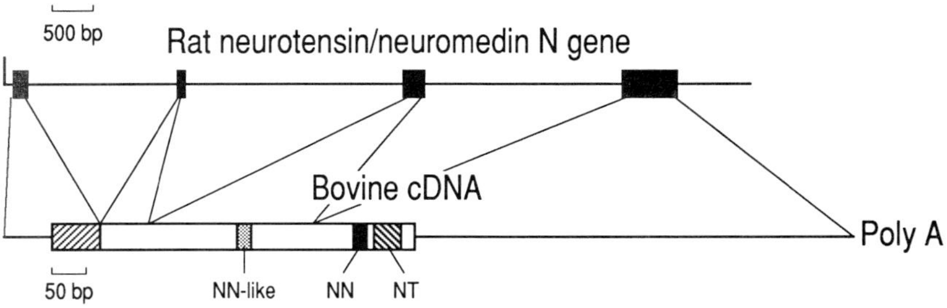

In the upper line, black boxes indicate the position of the four exons. The sequences encoded by the exons are projected on the bovine prepro-NT/NN cDNA.

response element (GRE) (Kislauskis *et al.*, 1988). The AP1 site and the CRE sequences were closely conserved between the rat and human genes (Bean *et al.*, 1992). Interestingly, one of the CREs was identical to a human c-*jun* gene autoregulatory site (Bean *et al.*, 1982). In contrast, the GRE identified in the rat gene was not present in the human (Bean *et al.*, 1992).

The regulation of the NT/NN gene expression by the putative inducible *cis*-regulatory sequences presented above has been investigated in PC12 phaeochromocytoma cells (Dobner *et al.*, 1988; Kislauskis and Dobner, 1991). This cell line has proved useful since the biosynthesis and secretion of NT could be induced by a combination of nerve growth factor, glucocorticoids and activators of adenylate cyclase (Tischler *et al.*, 1986). Such effects could be drastically enhanced by lithium (Dobner *et al.*, 1988). Interestingly, the gene expression did not appear in cells treated with the individual inducers and a combination of glucocorticoids and adenylate cyclase activators was also ineffective. It seemed therefore that the NT/NN gene expression required the co-operative actions of all the above inducers. Transfection experiments showed that the above regulation of NT/NN gene expression occurred at the transcriptional level (Kislauskis and Dobner, 1991). It was shown by progressive deletions of the 5′ flanking region of the gene that all the *cis*-regulatory sequences were borne by the −216/+56 fragment (Kislauskis and Dobner, 1991). However, the control of the NT/NN gene expression seemed very complex. Thus, mutational analysis of either the AP1 site or each of the two CREs, showed a drastic alteration of the gene expression increase elicited by different combinations of inducers (Kislauskis and Dobner, 1991). This seemed to indicate that these regulatory elements could be functionally interdependent. Whether it is also the case for the GRE remains unclear.

Modulation of prepro-NT/NN mRNA expression

As stated in the Introduction, there exists a dense network of evidence indicating a close interaction between NT and the brain dopaminergic systems. Particularly interesting were the reports showing that NT could mimic the pharmacological pattern elicited by neuroleptics which acted through dopaminergic circuits (Widerlov *et al.*, 1982*a*; Bissette and Nemeroff, 1985; Bissette *et al.*, 1988). Conversely, the neuroleptic haloperidol which antagonized dopamine action was shown to enhance the NT content in striatum (Govoni *et al.*, 1980; Kilts *et al.*, 1988). Such neuroleptic treatment also increased NT binding sites in the substantia nigra (Uhl and Kuhar, 1984). The antipsychotic-like profile of NT had led to the hypothesis that the peptide could act as a neuromodulator of the dopaminergic pathways and, thereby, could be involved in the aetiology of schizophrenia. In light of this hypothesis, it was reported that schizophrenic patients treated with antipsychotic drugs recovered the normal concentration of NT in their cerebrospinal fluid while the NT content was significantly lowered in drug-free schizophrenics (Widerlov *et al.*, 1982*b*). These observations led several groups to examine the putative influence of various drugs or pharmacological agents targeting the dopaminergic systems on the modulation of prepro-NT/NN mRNA content or gene expression in various brain regions (Table 3).

Three studies reported the influence of acute intraperitoneal administration of haloperidol on the brain nuclei content of mRNA (Williams *et al.*, 1990; Merchant *et al.*, 1991, 1992). All these works indicated that the prepro-NT/NN mRNA expression was increased in the dorsolateral part of the striatum. Haloperidol seemed to enhance the frequency and distribution of hybridization-positive neurons in striatum but also in septal nuclei (Williams *et al.*, 1990). The effect appeared to be transient. One hour after administration, haloperidol induced a 3-fold increase in the number of cells expressing the NT/NN mRNA (Merchant *et al.*, 1991). Maximal increase occurred after 7 h then the level significantly declined by 20 h (Merchant *et al.*, 1992). *In situ* hybridization analysis using an intron-derived probe indicated a transient increase in intron-containing NT/NN gene transcripts showing that haloperidol probably modulated the mRNA expression at the transcriptional level. Haloperidol seemed to exert a region-specific influence since the NT/NN mRNA expression appeared to be stable or only weakly increased in the ventro-tegmental area, nucleus accumbens and other parts of the midbrain (Table 3).

Interestingly, Merchant and his collaborators reported on a strikingly distinct pattern elicited by the atypical antipsychotic drug clozapine which did not affect NT/NN mRNA in the dorsolateral striatum (Merchant *et al.*, 1992). It is not yet clear whether these observations could explain the distinct pharmacological profile elicited by typical and atypical anti-

Table 3. Modulation of prepro-NT/NN mRNA expression by pharmacological agents related to dopamine function

Pharmacological agent	Site of administration	Modulation of mRNA expression		Reference
		Site	Variation	
Haloperidol	i.p.	Striatum and septal nuclei	Increase	Williams *et al.* (1990)
		Ventro-tegmental area	Modest increase	
		Other parts of midbrain	Stability	
Haloperidol	i.p.	Dorsolateral striatum	Increase	Merchant *et al.* (1991)
		Nucleus accumbens and septal nuclei	Stability	
Haloperidol	i.p.	Dorsolateral striatum	Increase	Merchant *et al.* (1992)
		Nucleus accumbens	Weak increase	
Clozapine	i.p.	Dorsolateral striatum	Stability	
		Nucleus accumbens	Weak increase	
SCH 23390 (D_1 antagonist)	i.p.	Dorsolateral striatum	Stability	Augood *et al.* (1991)
Raclopride (D_2 antagonist)	i.p.	Dorsolateral striatum	Increase	
BMY 14802	i.p.	Nucleus accumbens and dorsal striatum	Increase	Levant *et al.* (1992)
Reserpine	i.p.	Striatum	Increase	Bean and Hökfelt (1992)
Pertussis toxin	Striatum	Ipsilateral striatum	Increase	Augood and Emson (1992)
		Nucleus accumbens	Stability	

Abbreviations used: i.p., intraperitoneal.

psychotics. A recent report also indicated that BMY 14802, a potential antipsychotic drug which also interacts with δ-receptors, rapidly increased prepro-NT/NN expression in the dorsal striatum and in the nucleus accumbens (Levant *et al.*, 1992). Interestingly, Augood *et al.* showed that it was possible to delineate the regulation of NT/NN mRNA expression by dopaminergic D_1 and D_2 antagonists (Augood *et al.*, 1991). Thus, the D_2-antagonist raclopride increased the mRNA expression in the striatum while the D_1-blocking agent SCH 23390 failed to elicit any increase in the number of NT mRNA-containing cells (Augood *et al.*, 1991). Since the D_2 receptors are negatively coupled to adenylate cyclase by a G_i-protein, these authors examined the effect of pertussis toxin (Augood and Emson, 1992) which interacts with such a G_i protein, thereby lowering the receptor-mediated cyclic AMP inhibition. Intrastriatal administration of the toxin elicited an increase in the number of mRNA-containing cells in the ipsilateral striatum while no modifications were observed in the nucleus accumbens (Table 3). Finally, it was previously shown that reserpine could enhance NT content in the striatum with kinetics that paralleled those observed for dopamine depletion. This agent clearly increased the number of NT/NN mRNA-expressing cells (Bean and Hökfelt, 1992). Altogether, the above data confirmed that NT closely interacts with the dopaminergic systems and that modulation could be mediated at the transcriptional level. However, it is particularly important to emphasize the fact that in the ventral mesencephalon, the percentage of dopaminergic neurons expressing NT/NN mRNA did not seem to vary between schizophrenics and aged-matched patients (Bean *et al.*, 1992).

The influence of thyroid and adrenal hormone status on the NT/NN gene expression has been examined. Thyroidectomy was previously reported to reduce the NT content in the rat pituitary gland (Goedert *et al.*, 1982). This was further documented by the observation that a drop in the level of prepro-NT/NN mRNA was observed in the pituitary of hypothyroid animals (Table 4). Such a decrease affected the 1 kb mRNA species which is the virtually unique mRNA form found in the rat anterior pituitary (Jones *et al.*, 1989).

Jones and his co-workers (1990) examined the putative modification of the expression of NT mRNA in various animal models in which the corticosteroid hormone status was clearly affected (Table 4). The anterior pituitary NT/NN mRNA expression did not vary in castrated or adrenalectomized animals or when the effect of both surgical manipulations was examined (Jones *et al.*, 1990). These data were in agreement with a previous work showing that adrenalectomy or gonadectomy did not modify the NT content in the anterior pituitary (Goedert *et al.*, 1982). In contrast, a high dose of dexamethasone elicited a significant decrease of mRNA expression, representing about 70% of the control value and which seemed to particularly affect the 1 kb mRNA species (Jones *et al.*, 1990). It is not yet clear whether the decrease observed upon dexamethasone treatment directly derives from a modulation of endogenous steroid content.

Table 4. Modulation of prepro-NT/NN mRNA in experimental models

Experimental model	Modulation of mRNA expression: Site	Modulation of mRNA expression: Variation	Reference
Colchicine-treated by i.c.v. injection	Dorsomedial striatum and paraventricular and arcuate nuclei	Increase	Kiyama et *al*. (1991*b*)
	Medical preoptic area and lateral septum	Stability	
Ovariectomized rats	Pituitary	Increase	O'Halloran et *al*. (1990)
Oestrogen-treated ovariectomized rats	Pituitary	Reversal of increase	
Ovariectomized rats implanted subcutaneously with 17β-oestradiol	Anterior medial preoptic nucleus	Strong increase	Alexander et *al*. (1989*a*)
	Lateral septum, diagonal band of Broca, bed nucleus of stria terminalis and lateral preoptic area	Stability	
Hypothyroidism	Anterior pituitary	Decrease	Jones et *al*. (1989)
Castrated or ovariectomized rats	Anterior pituitary	Stability	Jones et *al*. (1990)
Daily chronic injection of dexamethasone	Anterior pituitary	Decrease	
Osmotic stimulation	Lateral hypothalamic area	Increase	Watts (1992)
	Central nucleus of the amygdala	Decrease	

Abbreviations used: i.c.v., intracerebro-ventricular.

Two papers reported on the effect of regulation of NT/NN mRNA expression in model animals displaying an altered oestrogen status (Alexander *et al.*, 1989*a*; O'Halloran *et al.*, 1990). In ovariectomized rats, supraphysiological concentrations of 17β-oestradiol markedly increased the NT/NN mRNA content in the preoptic region (Alexander *et al.*, 1989*a*). Particularly interesting was the 3 to 4-fold increase observed in the anterior medial preoptic nucleus (AMPN) and in the adjacent medial preoptic nucleus (MPN) whereas levels remained constant in the lateral septum, diagonal band of Broca, bed nucleus of the stria terminalis and lateral preoptic area (Alexander *et al.*, 1989*a*). The fact that AMPN and MPN are specifically affected in their mRNA content could reflect a role of NT and/or NN in the control of preovulatory secretion of luteinizing hormone (LH) which seems to be controlled by these two nuclei. This would be in keeping with previous observations indicating that the immuno-neutralization of NT in the preoptic region lowered LH secretion induced by oestrogen treatment in ovariectomized rats (Alexander *et al.*, 1989*b*). The relation between the NT/NN mRNA level in anterior pituitary and the oestrogen status was further described (O'Halloran *et al.*, 1990).

A recent report investigated the effect of osmotic stimulation on the NT/NN mRNA content in the lateral hypothalamic area (LHA) and the central nucleus of the amygdala (CNA) (Watts, 1992). Interestingly, mRNA levels increased in the LHA while a decrease of the mRNA content was observed in the CNA (Table 4). This suggested that the adaptative response to alterations of fluid homeostasis could involve NT and/or NN in the forebrain circuits.

Distribution of prepro-NT/NN mRNA

In the forebrain, the NT/NN mRNA was widely detectable in the septum, bed nucleus of the stria terminalis, the preoptic area, the amygdala, the hypothalamus, the nucleus accumbens, the striatum and the various cortical zones including the piriform and retrosplenial regions (Alexander *et al.*, 1989*c*). These observations were confirmed by a novel fluorescence technique using biotinylated oligonucleotide probes which allowed sensitive visualization of *in situ* hybridization (Williams *et al.*, 1990). This study demonstrated the occurrence of mRNA in the dorsomedial and ventrolateral parts of the caudate putamen and in the olfactory tubercles and septal nuclei of the ventral striatum (Williams *et al*, 1990). These authors extended their cartography to the midbrain and showed that the highest frequency of hybridization-positive neurons was located in the substantia nigra and in the superior colliculus. Intense staining was also observed in the dorsal and ventral periaqueductal grey, the oculomotor region and the medial longitudinal fasciculus while less prominent populations were detected in the ventral tegmental area and the dorsomedial mesencephalic nuclei (Williams *et al.*, 1990).

It was noticeable that the distribution of the prepro-NT/NN mRNA in the forebrain generally matched that of NT immunoreactivity. However, few discrepancies took place in the CA1 region of the hypothalamus and in the subiculum where mRNA was observed (Alexander *et al.*, 1989 *c*) while NT was never detected in such brain areas (Jennes *et al.*, 1982). Whether such discrepancies could reflect a tissue-specific maturation of the prepro-NT/NN precursor, yielding products other than NT and NN, or an efficient proteolytic mechanism leading to an underestimation of biosynthesized products is not clear and will be discussed later (see the Post-transcriptional Events section below).

Ontogeny of NT and NT/NN mRNA

Ontogeny of NT

Several studies have documented the ontogeny of NT and NT-containing pathways in the central nervous system. Hara *et al.* (1982) established the precise ontogeny of the neurotensinergic circuits in the forebrain and diencephalon. The NT-positive structures first appeared by day 16 of the fetal stage in the lateral preoptic area, lateral hypothalamus, olfactory bulb, anterior amygdala and piriform cortex. NT-bearing structures gradually increased and reached a maximum on postnatal day 7 (Hara *et al.*, 1982). Then, the number of NT-positive structures slightly decreased until adulthood although numerous structures remained positive for NT in adult rats (Hara *et al.*, 1982). The same group examined the ontogeny of the NT-containing neuron system in the lower brainstem (Minagawa *et al.*, 1983). Here again, a noticeable increase in the number of NT-positive structures occurred between day 19 of gestation and postnatal day 7 (Minagawa *et al.*, 1983). Most of the regions of the lower brainstem underwent a strong decrease in the number of NT-positive structures in parallel with rat growth. This was in agreement with another study demonstrating the transient increase from days 2 to 10 followed by a decrease between days 20 to 60 in the NT-like immunoreactivity observed in the brainstem (Bissette *et al.*, 1984). The only exception occurred in the nucleus tractus solitarii which maintained their NT immunoreactivity at rat adulthood (Minagawa *et al.*, 1983). Such transient expression of NT immunoreactivity was also documented in the inferior olive of the medulla oblongata in the human (Mailleux and Vanderhaeghen, 1988) and the cat (Mailleux *et al.*, 1989). Finally, the developmental changes in the distribution and abundance of NT-like immunoreactivity were examined in the mediodorsal thalamic nucleus and in the striatum. During the first postnatal week, an intense staining was observed in both medial and central parts of the mediodorsal thalamic nucleus. The density of the fibres thereafter decreased and reached adult status after 5 weeks where only sparse fibres appeared to be labelled

(Ray and Price, 1990). NT immunoreactivity was absent in the rostral striatum at birth (Zahm *et al.*, 1990) while few fibres were observed in the caudate putamen at this stage. The NT immunoreactive structures spread into the striatum between days 5 and 10 and decreased later (Zahm *et al.*, 1990).

All the above studies have clearly established that NT and its neuronal circuits are developmentally regulated. The development of prepro-NT/NN probes has proved useful for investigating whether the ontogenic modulation of the NT content was also accompanied by modifications at the transcriptional level.

Ontogeny of prepro-NT/NN mRNA

Central nervous system

The ontogeny of the NT/NN mRNA in the central nervous system has been very recently reviewed (Kiyama *et al.*, 1991 *b*). These authors developed an original non-radioactive *in situ* hybridization approach by means of an oligodeoxynucleotide probe labelled with alkaline phosphatase (Kiyama *et al.*, 1991 *a*, *c*). It was therefore possible to record and quantify mRNA signals by densitometry. They demonstrated that NT mRNA expression appeared transiently in several brain nuclei. As illustrated in Fig. 4, NT/NN mRNA-containing cells were observed on day 14 of the embryonic stage in the hypoglossal nucleus (Kiyama *et al.*, 1991 *a*) and in the mitral cells of the olfactory bulb (Kiyama *et al.*, 1991 *c*). In both brain nuclei, the mRNA signal subsequently decreased during development and reached a very low absorbance value at the adult stage (Fig. 4).

Extensive study of the postnatal expression of prepro-NT/NN mRNA in cells of the rat forebrain and midbrain revealed distinct types of ontogenic behaviours (Sato *et al.*, 1991) that can be roughly divided into two groups. One group of neurons exhibited the messenger around birth and reached a maximum during the postnatal period. These cells still expressed the mRNA at adulthood. The second type of mRNA-expressing cells also exhibited the mRNA during the postnatal period but the signal drastically diminished after the first week and was barely detectable a few weeks after birth (Sato *et al.*, 1991). Brain regions belonging to the first group included the piriform cortex, the diagonal band of Broca, the intermediate part of the lateral septal nucleus, the bed nucleus of the stria terminalis, the medial preoptic area, the lateral hypothalamus, the caudal part of the caudate putamen, the medial, cortical and central amygdaloid nuclei, the ventral tegmental area, the deep mesencephalic nucleus, the cuneiform nucleus, the dorsal raphe nucleus, the laterodorsal tegmental nucleus, the parabrachial nucleus, the oral part of the pontine reticular nucleus and the subiculum (Fig. 5). Cells belonging to the second type were observed in the anterior and posterior cingulate cortex, the mitral cell layer of the olfactory bulb and the rostral part of the caudate putamen (Sato *et al.*, 1991).

Fig. 4. **Ontogenic profile of prepro-NT/NN mRNA in the olfactory bulb and hypoglossal nucleus**

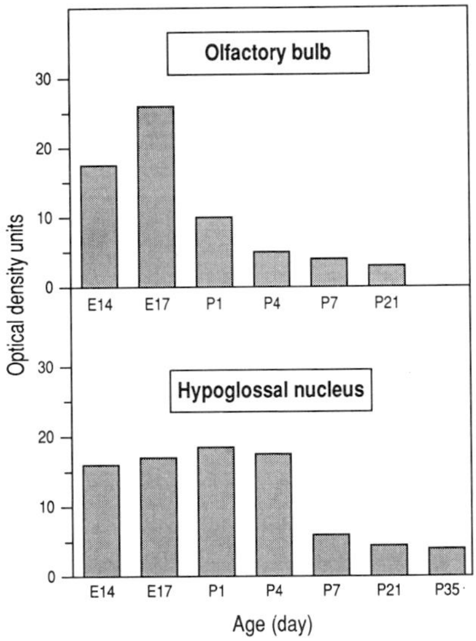

The ontogenic change of prepro-NT/NN mRNA in the mitral cells of the olfactory bulb and in the hypoglossal nucleus was recorded by microdensitometry by means of a non-radioactive in situ *hybridization technique using oligodeoxynucleotide probes labelled with alkaline phosphatase. (Data are from Kiyama* et al., *1991*a,b.*) Abbreviations used: E, embryonic day; P, postnatal day.*

Northern blot analysis and *in situ* hybridization experiments confirmed that the rat forebrain underwent important changes in the NT/NN mRNA between youth and adulthood (Nicot *et al.*, 1992). Whether the transient and region-specific expression of the NT/NN mRNA reflects an involvement of the two peptides in the development of brain nuclei remains to be established.

Gastrointestinal tract

The ontogeny of prepro-NT/NN mRNA in various segments of rat gastrointestinal tract has been examined. Here again, the ontogenic development appeared to vary according to the tissue source. Thus, the NT mRNA remained virtually constant between day 0 and 8 weeks in the rat jejunum and ileum. In contrast, after day 7, there existed a total absence of the message in the stomach, the duodenum and the colon (M. Oka, N. Khandon Nia and S. R. Bloom, personal communication). These data are in agreement with those obtained by Evers *et al.* (1993) showing that NT mRNA expres-

Fig. 5. Developmental changes in prepro-NT/NN mRNA in various brain regions

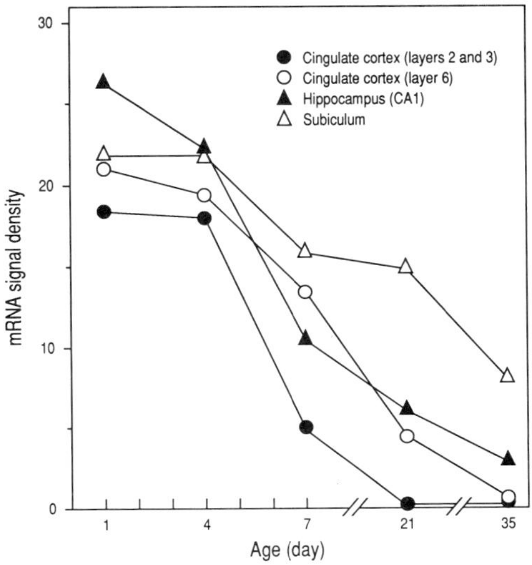

mRNA signal was densitometrically recorded as explained in the legend to Fig. 4. (Data are from Kiyama et al., *1991*b.*)*

sion rapidly increased after birth in rat ileum and jejunum. mRNA levels reached a maximum by day 21 after birth then remained constant in the ileum while the expression gradually decreased in the jejunum. The main mRNA species in these tissues was the 1 kb form (see the Xenopsin and NT/NN precursors section). In the colon, the small mRNA form was observed at the fetal stage, decreased after day 14 and was not detectable at adulthood (Evers *et al.*, 1993). In this tissue, the 1.5 kb mRNA appeared virtually absent. The profile of the prepro-NT/NN mRNA expression in these intestinal regions seemed to be similarly developmentally regulated in the human (Evers *et al.*, 1993). The reasons for the distinct mechanisms controlling the transcription of the prepro-NT/NN in a tissue-specific manner are not yet clear.

Post-transcriptional events

Processing of the NT/NN precursor

As stated in the section on Xenopsin and NT/NN precursors, there exist several doublets of basic amino acids in the NT precursor sequence which correspond to putative targets for processing enzymes. It was therefore possible that tissue or region-specific maturation processes led to varying concentrations of NT, NN or other related sequences. Such an hypothesis has led several authors to establish the contents of NT and NN-like immunoreactivities in brain and peripheral organs. By means of highly sensitive and specific radioimmunoassays, De Nadai *et al.* (1989) compared the distribution of the two parent peptides in several brain regions. It was established that the ratio of NT-like immunoreactivity (iNT) over that of NN (iNN) did not vary significantly according to the brain zone (Table 5). However, when such a study was performed with more discrete nuclei, there existed marked variations of the iNT/iNN ratio. For instance, the amount of iNT was about 4–6-fold higher than that of NN in the substantia nigra pars compacta, globus pallidus or posterior hypothalamus (Kitabgi *et al.*, 1991). Such a drastic difference in the content of NT versus NN was also reported recently in various ocular tissues (Hayes *et al.*, 1992). Altogether, this

Table 5. Immunoreactive NT and NN in brain and peripheral tissues

Tissue	iNT (pmol/g of tissue)	iNT/iNN
Brain		
Frontal cortex	2.65 ± 0.85	1.7
Hippocampus	9.82 ± 1.96	1.1
Striatum	6.8 ± 1.78	1.4
Thalamus	7.8 ± 1.65	1.3
Hypothalamus	45.9 ± 6.14	1.6
Mesencephalon	15.15 ± 0.98	1.1
Brainstem	13.06 ± 1.86	1.4
Cerebellum	0.61 ± 0.04	1.6
Peripheral organs		
Caecum	3.7 ± 1.3	1.4
Colon	0.28 ± 0.03	1.8
Duodenum	3.4 ± 1.2	0.8
Ileum	40.8 ± 3.5	2.1
Jejunum	40.8 ± 5.3	1.7
Stomach	4.3 ± 1.1	5.1
Heart	0.54	—
Kidneys	0.3	—
Pancreas	0.54	—
Ocular tissues		
Retina	0.53	0.31
Choroid/sclera	0.21	—
Iris/ciliary body	0.104	7.4

Data were taken from De Nadai et al. *(1989); Kitabgi* et al. *(1992) and Hayes* et al. *(1992).*

suggested the possibility that post-translational events could, in a tissue-specific manner, modulate the concentration of endogenous NT-related sequences.

Table 5 indicates that the iNT/iNN ratio also varied according to the rat peripheral tissue examined (Kitabgi *et al.*, 1992). The situation is particularly interesting in the intestine. Carraway and Mitra (1990*a*) showed that the canine intestinal prepro-NT/NN precursor underwent processing mechanisms clearly distinct from those occurring in the brain. Thus, the authors clearly established that more than 75% of the iNN present in the canine small intestine consisted of large molecular forms of NN (Carraway and Mitra, 1990*a*). Such differential processing of the NT/NN precursor between central and peripheral tissues was also reported in mice (Shaw *et al.*, 1990).

The processing of the precursor was also examined in the brain and intestine of other mammals including cat, guinea-pig, rat and dog. In these species, a large form of NN consisting of 55–91% of total iNN was detected in the ileum while this form only represented 2–8% of total iNN observed in the brain (Carraway *et al.*, 1991). This high molecular mass NN-like sequence was partially purified and characterized by means of antisera directed towards various regions of the precursor (Carraway *et al.*, 1991). The data indicated that the extended form of NN contained NN at its *C*-terminus and corresponded at least to the 48–148 sequence of the canine pre-cursor (Carraway *et al.*, 1991). The size of the high molecular weight NN sequence was further confirmed by Western blot analysis (Carraway *et al.*, 1991).

The large molecular form of NN occurring in the gastrointestinal tract was further characterized recently in the canine ileal mucosa (Carraway and Mitra, 1991). The purification and sequencing of the peptide indicated that it corresponded to the *N*-terminal extended form of NN yielded by a cleavage taking place at the Cys^{23}–Ser^{24} bond (end of the signal peptide) of the dog precursor. The large molecular form which extended from residues 24 to 148 was called NN 125 (Carraway and Mitra, 1991). It was very interesting to note that the partially purified preparation of this NN form was found to display an equipotent ability to contract guinea-pig ileum when compared with that of NN (Carraway *et al.*, 1991).

The above studies clearly indicated that the maturation of the NT precursor in the intestine led to formation of a minor amount of genuine NN. This could explain the striking lack of secretion of this peptide in a vascularly perfused model of isolated rat ileum (Cuber *et al.*, 1990). Thus, the authors established that the luminal infusion of various nutrients led to a transient secretion of NT (600% above the basal circulating concentration) while NN did not seem to be co-released with its parent peptide (Cuber *et al.*, 1990). These data corroborated the hypothesis of a novel fate of the prepro-NT/NN precursor in the ileal mucosa. Whether the extended forms of NN could have contributed to the pharmacological profile generally ascribed to NN in the gastrointestinal tract remains speculative.

Recently, Bidard *et al.* (1992) reported the characterization of prepro-NT/NN processing in a cell line derived from a rat medullary thyroid carcinoma rMTC 6–23. The results showed that dexamethasone-treated rMTC 6–23 cells produced equimolar amounts of NT, NN and the *N*-terminal counterpart of the precursor with the signal peptide deleted (sequence 23–139).

Proteolytic inactivation of NT and NN

Differential maturation processes can lead to marked variations of the NT and NN content. However, besides such post-transcriptional events and downstream to them, proteolytic mechanisms occur that probably contribute to the modulation of peptide concentration, thereby interrupting the peptidergic signal.

We previously studied the metabolism of NT in various tissue preparations and cell lines from various neural and extraneural origins that were previously shown to bear specific NT receptors likely to be involved in the mediation of the peptidergic message. We showed that NT underwent proteolytic cleavages by various endopeptidases that acted at the *C*-terminal end of NT which sustains the whole biological activity of this neuropeptide (Checler *et al.*, 1983*a*,*b*, 1984, 1985, 1986*a*,*b*, 1987*a*,*b*). We detected (Checler *et al.*, 1983*a*) and purified a novel metallo-endopeptidase (Checler *et al.*, 1986*d*; Barelli *et al.*, 1988, 1993*b*) now called endopeptidase 3.4.24.16 which was the only peptidase that ubiquitously participated in the inactivating mechanisms of NT in these NT-receptor-bearing preparations (Checler *et al.*, 1988*a*).

In whole rat brain, the three major endopeptidases inactivating NT were endopeptidases 24.11, 24.15 and 24.16 (Fig. 6). The participation of endopeptidase 24.11 (Coquerel *et al.*, 1986) and endopeptidase 24.16 (F. Checler, H. Barelli and J. P. Vincent, unpublished work) in the metabolism of NT *in vivo*, was further supported by the strong potentiation of the hypothermic-induced effect of NT (following intra-cerebro-ventricular administration) by specific inhibitors of endopeptidase 24.11 (Coquerel *et al.*, 1986) and endopeptidase 24.16 (Barelli *et al.*, 1992). In the periphery, in a model of vascularly perfused dog ileum, we clearly demonstrated the involvement of endopeptidase 24.11 (Checler *et al.*, 1988*b*) and endopeptidase 24.16 (Barelli *et al.*, 1994) in NT catabolism.

There existed a regionalization of NT degradative mechanisms in brain. Davis *et al.* (1992) clearly demonstrated by means of regional rat brain slices that, although endopeptidase 24.15 appeared to be of major importance in all the brain regions studied, additional contributions of angiotensin-converting enzyme and endopeptidase 24.11 were markedly elevated in the caudate putamen as compared with hippocampus and nucleus accumbens.

Interestingly, the pattern of hydrolysis can also vary according to the tissue layer examined. Thus, we examined the NT inactivating mechanisms taking place in the various muscular, nervous and mucosal layers of

Fig. 6. Peptidases involved in NT and NN inactivation in rat brain and canine mucosa

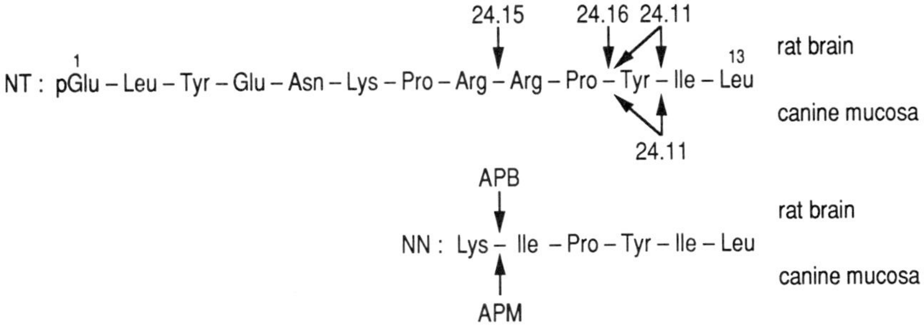

Arrows indicate the primary sites of hydrolysis inactivating the two peptides and the peptidases responsible for these cleavages. These are endopeptidases 24.11, 24.15 and 24.16 as well as aminopeptidase B (APB) and aminopeptidase M (APM).

canine ileum (Checler *et al.*, 1987*a*; Barelli *et al.*, 1989). The degradation products and the peptidases responsible for their formation varied according to the muscular or nervous origin of the tissue (Checler *et al.*, 1987*a*; Barelli *et al.*, 1989; 1993*a*). Furthermore, in the dog intestinal mucosa, endopeptidase 24.11 contributed to NT breakdown while endopeptidases 24.15 and 24.16 were absent and did not contribute to the hydrolysis (Fig. 6) (Barelli *et al.*, 1993*a*). It can be derived from the above observations that the qualitative pattern of inactivation of neuropeptides can vary according to the set of neuropeptidases constituting the enzymic machinery that equipped a given area or tissue layer. Such a statement could also stand when the proteolytic mechanisms leading to the clearance of related peptides are examined. Thus, we demonstrated that NN was particularly susceptible to degradation by amino-exopeptidases in rat brain synaptic membranes (Checler *et al.*, 1986*c*). The main degradation product generated by these membranes corresponded to des-Lys[1]-NN (i.e. resulting from the release of the *N*-terminal lysyl residue of NN) (Fig. 7). This cleavage appeared to be elicited by distinct aminopeptidases (Fig. 6) in rat brain (Checler *et al.*, 1991) and in canine mucosa (Barelli *et al.*, 1993*a*). The involvement of aminopeptidase(s) in the inactivation of NN was confirmed by the potentiation of the hypothermic and analgesic effect of NN by the general aminopeptidase inhibitor bestatin (Coquerel *et al.*, 1988; Dubuc *et al.*, 1988). We also demonstrated that bestatin and arphamenine B (aminopeptidase B inhibitor) increased the recovery of NN after infusion in dog ileum (H. Barelli, E. E. Daniel, J. P. Vincent and F. Checler, unpublished work).

It should be noted that such *N*-terminal attack was totally prevented by the pyroglutamate residue blocking the *N*-terminus of the NT sequence. Therefore, it appears that the set of proteolytic activities together with the various susceptibilities to proteolytic attack (conferred by the struc-

Fig. 7. Kinetics of NN inactivation by rat brain synaptic membranes

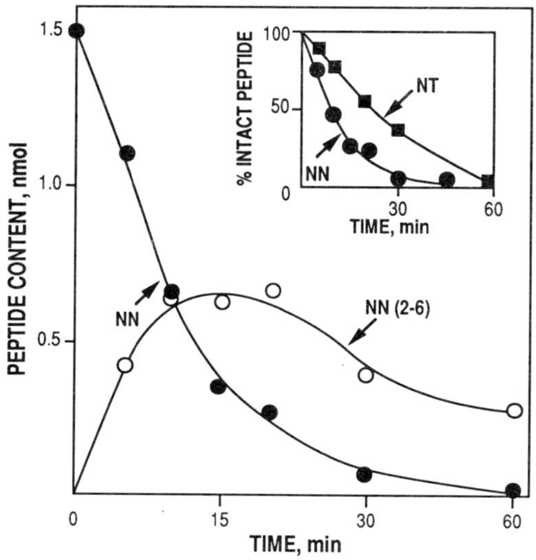

NN was incubated for various times at 37 °C and degradation was monitored by h.p.l.c. Neuromedin N (●) and des-Lys[1] neuromedin N (NN 2–6) (○) were quantified by comparison with known amounts of corresponding synthetic peptides. Insert compares the rates of inactivation of NT and NN by rat brain synaptic membranes.

Table 6. Comparison of initial rates of NN and NT hydrolysis in rat brain and canine mucosa

	NT	NN	NN/NT
Rat brain			
Nucleus accumbens	17.2 ± 1.6	48.7 ± 4.9	2.8
Ventro-tegmental area	10.9 ± 1.2	37.7 ± 3.2	3.5
Canine mucosa			
Mucosa-P_2	430 ± 30	7070 ± 490	16.5
Mucosa-Mic_{II}	225 ± 19	5340 ± 678	23.7

Data are expressed as nmol/mg of protein per h and were taken from Checler et al. *(1991) and Barelli* et al. *(1993a).*

tural features) might be responsible for clearly distinct 'qualitative' patterns of degradation for two parent peptides. It should be added that this can also lead to drastically different 'quantitative' rates of hydrolysis. As illustrated in Table 6, we showed that in synaptic membranes prepared from whole brain

(Checler *et al.*, 1986*c*) and in various brain areas (Checler *et al.*, 1991), NN was degraded more rapidly than NT. This feature appeared much more pronounced in the mucosa where NN was degraded at 17–24-fold higher rates than NT (Table 6).

NT and NN (and also probably structurally related peptides belonging to other families) can be qualitatively and quantitatively susceptible to differential metabolic inactivation. If the processing of the precursor leads to the concomitant biosynthesis of the two related peptides (as seems, at least, to be the case in most brain areas), metabolic processes could be the way to balance between long-acting or short-lived peptides, thereby directly influencing the neuromodulatory properties of the two related peptides. In the periphery, distinct patterns of hydrolysis could dictate the paracrine or endocrine function of NT and NN. It would be tempting to speculate that the first role of the *N*-terminal extended region of the large form of NN recovered as the major form in the intestine would be to protect the NN sequence from aminopeptidase attack, thereby leading to a biologically active peptide with prolonged lifetime in the gastrointestinal tract.

The author wishes to gratefully acknowledge V. Chéron for expert secretarial assistance and F. Aguila for excellent artwork. This work was supported by the Centre National de la Recherche Scientifique, the Institut National de la Santé et de la Recherche Médicale and the Fondation pour la Recherche Médicale.

References

Alexander, M. J., Dobner, P. R., Miller, M. A., Bullock, B. P., Dorsa, D. M. and Leeman, S. E. (1989*a*) Estrogen induces neurotensin/neuromedin N messenger ribonucleic acid in a preoptic nucleus essential for the preovulatory surge of luteinizing hormone in the rat. Endocrinology **125**, 2111–2117

Alexander, M. J., Mahoney, P. D., Ferris, C. F., Carraway, R. E. and Leeman, S. E. (1989*b*) Evidence that neurotensin participates in the central regulation of preovulatory surge of luteinizing hormone in the rat. Endocrinology **124**, 783–788

Alexander, M. J., Miller, M. A., Dorsa, D. M., Bullock, B. P., Melloni, R. H., Dobner, P. R. and Leeman, S. E. (1989*c*) Distribution of neurotensin/neuromedin N mRNA in rat forebrain: unexpected abundance in hippocampus and subiculum. Proc. Natl. Acad. Sci. U.S.A. **86**, 5202–5206

Araki, K., Takibana, S., Ushiyama, M., Nakajima, T. and Yashuhara, T. (1975) Isolation and structure of a new active peptide xenopsin on rat stomach and some biogenic amines in the skin of *Xenopus laevis*. Chem. Pharm. Bull. **23**, 3132–3140

Araki, K., Tachibana, S., Kato, Y. and Tajima, T. (1979) Comparative studies of xenopsin and neurotensin and some biological activities. Yakugaku Zasshi **99**, 466–470

Augood, S. J. and Emson, P. C. (1992) Pertussis toxin administration increases the expression of proneurotensin and preproenkephalin A mRNAs in rat striatum. Neuroscience **47**, 317–324

Augood, S. J., Kiyama, H., Faull, R. L. M. and Emson, P. C. (1991) Differential effects of acute dopaminergic D_1 and D_2 receptor antagonists on proneurotensin mRNA expression in rat striatum. Mol. Brain Res. **9**, 341–346

Barelli, H., Vincent, J. P. and Checler, F. (1988) Peripheral inactivation of neurotensin: isolation and characterization of a metallopeptidase from rat ileum. Eur. J. Biochem. **175**, 481–489

Barelli, H., Ahmad, S., Kostka, P., Fox, J.-A. E. T., Daniel, E. E., Vincent, J. P. and Checler, F. (1989) Neuropeptide-hydrolysing activities in synaptosomal fractions from dog ileum myenteric, deep muscular and submucous plexi. Their participation in neurotensin inactivation. Peptides **10**, 1055–1061

Barelli, H., Dive, V., Yiotakis, A., Vincent, J. P. and Checler, F. (1992) Potent inhibition of endopeptidase 24.16 and endopeptidase 24.15 by the phosphonamide peptide *N*-(phenylethylphosphonyl)-Gly-L-Pro-L-aminohexanoic acid. Biochem. J. **287**, 621–625

Barelli, H., Mao, Y. K., Vincent, B., Daniel, E. E., Vincent, J. P. and Checler, F. (1993*a*) Differential catabolic fate of neuromedin N and neurotensin in the canine intestinal mucosa, Peptides **14**, 457–463

Barelli, H., Vincent, J. P. and Checler, F. (1993*b*) Rat kidney endopeptidase 24.16: purification, physico-chemical characteristics and specificity towards opiates, tachykinins and neurotensin-related peptides. Eur. J. Biochem., **211**, 79–90

Barelli, H., Fox-Threlkeld, J. E. T., Dive, V., Daniel, E. E., Vincent, J. P. and Checler, F. (1994) Role of endopeptidase-3.4.24.16 in the catabolism of neurotensin, *in vivo*, in the vascularly perfused dog ileum. Br. J. Pharmacol., in the press

Bean, A. J. and Hökfelt, T. (1992) Reserpine increases striatal neurotensin mRNA levels. Mol. Brain Res. **12**, 345–348

Bean, A. J., Dagerlind, A., Hökfelt, T. and Dobner, P. R. (1992) Cloning of human neurotensin/neuromedin N genomic sequences and expression in the ventral mesencephalon of schizophrenics and age/sex matched controls. Neuroscience **50**, 259–260

Bidard, J. N., De Nadai, F., Rovère, C., Laur, J., Martinez, J., Cuber, J. C. and Kitabgi, P. (1993) Immunological and biochemical characterization of processing products from the neurotensin/neuromedin N precursor in the rat medullary thyroid carcinoma 6–23 cell line, Biochem. J. **291**, 225–233

Bissette, G. and Nemeroff, C. B. (1985) Do neuropeptide systems mediate some of the effects of antipsychotic drugs? In Endocoids, Progress in Clinical Biology Research (Lal, H., Labella, F. and Lane, J., eds.), pp. 349–353, 192, Liss, New York

Bissette, G., Richardson, C., Kizer, J. S. and Nemeroff, C. B. (1984) Ontogeny of brain neurotensin in the rat: a radioimmunoassay study. J. Neurochem. **43**, 283–287

Bissette, G., Dauer, W. T., Kilts, C. D., O'Connor, L. and Nemeroff, C. B. (1988) The effect of the stereoisomers of butachamol on neurotensin content in discrete regions of the rat brain. Neuropsychopharmacology **1**, 329–335

Carraway, R. and Bhatnagar, Y. M. (1980) Immunochemical characterization of neurotensin-like peptides in chicken. Peptides **1**, 159–165

Carraway, R. E. and Ferris, C. F. (1983) Isolation, biological and chemical characterization, and synthesis of a neurotensin-related hexapeptide from chicken intestine. J. Biol. Chem. **258**, 2475–2479

Carraway, R. E. and Leeman, S. E. (1973) The isolation of a new hypotensive peptide, neurotensin, from bovine hypothalami. J. Biol. Chem. **248**, 6854–6861

Carraway, R. E. and Leeman, S. E. (1976) Characterization of radioimmunoassayable neurotensin in the rat. Its differential distribution in the central nervous system, small intestine and stomach. J. Biol. Chem. **251**, 7045–7052

Carraway, R. E. and Mitra, S. P. (1990*a*) Differential processing of neurotensin/neuromedin N precursor(s) in canine brain and intestine. J. Biol. Chem. **265**, 8627–8631

Carraway, R. E. and Mitra, S. P. (1990*b*) Isolation and sequence of canine xenopsin and an extended fragment from its precursor. Peptides **11**, 747–752

Carraway, R. E. and Mitra, S. P. (1991) Purification of large neuromedin N (NMN) from canine intestine and its identification as NMN-125. Biochem. Biophys. Res. Commun. **179**, 301–308

Carraway, R. E., Mitra, S. P. and Cochrane, D. E. (1987*a*) Structure of a biologically active neurotensin-related peptide obtained from pepsin-treated albumin(s). J. Biol. Chem. **262**, 5968–5973

Carraway, R. E., Cochrane, D. E. and Ruane, S. E. (1987*b*) Isolation, structures, and biological activity of neurotensin-related peptides generated in extracts of avian tissue. J. Biol. Chem. **262**, 15886–15889

Carraway, R. E., Cochrane, D. E. and Mitra, S. P. (1988) Xenopsin-related peptide generated in avian gastric extracts. Regul. Peptides, **22**, 303–314

Carraway, R. E., Mitra, S. P. and Muraki, K. (1990) Isolation and structures of xenopsin-related peptides from rat stomach, liver and brain. Regul. Peptides **29**, 229–239

Carraway, R. E., Mitra, S. P. and Paradise, C. (1991) Characterization of large neuromedin-N using antisera towards regions of the neurotensin/neuromedin-N precursor. Peptides **12**, 601–607

Checler, F., Vincent, J. P. and Kitabgi, P. (1983*a*) Degradation of neurotensin by rat brain synaptic membranes. Involvement of a thermolysin-like metalloendopeptidase (enkephalinase), angiotensin-converting enzyme and other unidentified peptidases. J. Neurochem. **41**, 375–384

Checler, F., Vincent, J. P. and Kitabgi, P. (1983*b*) Neurotensin analogs D-Tyr^{11} and D-Phe^{11} neurotensin resist degradation by brain peptidases *in vitro* and *in vivo*. J. Pharmacol. Exp. Ther. **227**, 743–748

Checler, F., Emson, P. C., Vincent, J. P. and Kitabgi, P. (1984) Inactivation of neurotensin by rat brain synaptic membranes. Cleavage at the Pro^{10}-Tyr^{11} bond by endopeptidase 24.11 (enkephalinase) and a peptidase distinct from proline endopeptidase. J. Neurochem. **43**, 1295–1301

Checler, F., Vincent, J. P. and Kitabgi, P. (1985) Inactivation of neurotensin by rat brain synaptic membranes partly occurs through cleavage at the Arg^{8}–Arg^{9} peptide bond by a metalloendopeptidase. J. Neurochem. **45**, 1509–1513

Checler, F., Amar, S., Kitabgi, P. and Vincent, J. P. (1986*a*) Catabolism of neurotensin by neural (neuroblastoma clone N1E115) and extraneural (HT29) cell lines. Peptides **7**, 1071–1077

Checler, F., Mazella, J., Kitabgi, P. and Vincent, J. P. (1986*b*) High affinity receptor sites and rapid proteolytic inactivation of neurotensin in primary cultured neurons. J. Neurochem. **47**, 1742–1748

Checler, F., Vincent, J. P. and Kitabgi, P. (1986*c*) Neuromedin N: high affinity interaction with brain neurotensin receptors and rapid inactivation by brain synaptic peptidases. Eur. J. Pharmacol. **126**, 239–244

Checler, F., Vincent, J. P. and Kitabgi, P. (1986*d*) Purification and characterization of a novel neurotensin degrading peptidase from rat brain synaptic membranes. J. Biol. Chem. **261**, 11274–11281

Checler, F., Ahmad, S., Kostka, P., Barelli, H., Kitabgi, P., Fox, J. E. T., Kwan, C. Y., Daniel, E. E. and Vincent, J. P. (1987*a*) Peptidases in dog ileum circular and longitudinal smooth muscle plasma membranes: their relative contribution to the metabolism of neurotensin. Eur. J. Biochem. **166**, 461–468

Checler, F., Barelli, H., Kwan, C. Y. and Vincent, J. P. (1987*b*) Neurotensin-metabolizing peptidases in rat fundus plasma membranes. J. Neurochem. **49**, 507–512

Checler, F., Barelli, H., Kitabgi, P. and Vincent, J. P. (1988*a*) Neurotensin metabolism in various tissues from central and peripheral origins. Ubiquitous involvement of a novel neurotensin degrading metalloendopeptidase. Biochimie **70**, 75–82

Checler, F., Kostolonsko, B. and Fox, J. A. (1988*b*) *In vivo* inactivation of neurotensin in dog ileum: major involvement of endopeptide 24.11. J. Pharmacol. Exp. Ther. **244**, 1040–1141

Checler, F., Dauch, P., Masuo, Y. and Vincent, J. P. (1991) Neurotensin and neuromedin N are differently metabolized in ventral tegmental area and nucleus accumbens. J. Neurochem. **56**, 1320–1328

Coquerel, A., Dubuc, I., Menard, J. F., Kitabgi, P. and Costentin, J. (1986) Naloxone-insensitive potentiation of neurotensin hypothermic effect by the enkephalinase inhibitor thiorphan. Brain Res. **398**, 386–389

Coquerel A., Dubuc, I., Kitabgi, P. and Costentin, P. (1988) Potentiation by thiorphan and bestatin of the naloxone-insensitive analgesic effects of neurotensin and neuromedin N. Neurochem. Int. **12**, 361–366

Cuber, J. C., Herrmann, C., Kitabgi, P., Bosshard, A., Bernard, C., De Nadai, F. and Chayvialle, J. A. (1990) Neuromedin-N is not released with neurotensin from rat ileum. Endocrinology **126**, 1584–1592

Davis, T. P., Gillespie, T. J. and Konings, P. N. M. (1992) Specificity of neurotensin metabolism by regional rat brain slices. J. Neurochem. **58**, 608–617

De Nadai, F., Cuber, J. C. and Kitabgi, P. (1989) The characterization and regional distribution of neuromedin N-like immunoreactivity in rat brain using a highly sensitive and specific radioimmunoassay. Comparison with the distribution of neurotensin. Brain Res. **500**, 193–198

Dobner, P. R., Barber, D. L., Villa-Komaroff, L. and McKierman, C. (1987) Cloning and sequence analysis of cDNA for the canine neurotensin/neuromedin N precursor. Proc. Natl. Acad. Sci. U.S.A. **84**, 3516–3520

Dobner, P. R., Tischler, A. S., Lee, Y. C., Bloom, S. R. and Donahue, S. R. (1988) Lithium dramatically potentiates neurotensin/neuromedin N gene expression. J. Biol. Chem. **263**, 13983–13986

Dubuc, I., Nouel, D., Coquerel, A., Menard, J. F., Kitabgi, P. and Costentin, J. (1988) Hypothermic effect of neuromedin N in mice and its potentiation by peptidase inhibitors. Eur. J. Pharmacol. **151**, 117–121

Evers, B. M., Rajamaran, S., Chung, P. H., Townshend, E. P., Townsend, C. M. Jr., Wang, X., Graves, K. and Thompson, C. M. Jr. (1993) Differential expression of the neurotensin gene in the developing rat and human gastrointestinal tract. Am. J. Physiol. **265**, G482–G590.

Goedert, M., Lightman, S. L., Nagy, J. I., Marley, P. D. and Emson, P. C. (1982) Neurotensin in the rat anterior pituitary gland. Nature (London) **298**, 163–165

Govoni, S., Hong, J. S., Yang, H. Y. T. and Costa, E. (1980) Increase of neurotensin content elicited by neuroleptics in nucleus accumbens. J. Pharmacol. Exp. Ther. **215**, 413–419

Hammer, R. A., Carraway, R. E. and Leeman, S. E. (1982) Elevation of plasma neurotensin-like immunoreactivity after a meal. Characterization of the elevated components. J. Clin. Invest. **70**, 74–81

Hara, Y., Shiosaka, S., Senba, E., Sakanaka, M., Inagaki, S., Takagi, H., Kawai, Y., Takatsuki, K., Matsuzaki, T. and Tohyama, M. (1982) Ontogeny of the neurotensin-containing neuron system of the rat: immunohistochemical analysis. I. Forebrain and diencephalon. J. Comp. Neurol. **208**, 177–195

Hayes, R. G. J., Shaw, C., Kitabgi, P. and Buchanan, K. D. (1992) Different relative abundance of neurotensin and neuromedin N in bovine ocular tissues. Regul. Peptides **39**, 147–155

Hirsch Fernstrom, M., Carraway, R. E. and Leeman, S. E. (1980) Neurotensin. In Frontiers in Neuroendocrinology (Martini, L. and Ganong, W. F., eds.), pp. 103–127, Raven Press, New York

Hoffman, E., Richter, K. and Kreil, G. (1983) A novel peptide designated PYL[a] and its precursor as predicted from cloned mRNA of *Xenopus laevis* skin. EMBO J. **2**, 711–714

Holzer, P. A., Bucsics, A., Saria, A. and Lembeck, F. (1982) A study of the concentrations of substance P and neurotensin in the gastrointestinal tract of various mammals. Neuroscience **7**, 2919–2924

Jennes, L., Stumpf, W. E. and Kalivas, P. W. (1982) Neurotensin: topographical distribution in rat brain by immunohistochemistry. J. Comp. Neurol. **210**, 211–224

Jones, P. M., Ghatei, M. A., Steel, J., O'Halloran, D., Gon, G., Legon, S., Burrin, J. M., Leonhardt, U., Polak, J. M. and Bloom, S. R. (1989) Evidence for neuropeptide Y synthesis in the rat anterior pituitary and the influence of thyroid hormone status: comparison with vasoactive intestinal peptide, substance P and neurotensin. Endocrinology **125**, 334–341

Jones, P. M., O'Halloran, D. J., Ghatei, M. A., Domin, J. and Bloom, S. R. (1990) The influence of adrenal hormone status on neuroendocrine peptides in the rat anterior pituitary gland. J. Endocrinol. **127**, 437–444

Kasckow, J. and Nemeroff, C. B. (1991) The neurobiology of neurotensin: focus on neurotensin–dopamine interactions. Regul. Peptides **36**, 153–164

Kilts, C. D., Anderson, C. M., Bissette, G., Ely, T. D. and Nemeroff, C. B. (1988) Differential effects of antipsychotic drugs on the neurotensin concentration of discrete rat brain nuclei. Biochem. Pharmacol. **37**, 1547–1554

Kislauskis, E. and Dobner, P. R. (1991) Mutually dependent response elements in the *cis*-regulatory region of the neurotensin/neuromedin N gene integrate environmental stimuli in PC12 cells. Neuron **4**, 783–795

Kislauskis, E., Bullock, B., McNeil, S. and Dobner, P. R. (1988) The rat gene encoding neurotensin and neuromedin N. Structure, tissue-specific expression, and evolution of exon sequences. J. Biol. Chem. **263**, 4963–4968

Kitabgi, P. (1982) Effects of neurotensin on intestinal smooth muscle: application to the study of structure–activity relationships. Ann. N.Y. Acad. Sci. **400**, 37–55

Kitabgi, P. (1989) Neurotensin modulates dopamine neurotransmission at several levels along brain dopaminergic pathways. Neurochem. Int. **14**, 111–119

Kitabgi, P., Carraway, R. E. and Leeman, S. E. (1976) Isolation of a tridecapeptide from bovine intestinal tissue and its partial characterization as neurotensin. J. Biol. Chem. **251**, 7053–7058

Kitabgi, P., Carraway, R. E., Van Rietschoten, J., Granier, C., Morgat, J. L., Menez, A., Leeman, S. and Freychet, P. (1977) Neurotensin: specific binding to synaptic membranes from rat brain. Proc. Natl. Acad. Sci. U.S.A. **74**, 1846–1850

Kitabgi, P., Checler, F., Mazella, J. and Vincent, J. P. (1985) Pharmacology and biochemistry of neurotensin receptors. Rev. Basic Clin. Pharmacol. **5**, 397–484

Kitabgi, P., Masuo, Y., Nicot, A., Berod, A., Cuber, J. C. and Rostène, W. (1991) Marked variations of the relative distributions of neurotensin and neuromedin N in micropunched rat brain areas suggest differential processing of their common precursor. Neurosci. Lett. **124**, 9–12

Kitabgi, P., De Nadai, F., Rovère, C. and Bidard, J. N. (1993) Biosynthesis, maturation, release and degradation of neurotensin and neuromedin N. Ann. N.Y. Acad. Sci., **668**, 30–42

Kiyama, H., Emson, P. C., Sato, M. and Tohyama, M. (1991 *a*) The transient appearance of proneurotensin mRNA in the rat hypoglossal nucleus during development. Dev. Brain Res. **58**, 293–296

Kiyama, H., Sato, M. and Emson, P. C. (1991 *b*) Ontogeny of neurotensin immunoreactivity and mRNA in the rat central nervous system, in Handbook of Chemical Neuroanatomy (Björklund, A., Hökfelt, T. and Tohyama, M., eds.), Vol. 10, pp. 399–431, Elsevier, Amsterdam

Kiyama, H., Sato, M., Emson, P. C. and Tohyama, M. (1991 *c*) Transient expression of neurotensin mRNA in the mitral cells of rat olfactory bulb during development. Neurosci. Lett. **128**, 85–89

Lazarus, L. H., Brown, M. R. and Perrin, M. H. (1977) Distribution, localization and characteristics of neurotensin binding sites in the rat brain. Neuropharmacology **16**, 625–629

Levant, B., Merchant, K. M., Dorsa, D. M. and Nemeroff, C. B. (1992) BMY 14802, a potential antipsychotic drug, increases expression of proneurotensin mRNA in the rat striatum. Mol. Brain Res. **12**, 279–284

Mailleux, P. and Vanderhaeghen, J. J. (1988) Transient neurotensin in the human inferior olive during development. Brain Res. **456**, 199–203

Mailleux, P., Schiffmann, S. N., Halleux, P., Lotstra, F. and Vanderhaeghen, J. J. (1989) Transient neurotensin in the cat inferior olive during development. Neurochem. Int. **14**, 159–161

Mashford, M. L., Nilsson, G., Rokaeus, A. and Rosell, S. (1978) The effect of food ingestion on circulating neurotensin-like immunoreactivity (NTLI) in the human. Acta Physiol. Scand. **104**, 244–246

Mazella, J., Poustis, C., Labbé, C., Checler, F., Kitabgi, P., Granier, C., Van Rietschoten, J. and Vincent, J. P. (1983) Monoiodo-Trp11 neurotensin, a highly radioactive ligand of neurotensin receptors. Preparation, biological activity and binding properties to rat brain synaptic membranes. J. Biol. Chem. **258**, 3476–3481

Mazella, J., Kitabgi, P. and Vincent, J. P. (1985*a*) Molecular properties of neurotensin receptors in rat brain. Identification of subunits by covalent labeling. J. Biol. Chem. **260**, 508–514

Mazella, J., Kwan, C. Y., Kitabgi, P. and Vincent, J. P. (1985*b*) Covalent labeling of neurotensin receptors in rat gastric fundus plasma membranes. Peptides **6**, 1137–1141

Mazella, J., Chabry, J., Zsürger, N. and Vincent, J. P. (1989) Purification of the neurotensin receptor from mouse brain by affinity chromatography. J. Biol. Chem. **264**, 5559–5563

Merchant, K. M., Miller, M. A., Ashleigh, E. A. and Dorsa, D. M. (1991) Haloperidol rapidly increases the number of neurotensin mRNA-expressing neurons in neostriatum of the rat brain. Brain Res. **540**, 311–314

Merchant, K. M., Dobner, P. R. and Dorsa, D. M. (1992) Differential effects of haloperidol and clozapine on neurotensin gene transcription in rat neostriatum. J. Neurosci. **12**, 652–663

Mills, A., Demoliou-Mason, C. D. and Bernard, E. A. (1988) Purification of the neurotensin receptor from bovine brain. J. Biol. Chem. **263**, 13–16

Minagawa, H., Shiosaka, S., Inagaki, S., Sakanaka, M., Takatsuki, K., Ishimoto, E., Senba, Y., Kawai, Y., Hara, Y., Matsuzaki, T., *et al.* (1983) Ontogeny of neurotensin-containing neuron system of the rat: immunohistochemical analysis — II. Lower brain stem. Neuroscience **8**, 467–486

Minamino, N., Kangawa, K. and Matsuo, H. (1984) Neuromedin N: a novel neurotensin-like peptide identified in porcine spinal cord. Biochem. Biophys. Res. Commun. **122**, 542–549

Miyamoto-Lee, Y., Shiosaka, S. and Tohyama, M. (1991) Purification and characterization of neurotensin receptor from rat brain with special reference to comparison between newborn and adult age rats. Peptides **12**, 1001–1006

Mogard, M. H., Kobayashi, R., Chen, C. F., Lee, T. D., Reeve, J. R., Shively, J. E. and Walsh, J. H. (1986) The amino acid sequence of kinetensin, a novel peptide isolated from pepsin-treated human plasma: homology with human serum albumin, neurotensin and angiotensin. Biochem. Biophys. Res. Commun. **136**, 983–988

Nemeroff, C. B., Luttinger, D. and Prange, A. J. (1980) Neurotensin: central nervous system effects of a neuropeptide. Trends Neurosci. **3**, 212–215

Nemeroff, C. B., Luttinger, D. and Prange, A. J. (1983) Neurotensin and Bombesin, in Handbook of Psychopharmacology (Iversen, S. D., Iversen, L. and Snyder, S., eds.), Vol. 18, pp. 363–466, Plenum, New York

Nicot, A., Berod, A. and Rostène, W. (1992) Distribution of prepro-neurotensin/neuromedin N mRNA in the young and adult rat forebrain, in The Neurobiology of Neurotensin (Kitabgi, P. and Nemeroff, C. B., eds.), Ann. N.Y. Acad. Sci. 361–363

O'Halloran, D. J., Jones, P. M., Ghatei, M. A., Domin, J. and Bloom, S. R. (1990) The regulation of neuropeptide expression in rat anterior pituitary following chronic manipulation of estrogen status: a comparison between substance, neuropeptide Y, neurotensin and vasoactive intestinal peptide. Endocrinology **127**, 1463–1469

Orci, L., Baetens, O., Rufener, C., Brown, M., Vale, W. and Guillemin, R. (1976) Evidence for immunoreactive neurotensin in dog intestinal mucosa. Life Sci. **19**, 559–562

Ray, K. and Price, J. L. (1990) Postnatal changes in the density and distribution of neurotensin-like immunoreactive fibers in the mediodorsal nucleus of the thalamus in the rat. J. Comp. Neurol. **292**, 269–282

Rodriguez-Bello, A., Kah, O., Tramu, G. and Conlon, J. M. (1993) Purification and final structure and alligator neurotensin. Peptides **14**, 1055–1058

Rosell, S. and Rökaeus, A. (1979) The effect of ingestion of amino acids, glucose and fat on circulating neurotensin-like immunoreactivity in man. Acta Physiol. Scand. **107**, 263–267

Sadoul, J. L., Kitabgi, P., Rostène, W., Javoy-Agid, F., Agid, Y. and Vincent, J. P. (1984) Characterization and visualization of neurotensin binding to receptor sites in human brain. Biochem. Biophys. Res. Commun. **120**, 206–213

Saint-Pierre, S., Krouac, R., Quirion, R., Jolicoeur, F. B. and Rioux, F. (1984) Neurotensin. Peptide Protein Rev. **2**, 83–171

Sato, M., Kiyama, H., Yoshida, S., Saika, T. and Tohyama, M. (1991) Postnatal ontogeny of cells expressing prepro-neurotensin/neuromedin N mRNA in the rat forebrain and midbrain: a hybridization histochemical study involving isotope-labeled and enzyme-labeled probes. J. Comp. Neurol. **310**, 300–315

Schotte, A., Leysen, J. E. and Laduron, P. M. (1986) Evidence for a displaceable non-specific 3H neurotensin binding site in rat brain. Naunyn-Schmiedeberg's Arch. Pharmacol. **333**, 400–405

Shaw, C., Thim, L. and Conlon, J. M. (1986) (Ser7)-neurotensin: isolation from guinea-pig intestine. FEBS Lett. **202**, 187–192

Shaw, C., McKay, D., Johnston, C. F., Halton, D. W., Fairweather, I., Kitabgi, P. and Buchanan, K. D. (1990) Differential processing of the neurotensin/neuromedin N precursor in the mouse. Peptides **11**, 227–236

Shaw, C., Murphy, R., Thim, L., Furness, J. B. and Buchanan, K. D. (1991) Marsupial possum neurotensin: a unique mammalian regulatory peptide exhibiting structural homology to the avian analogue. Regul. Peptides **35**, 49–57

Shaw, C., McKay, D. M., Halton, D. W., Thim, L. and Buchanan, K. D. (1992) Isolation and primary structure of an amphibian neurotensin. Regul. Peptides **38**, 23–31

Sundler, F., Hakansson, R. and Hammer, R. A. (1977) Immunohistochemical localization of neurotensin in endocrine cells of the gut. Cell Tissue Res. **178**, 313–321

Sures, I. and Grippa, M. (1984) Xenopsin: the neurotensin-like octapeptide from *Xenopus* skin at the carboxyl terminus of its precursor. Proc. Natl. Acad. Sci. U.S.A. **81**, 380–384

Tanaka, K., Masu, M. and Nakanishi, S. (1990) Structure and functional expression of the cloned rat neurotensin receptor. Neuron **4**, 847–854

Tang, Y. Q., Tian, S. H., Hua, J. C., Wu, S. X., Zou, G., Wu, G. F. and Zhao, E. M. (1990) Isolation, chemical and biological characterization of margaratensin, a neurotensin-related peptide from the skin of *Rana margaratae*. Sci. China **33**, 828–834

Tischler, A. S., Lee, Y. C., Costopoulos, D., Slayton, V. W., Jason, W. J. and Bloom, S. R. (1986) Cooperative regulation of neurotensin content in PC12 pheochromocytoma cell cultures: effects of nerve growth factor, dexamethasone and activators of adenylate cyclase. J. Neurosci. **6**, 1719–1725

Uhl, G. R. and Kuhar, M. J. (1984) Chronic neuroleptic treatment enhances neurotensin receptor binding in human and rat substantia nigra. Nature (London) **309**, 350–352

Vincent, J. P., Mazella, J., Chabry, J. and Zsürger, N. (1990) The neurotensin receptor from mammalian brain: solubilization and purification by affinity chromatography, in Receptor Purification (Litwack, G., ed.), Vol. 1, pp. 131–145, Humana Press, Clifton, NJ, USA

Watts, A. G. (1992) Osmotic stimulation differentially affects cellular levels of corticotropin-releasing hormone and neurotensin/neuromedin N mRNAs in the lateral hypothalamic area and central nucleus of the amygdala. Brain Res. **581**, 208–216

Widerlov, E., Kilts, C. D., Mailman, R. B., Nemeroff, C. B., Prange, A. J. and Breese, G. R. (1982*a*) Increase in dopamine metabolites in rat brain by neurotensin. J. Pharmacol. Exp. Ther. **223**, 1–6

Widerlov, E., Lindstrom, L. H., Besev, G., Manberg, P. J., Nemeroff, C. B., Bresse, G. R., Kizer, J. S. and Prange, A. J. J. (1982*b*) Subnormal CSF levels of neurotensin in a subgroup of schizophrenic patients: normalization after neuroleptic treatment. Am. J. Psychiatry **139**, 1122–1126

Williams, F. G., Murtaugh, M. P. and Beitz, A. J. (1990) The effect of acute haloperidol treatment on brain proneurotensin mRNA: *in situ* hybridization analyses using a novel fluorescence detection procedure. Mol. Brain Res. **7** 347–358

Zahm, D. S., Eggerman, K. W., Sprung, R. F., Wesche, D. E. and Payne, E. (1990) Postnatal development of striatal neurotensin immunoreactivity in relation to clusters of substance P. Immunoreactive neurons and the 'dopamine islands' in the rat. J Comp. Neurol. **296**, 403–414

Mammalian KEX2/subtilisin-like paired basic amino acid convertases of prohormones and proproteins

Nabil G. Seidah, †Robert Day and Michel Chrétien*

J. A. DeSève Laboratory of Biochemical and *Molecular Neuroendocrinology, Clinical Research Institute of Montreal (Affiliated to l'Université de Montréal), 110 Pine Ave West, Montreal, Quebec, Canada H2W 1R7

Introduction

The hypothesis of the existence of an inactive prohormone precursor was formulated in the 1960s. Over the years this concept of a precursor which undergoes limited proteolysis to release the active component(s) was found to be applicable not only to prohormones but also to hormonal receptors, viral surface glycoproteins, zymogens and many proteins performing a variety of biological functions. In 1964, Sachs and Takabatake advanced the hypothesis that vasopressin could be synthesized in the form of a precursor which would be transformed into the active nonapeptide vasopressin in little more than 1 h after the start of pulse-labelling of hypothalamic tissue (Sachs and Takabatake, 1964). However, these authors did not characterize molecularly the precursor form of vasopressin.

In 1967, two laboratories using completely different approaches, also advanced the hypothesis that prohormones existed and needed to be cleaved post-translationally in order to release bioactive end-products. Steiner and Oyer (1967) and Steiner *et al.* (1967, 1969) reached this summation by a series of elegant pulse-chase studies dealing with the conversion of proinsulin to insulin. Such pulse-chase studies showed that once the labelling reaction was initiated, there was a delay of about 40 min before significant amounts of intracellular insulin began to accumulate (Steiner *et al.*, 1969). Based on their comparison of the amino acid sequences of β-lipotropic hormone (LPH), γ-LPH and β-melanin-stimulating hormone (MSH), Chrétien and Li (1967) also proposed the existence of prohormones and provided evidence that prohormone processing occurs at pairs of basic

†To whom correspondence should be addressed.

residues of the type Lys-Lys and Lys-Arg. When the sequence of proinsulin was determined (Chance *et al.*, 1968), pairs of basic residues were also found at the site of activation of this precursor. Since the recognition of the pairs of basic residues as a general motif at the site of prohormone precursor processing in proinsulin and in β-LPH, it has been proposed that a trypsin-like/carboxypeptidase B-like enzyme combination might be responsible for the endogenous processing of prohormones and can convert proinsulin into native insulin (Kemmler *et al.*, 1971). Subsequently, it became apparent that the trypsin-like limited proteolysis of precursors at specific pairs of basic residues and/or at single basic amino acids is a widespread mechanism by which the cell expresses a repertoire of biologically active proteins and peptides (Lazure *et al.*, 1983; Gainer *et al.*, 1985; Mains *et al.*, 1990). However, until recently, the molecular identification of the tissue-specific and developmentally regulated proteinases responsible for such conversions proved to be a very difficult task.

The molecular characterization of the proteolytic enzyme participating in the biosynthesis of yeast pheromones as a subtilisin-like proteinase, was the major advance which led to the subsequent identification of the mammalian homologues. The cloning of the KEX2 gene (Julius *et al.*, 1984) and cDNA (Mizuno *et al.*, 1988) and the cellular expression of the yeast KEX2 gene product (also called Kex2 or kexin) demonstrated that this enzyme belonged to the Ca^{2+}-dependent subtilisin family of serine proteinases (Fuller *et al.*, 1989*a*), and that it exhibited exquisite selectivity of cleavage after pairs of basic residues in a number of yeast ((Fuller *et al.*, 1988) and mammalian (Bathurst *et al.*, 1987; Thomas *et al.*, 1988; Foster *et al.*, 1991) precursors. Furthermore, the ability of a yeast enzyme to cleave mammalian precursors both *in vitro* and *in vivo* reinforced the hypothesis that Kex2 represents a prototype for a mammalian subtilisin-like proteinase(s) physiologically responsible for the cleavage of proproteins at selected pairs of basic residues.

Discovery of the mammalian proprotein convertases furin, PC1 and PC2

Computer alignment of the amino acid sequences surrounding the active site Ser* and the catalytically important Asn* (Bryan *et al.*, 1986) of Kex2 against protein sequences found in databases, revealed a partial sequence, of unknown function, for a human gene product homologous to Kex2 (Fuller *et al.*, 1989*b*). The corresponding human gene (*fur*) lies immediately upstream of the tyrosine kinase *fps/fes* oncogene (Roebroek *et al.*, 1986). Further confirmation of the homology between Kex2 and the *fur*-product known as furin, was obtained from the complete cDNA sequence of furin (Van den Ouweland *et al.*, 1990; Wise *et al.*, 1990).

Based on the concept of sequence conservation around the active sites of serine proteinases, reverse transcriptase-polymerase chain reaction (RT-PCR) applied to mRNA amplification allowed two laboratories to simultaneously isolate other mammalian homologues of Kex2, known as PC1/PC3 (Seidah *et al.*, 1990, 1991*a*, 1992*a*; Smeekens *et al.*, 1991) and PC2 (Smeekens and Steiner, 1990; Seidah *et al.*, 1990), representing the first endocrine and neuroendocrine processing enzymes molecularly characterized in mammalian tissues. PCR amplification of a cDNA synthesized from a human insulinoma total RNA, using degenerate oligonucleotides coding for the consensus sequence surrounding the active sites Asp* and His* in Kex2 and related subtilisins, gave a 150 bp probe. The latter was used to screen a human insulinoma library and to isolate a full length cDNA coding for a novel proteinase called PC2 (Smeekens and Steiner, 1990). Independently, PCR on cDNA obtained from mouse pituitary total RNA using oligonucleotides coding for the sequence around the catalytically important Asn* and the active site Ser* of human furin (Seidah *et al.*, 1990, 1991*a*) allowed a 260 bp probe to be obtained. Screening mouse pituitary and mouse insulinoma libraries with this probe allowed the isolation of full length cDNA clones coding for mouse PC2 (Seidah *et al.*, 1990) and also for another proteinase which was called PC1 (Seidah *et al.*, 1990, 1991*a*; also called PC3 in Smeekens *et al.*, 1991). Northern blot analysis demonstrated the presence of two transcripts for each putative enzyme with apparent molecular masses of about 3 and 5 kb (Seidah *et al.*, 1990, 1991*a*; Smeekens and Steiner, 1990; Smeekens *et al.*, 1991). Recent data obtained from the sequence of rat PC1 and PC2 (Bloomquist *et al.*, 1991; Hakes *et al.*, 1991), demonstrated that the 5 kb form of rPC1 is a 3′ extended form of the 3 kb form, probably due to the utilization of an alternative polyadenylation site (Hakes *et al.*, 1991). Interestingly, human PC1 mRNA is found mostly as a band of 6.2 kb in length which has an extended 3′-end sequence which is species specific (Seidah *et al.*, 1992*a*).

Cellular and tissue distribution of furin, PC1 and PC2

Northern and *in situ* hybridization analyses demonstrated a tissue and cellular specificity of expression of PC1 and PC2 which were only found within endocrine and neuroendocrine cells (Seidah *et al.*, 1990, 1991*a*; Cullinan *et al.*, 1991). By contrast to PC1 and PC2, Northern blots of furin revealed a rather ubiquitous distribution in many tissues and cells with important enrichment in liver and kidney (Schalken *et al.*, 1987; Hatsuzawa *et al.*, 1991). *In situ* hybridization results demonstrated a distinct localization of the PC1 and PC2 transcripts in mouse and rat pituitary and brain. Northern blot analysis demonstrated that whereas both PC1 and PC2 are found in the intermediate lobe of the pituitary with an estimated PC1/PC2 ratio of about 1 to 15, within the diverse cells of the anterior pituitary this ratio is about 3–5 to 1 (Seidah *et al.*, 1990, 1991*a*; Benjannet *et al.*, 1991; Day *et al.*, 1992). By *in situ* hybridization, we examined the specific distribution of PC1 and PC2 in the anterior lobe corticotrophs. Double labelling co-localization studies indicated the presence of PC1 in pro-opiomelanocortin (POMC)-containing cells, whereas very little PC2 mRNA could be detected in corticotrophs (Day *et al.*, 1992).

The reported cellular distribution of furin suggested a rather ubiquitous distribution in most cell lines analysed (Hatsuzawa *et al.*, 1991). However, so far PC1 and PC2 transcripts have only been found in cells of endocrine or neuroendocrine origin at variable relative proportions. For example, PC1 and PC2 are found in the corticotroph AtT-20 cells at an estimated mRNA ratio of 20 to 1. No transcripts of PC1 or PC2 can be detected in the phaeochromocytoma PC12 cells (Seidah *et al.*, 1990, 1991*a*; Benjannet *et al.*, 1991). The somatomammotroph GH3 cells do not contain PC1 transcripts, but show the presence of PC2 mRNA whose level is up-regulated by hormonal treatments known to stimulate the amounts of secretory granules (Seidah *et al.*, 1990). The mouse insulinoma cell line βT3 contains predominantly PC2 transcripts (Seidah *et al.*, 1990). No transcripts of either PC1 or PC2 can be detected within constitutively secreting cells such as HepG2, BSC-40 or Ltk-cells (Seidah *et al.*, 1990; Benjannet *et al.*, 1991).

Comparative brain distribution of PC1 and PC2

In situ hybridization data showed that the distributions of PC1 and PC2 transcripts are distinct and sometimes overlapping both in mouse and rat brain (Seidah *et al.*, 1990, 1991*a*; Cullinan *et al.*, 1991). Both mRNAs are widely distributed throughout the central nervous system (CNS) neurons, from the olfactory bulb to the spinal cord. In general, PC2 is more widely distributed than PC1 both in terms of the structures examined and the number of positive neurons. The most abundant PC1 expression is noted in the supraoptic and the paraventricular nuclei of the hypothalamus. High levels of PC1 gene expression are also noted in large neurons of the lateral hypothalamus. A moderate abundance of PC1 mRNA is detected in the arcuate nucleus. PC2 mRNA is also observed in these areas with moderate abundance but additionally we note high levels of PC2 in the medial preoptic area, and the mammillary bodies. In other CNS regions, we note a high abundance of PC2 in various thalamic nuclei, in the hippocampus, in deep and superficial layers of the cortex, in the amygdala, and in the striatum. As for PC1, we note only a low or moderate abundance in these areas. As an example of the distinct, yet overlapping distribution of PC1 and PC2, we note that in the cortex the distribution of PC1 and PC2 overlaps in the deeper layers, while PC1 predominates in the medium layers. Other examples of the overlapping distribution of PC1 and PC2 include the hippocampal formation and the substantia nigra. These unique localization patterns of PC1 and PC2 indicate the differential and possible co-operative roles of these convertases in the brain, and may have some significance in terms of our understanding of the 'tissue-specific post-translational processing' of proproteins (Seidah and Chrétien, 1992*b*; Seidah *et al.*, 1991*b*).

PC4 and PACE4 are new members of the family of subtilisin-like convertases

Based on the conserved segments found in the catalytic region of PC1, PC2 and furin, degenerate oligonucleotides were designed and used to screen (by RT-PCR analysis) a number of tissues for the expression of other members of this growing family of subtilisin-like convertases. Recently, two new members were identified, namely, mouse (m) (Nakayama *et al.*, 1992) and rat (r) (Seidah *et al.*, 1992*c*) PC4 and human (h) PACE4 (Kiefer *et al.*, 1991).

Northern blot analysis of rPC4 which was originally derived from rat testis RNA, showed that the 2.8–3.0 kb rPC4 mRNA is expressed almost exclusively in the male germ cells of the testis. Cellular localization permitted the identification of the spermatocytes and round spermatids as the site of synthesis of rPC4 (Seidah *et al.*, 1992*c*). In females, rPC4 is exclusively expressed in the ovaries, and Northern blot data using RNA isolated from the ovaries of immature female rats primed with pregnant mare serum gonadotrophin (PMSG), revealed that PC4 mRNA levels were significantly increased. Further treatment of the female rats with chorionic gonadotrophin, resulting in pseudo-pregnancy, produced lower levels of ovarian PC4 as compared with the PMSG-treated rats (R. Day and N. G. Seidah, unpublished work). Therefore, the expression of PC4 may not be completely restricted to the testis but may also occur within hormonally stimulated ovaries. The rat cDNA sequence of PC4 demonstrated the presence of at least three rPC4 mRNAs resulting in the production of rPC4-A (654 amino acids), rPC4-B (619 amino acids) and rPC4-C (609 amino acids) with different *C*-terminal sequences. Analogous to rPC4, three cDNAs were also found for the mouse PC4 (Seidah *et al.*, 1992*c*). The observed molecular diversity of PC4 mRNA, possibly results from the differential splicing and/or exon skipping of the parent gene. PC4 mRNA, with a major form at 2.8 kb, was highly abundant in the rat testis but could not be detected, by Northern analysis, in any other tissues including the CNS and peripheral tissues. Testicular cell separation studies combined with Northern analysis indicated the high expression levels of PC4 in germ cells but not in Leydig, Sertoli or peritubular cells. *In situ* hybridization histochemistry confirmed the sites of PC4 gene expression as the pachytene spermatocytes and the round spermatids but not the elongating spermatids.

We further demonstrated the co-localization of PC4 with proenkephalin in testicular germ cells by *in situ* hybridization. A study of the ontogeny of PC4 indicated that its mRNA was first expressed postnatally between days 19 and 22, coinciding with the first stages of spermiogenesis. The stage-specific expression of PC4 in testis indicates its potential role in the developmental maturation of germ cells; this convertase may play a specific physiological function in reproduction. The endogenous substrate(s) of PC4 are not yet known. The co-localization of PC4 with proenkephalin within both meiotic and postmeiotic germ cells (Seidah *et al.*, 1992*c*) suggests that PC4 may be involved in the limited cleavage of this precursor to the incompletely processed enkephalin-containing proteins reported in this tissue. The function of these larger enkephalin-containing peptides is unknown.

Germ cells also synthesize a number of other precursors in which the sequences of the active fragment(s) are sandwiched between pairs of basic residues, making them potential PC4 substrates. These include POMC, growth factors such as nerve growth factor (NGF) and an NGF-like protein, and several mitogenic factors that have been reported to be secreted from germ cells (for review see Ackland *et al.*, 1992). Autocrine or paracrine effects of these factors in cell-to-cell communication have been postulated and evidence is now available to suggest that specific proteins produced by germ cells may regulate Sertoli cell function. For example, germ cells or their proteins released into culture media have been shown to stimulate Sertoli cell secretion of androgen-binding protein, inhibin and transferrin, as well as inhibiting the secretion of 17β-oestradiol.

Northern blot analysis demonstrated that the 4.4 kb transcript of human PACE4 (Kiefer *et al.*, 1991) was expressed in every tissue analysed, with the liver being the major tissue containing this putative convertase. Interestingly, the gene coding for hPACE4 was found within 5 Mb from the furin gene on the human chromosome 15 (Kiefer *et al.*, 1991). Extensive studies will be needed to study the substrate specificity of these novel putative convertases and to identify their physiological substrates.

Structural comparison of subtilisin, PC1, PC2, furin, PC4, PACE4 and Kex2

The comparative architectural features of the eukaryotic subtilisin-like pro-protein convertases and those of subtilisin BPN′ are presented in Fig. 1. It becomes apparent that all mammalian convertases of the subtilisin/Kex2-type contain a signal peptide followed by a prosegment and then a catalytic domain and a so-called P-domain which has been reported to be important for the folding of the proteinase and the elaboration of the full enzymic activity of Kex2 (Wilcox and Fuller, 1991) and furin (Hatsuzawa *et al.*, 1992). All the structures vary most at their *C*-terminus, thus, in contrast to Kex2 and furin, the structures of PC1, PC2, PC4 and PACE4 do not predict the presence of a transmembrane domain. The *C*-terminal segments of PC1 and

Fig. 1. **Schematic diagram depicting the various domains of the Kex2/subtilisin-like convertases and a comparison with the primitive subtilisin (subt) BPN′ enzyme**

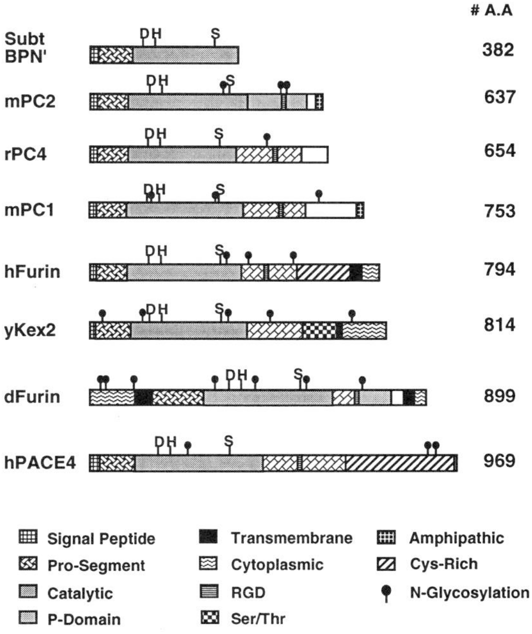

The number of amino acids (# A.A) in each primary translation product is shown on the right. The diagram emphasizes the various domains in each convertase and points out the positions of the catalytically important residues found in the active site of each enzyme, including the catalytic triad of Asp(D), His*(H), Ser*(S). Abbreviations used: m, mouse; r, rat; h, human; y, yeast; d,* Drosophila.

PC2 code for an amphipathic structure which could interact with membranes in a pH-dependent manner, in a similar fashion to the properties of the soluble carboxypeptidase E (Fricker *et al.*, 1990). Whereas Kex2 contains an *O*-glycosylation-rich segment, furin and PACE4 both contain a Cys-rich segment of different lengths. The functions of these domains are still obscure. The glycosylation pattern in the mammalian members shows that they each contain three *N*-glycosylation sites (except for PC4 which contains only one site) and, unlike Kex2, none appears within the proseg-ment.

Fig. 2 depicts the conservation of sequences around the active sites Asp*, His* and Ser*, as well as the catalytically important Asn* residues of some subtilisins and of all the members of the subtilisin family reported so

Fig. 2. Conservation of the sequence around the Asp*, His* and Ser* of the active sites found in mouse mPC1 (Seidah *et al.*, 1990, 1991*a*), mouse mPC2 (Seidah *et al.*, 1990), human hfurin (Van den Ouweland *et al.*, 1990), human hPACE4 (Kiefer *et al.*, 1991), rat rPC4 (Seidah *et al.*, 1992*c*), *Xenopus* furin Xen-14 and Xen-18 (Korner *et al.*, 1991), *Drosophila* Dfurin1 (Roebroek *et al.*, 1991), *Drosophila* Dfurin2 (Roebroek *et al.*, 1992), yeast Kex2 (Mizuno *et al.*, 1988), human tripeptidyl peptidase II (hTPP II; Tomkinson and Jonsson, 1991) and a number of subtilisins

Conservation of active site residues between Kex2/subtilisin-like convertases

	Asp*	His*	Asn*	Ser*
mPC1	T V L D **D** G L E	N E N K H G T **R** C A G **E**	S I F V W A **S** G N **G** G	E T **H** T G T S **A** S A P L A A G
mPC2	G I M D **D** G I D	W F N S H G T **R** C A G **E**	S I Y V W A **S** G **D** **G** G	L R **H** S G T S **A** A A P R Q L G
hFurin	S I L D **D** G I E	N D N R H G T **R** C A G **E**	I F V V W A **S** G N **G** G	E S **H** T G T S **A** S A P L A A G
hPACE4	T I L D **D** G I E	N E N K H G T **R** C A G **E**	S I F V W A **S** G N **G** G	D G **H** T G T S V S A P M V A G
rPC4	S I L D **D** G I E	D E N R H G T **R** C A G **E**	T L F I W A **S** G N **G** G	D K **H** T G T S **A** S A P L A A G
Xen-14	S I L D **D** G I E	N D N R H G T **R** C A G **E**	S I Y V W A **S** G N **G** G	D S **H** T G T S **A** S A P L A A G
Xen-18	S I L D **D** G I E	N D N R H G T **R** C A G **E**	S I Y V W A **S** G N **G** G	D S **H** T G T S **A** S A P L A A G
Dfurin1	T I L D **D** G L E	D S N R H G T **R** C A G **E**	S I F I W A **S** G N **G** G	V S **H** T G T S **A** S A P L A A G
Dfurin2	S I L D **D** G I Q	G D N K H G T **R** C A G **E**	S I F V W A **S** G N **G** G	V E **H** T G T S **A** S A P L A A G
yKex2	A I V D **D** G L D	S D D Y H G T **R** C A G **E**	I Y V V F A **S** G N **G** G	N S **H** G G T S **A** A A P L A A G
hTPP II	A C I D S N E D	S G G A H G T H V A S **E**	I I Y V S S A G N N G	Q L M N G T S M S S P N A C G
Ther	A I V D T G V Q	N G N G H G T H C A G I	S V V V A A A G N A G	A S L S G T S M A T P H V A G
Prok	Y V I D T G I E	D G N G H G T H C A G T	V M V A V A A G N N -	R S I S G T S M A T P H V A G
Prbl	Y V I D T G V N	D G N G H G T H C A G T	I H F A V A A G N E -	A T L S G T S M A S P H V A G
Subt(1)	A V L D T G V Q	D G N G H G T H V A G T	V V V V A A A G N S G	A T L N G T S M A S P H V A G
Subt(2)	G I I D T G I A	D G N G H G T H V A G T	I V V V A A A G N S G	T S L N G T S M A S P H V A G
Subt(3)	A V I D S G I D	D N N S H G T H V A G T	V V V V A A A G N E G	G A Y N G T S M A S P H V A G
Subt(4)	A V I D S G I D	D G S S H G T H V A G T	I V V A A A A G N E G	G A Y N G T S M A T P H V A G
Subt(5)	A V L D T G C D	D Y N G H G T H V A G T	V L V V C A A G N E G	G K L T G T S M A A P H V S G

Boxed amino acids, inverted bold triangles and bold residues represent invariant residues, amino acids found only in the eukaryotic members of the family, and conserved amino acids found only in the Kex2-like members of the eukaryotic subfamily, respectively. Abbreviations used: mPC1, mouse proprotein convertase 1; mPC2, mouse proprotein convertase 2; hFurin, human furin; hPACE4, human paired basic amino acids convertase 4; rPC4, rat proprotein convertase 4; Xen-14, Xenopus *furin-like-14; Xen-18,* Xenopus, *furin-like-18; dFurin1,* Drosophila *furin-like 1; dFurin2,* Drosophila *furin-like 2; yKex2, yeast Kex2; hTPP II, human tripeptidyl peptidase II; Ther,* Thermoactinomyces vulgaris *thermitase; Prok,* Tritirachium album Limber *proteinase K; Prbl,* S. cerevisiae *vacuolar proteinase B; Subt(1),* Bacillus subtilis *subtilisin Carlsberg; Subt(2),* B. subtilis *subtilisin DY; Subt(3),* B. amyloliquefaciens *subtilisin BPN′; Subt(4),* B. subtilis var. amylo-sacchariticus *subtilisin S; Subt(5), intracellular subtilisin from* B. subtilis *A-50.*

far. Aside from the conserved residues in the family (boxed in Fig. 2), we also noticed that certain residues (inverted triangle) are found only in the eukaryotic subtilisin-like family (bold), and within these, six of them (namely Asp, Arg, Ser, Gly, His and Ala, shown in bold and under the inverted triangle) are found only in the Kex2-like convertases, even in the recently reported *Drosophila* furin 1 (Dfurin1) (Roebroek *et al.*, 1991) and *Drosophila* furin 2 (Dfurin2) (Roebroek *et al.*, 1992), but not in the structure reported for the human tripeptidyl peptidase II (hTPP II) (Tomkinson and Jonsson, 1991). Finally, only in PC2 do we find Asp* instead of the usual Asn* residue. This interesting single point mutation of PC2 (but not PC1, or furin), may have an important bearing on the catalytic efficiency ($k_{cat.}/K_m$) of this proteinase, since it has been reported that this Asn residue forms a hydrogen bond with the developing substrate carbonyl oxygen anion in the transition state complex. This interaction was shown to be important, but not essential, for the stabilization of this oxyanion hole in subtilisins (Bryan *et al.*, 1986). From these data and the sequence similarity of the catalytic domains in these different enzymes, it becomes apparent that hTPP II is much closer to subtilisin BPN′ than to any other member of the Kex2-like proteinases, suggesting a possible classification of hTPP II as a mammalian degradative enzyme, rather than a Kex2-like specific proprotein convertase, in agreement with its observed broad substrate specificity (Bålöw *et al.*, 1986).

As shown in Figs. 1 and 3, primary sequence analysis of the subtilisin-like convertases revealed the presence of a hydrophilic Arg-Arg-Gly-Asp-Leu sequence in PC1, PC2, furin, PACE4 and PC4 but not in Kex2 nor in subtilisin BPN′. Notice that the *Xenopus* furin-like enzymes Xen-14 and Xen-18 contain a Cys-Arg-Gly-Asp-Leu sequence (Korner *et al.*, 1991), instead of the mammalian Arg-Arg-Gly-Asp-Leu structure. The tripeptide Arg-Gly-Asp (RGD) sequence is found in a number of extracellular matrix proteins and is a key recognition sequence by cell surface integrins implicated in cell adhesion (Ruoslahti, 1988). It would be interesting to verify whether this sequence is involved in the intracellular sorting of the convertases PC1, PC2, PC4, PACE4 and furin, via their interaction with integrins during their migration in the endoplasmic reticulum and/or in the Golgi apparatus.

Intracellular processing of PC1 and PC2

Confirming previous predictions which were based on the comparison of the structures of PC1 and PC2 with prosubtilisins (Seidah and Chrétien, 1992*b*), it has recently been reported that the purified convertases PC1 (Christie *et al.*, 1991) and PC2 (Christie *et al.*, 1991; Mackin *et al.*, 1991) from either bovine adrenal (Christie *et al.*, 1991) or anglerfish (Mackin *et al.*, 1991) secretory granules, lack the *N*-terminal prosegment. Such a prosegment, which in subtilisins was called an 'intramolecular chaperone' (Ohta *et al.*, 1991), was shown to be essential for the folding of the protein into the active enzyme, which then autoprocessed the prosegment intramolecularly (Zhu *et al.*, 1989; Ohta *et al.*, 1991).

Fig. 3. **Comparison of the amino acids surrounding the Arg-Gly-Asp (RGD) sequence found in most of the eukaryotic members of paired basic amino acid convertases with those found in fibronectin, chromogranin A, protein C, tyrosine mono-oxygenase (TyrMonoO), mouse transforming growth factor-β (mTGF-β), mouse interleukin-6 (mIL-6) and γ-crystallin**

Conservation of the "Arg-Gly-Asp" Sequence

hPC1	Tyr	Ser	Arg	**Arg**	**Gly**	**Asp**	Leu	His	Val
hPC2	Ala	Thr	Arg	**Arg**	**Gly**	**Asp**	Leu	Asn	Ile
hFurin	Tyr	Asn	Arg	**Arg**	**Gly**	**Asp**	Leu	Ala	Ile
hPACE4	His	Pro	Arg	**Arg**	**Gly**	**Asp**	Leu	Gln	Ile
rPC4	Tyr	Ser	Arg	**Arg**	**Gly**	**Asp**	Leu	Glu	Ile
dFurin1	Ser	Gln	Arg	**Arg**	**Gly**	**Asp**	Ile	Gln	Leu
Xen-14	Tyr	Asn	Cys	**Arg**	**Gly**	**Asp**	Leu	Ala	Ile
Xen-18	Tyr	Asn	Cys	**Arg**	**Gly**	**Asp**	Leu	Ala	Ile
Kex2	Thr	Glu	Ile	**Arg**	**Gly**	Thr	Thr	Thr	Val
Fibronectin	Val	Thr	Gly	**Arg**	**Gly**	**Asp**	Ser	Pro	Ala
Chromogranin A	Glu	Thr	Leu	**Arg**	**Gly**	**Asp**	Glu	Arg	Ile
Prot.C	Met	Thr	Arg	**Arg**	**Gly**	**Asp**	Ser	Pro	Trp
TyrMonoO	Glu	Val	Arg	**Arg**	**Gly**	**Asp**	Leu	Ala	Ala
mTGF-β	Thr	Gly	Arg	**Arg**	**Gly**	**Asp**	Leu	Ala	Thr
mIL-6	Gln	Val	Arg	**Arg**	**Gly**	**Asp**	Phe	Thr	Glu
γ-Cryst	Phe	Leu	Arg	**Arg**	**Gly**	**Asp**	Tyr	Pro	Asp

Abbreviations used: Prot. C, protein C; γ-Cryst., γ-crystallin.

Using pulse and pulse-chase analysis of cells expressing PC1 and PC2, we have undertaken the study of the biosynthesis of PC1 and PC2 in both regulated and constitutive cell lines and have characterized by amino acid sequence both the intracellular and secreted molecular forms of PC1 and PC2. Pulse analysis (Benjannet *et al.*, 1992*a*) revealed that intracellularly, PC1 is rapidly processed with the removal of the *N*-terminal 83 amino acid prosegment, in cells containing (regulated) or lacking (constitutive) secretory granules. These data suggested that the intracellular cleavage of pro-PC1 at the Arg-Ser-Lys-Arg_{83}↓Ser-Val sequence was highly efficient. Furthermore, we also showed that the resulting 80 kDa PC1 was secretable from either cell type. In contrast, the major intracellular form of PC2 had a molecular mass of 65 kDa. The microsequence analysis of the PC2 forms demonstrated that the 75 kDa form contained the 84 amino acid prosegment at its *N*-terminus, whereas this prosegment was removed in the 65 kDa form, implying that the secretable 65 kDa form of PC2 was obtained through the cleavage of pro-

Fig. 4. Pulse-chase analysis of the biosynthesis of mPC1 and mPC2 in GH4C1 cells

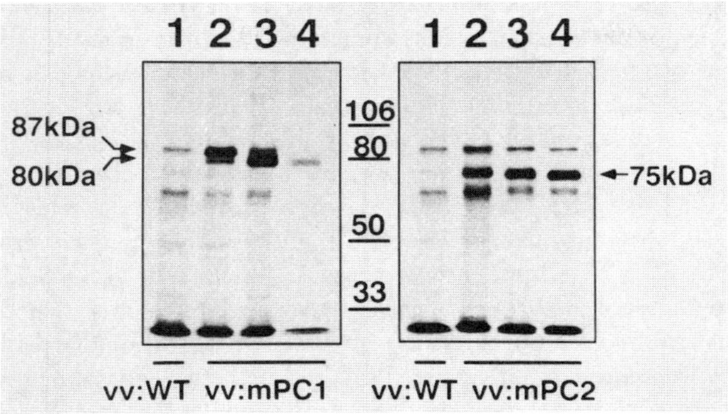

GH_4C_1 cells which were infected with the control vaccinia wild-type virus VV:WT (lane 1, a and b), vaccinia virus recombinant VV:mPC1 (a), or vaccinia virus recombinant VV:mPC2 (b), were pulse-labelled with [^{35}S]methionine for 5 min (lanes 1 and 2) and then chased with unlabelled methionine for 0 min (lane 2), 20 min (lane 3) and 60 min (lane 4). The total of the cell extracts (Benjannet et al., 1992b) was immunoprecipitated with AbC-mPC1 (a), or AbC-mPC2 (b), antibodies specific to the C-terminal segments of mPC1 and mPC2 (Benjannet et al., 1992a), and an aliquot representing 3% (cell extract) or 20% (medium) of each total immunoprecipitate was separated on an 8% sodium dodecyl sulphate–polyacrylamide gel and fluorographed as described previously (Benjannet et al., 1991). The resulting autoradiogram also depicts the estimated molecular mass of the intracellular forms of PC1 (a), and PC2 (b), based on the migration position of the molecular mass standards obtained from Biorad. The secreted PC1 and PC2 migrated with apparent molecular masses of 80 and 65 kDa, respectively (Benjannet et al., 1992b)

PC2 at the Arg-Lys-Lys-Arg$_{84}$↓Gly-Tyr sequence (Seidah *et al.*, 1990, 1991*a*).

The pulse-chase analysis (S. Benjannet, N. Rondeau, C. Lazure, M. Chrétien and N. G. Seidah, unpublished work) demonstrated that in cells devoid of or containing secretory granules, pro-PC1 was cleaved into PC1 at a much faster rate than the conversion of pro-PC2 into PC2 (Fig. 4). Since the *N*-terminal prosegment of mPC1 was excised within the first 20 min of chase (Fig. 4A, lane 3), and since at least 10% of pro-mPC1 was cleaved within the first 5 min of pulse (Fig. 4A, lane 2), this suggested that the activation of PC1 occurred very early during the biosynthesis of this enzyme. This time scale of pro-PC1 to PC1 conversion suggested that such an activation could begin to take place within either the endoplasmic reticulum or early Golgi cisternae (such as in the *cis*- or medial-Golgi) (Shields, 1992). In addition, the results of pulse-labelling experiments in a number of cells, including the pituitary pars intermedia and mouse insulinoma βTC3 cells, showed that pro-PC2 predominated intracellularly, as compared with PC2. We also demonstrated that in cells containing secretory granules, PC1 and PC2 were secreted while pro-PC1 and pro-PC2 were selectively retained within particulate intracellular fractions (S. Benjannet, N. Rondeau, C. Lazure, M. Chrétien and N. G. Seidah, unpublished work).

Evidence obtained in our laboratory has shown that the pro-PC1 and pro-PC2 enzymes, when isolated from insect Sf9 cells infected with baculovirus recombinants of PC1 or PC2, are essentially inactive zymogens which keep their *N*-terminal prosegment (A. Boudreault, N. G. Seidah and C. Lazure, unpublished work). These data suggested either that another enzyme is required for the activation of these zymogens, or that within the cellular environment of Sf9 cells, PC1 and PC2 are not able to excise efficiently their own *N*-terminal prosegment. Since both Kex2 (Wilcox and Fuller, 1991) and furin (Leduc *et al.*, 1992) have been proven to remove their prosegments autocatalytically by an intramolecular mechanism, it is likely that a similar auto-activation also applies to the zymogen activation of PC1 and PC2. However, since the mechanism of the prosegment removal of PC1 and PC2 was not known, it was still possible that the Golgi-associated furin (Misumi *et al.*, 1991) might participate in this process. This idea was plausible since the motif Arg_{-4}-Xaa-Lys/Arg-Arg_{-1} suggested to be a signal for precursor cleavage by furin within the constitutive pathway of secretion (Burgess and Kelly, 1987; Hosaka *et al.*, 1991), is also present at the furin, PC1 and PC2 prosegment cleavage sites. However, the cellular co-expression of furin with pro-PC1 or pro-PC2 demonstrated the inability of furin to cleave either zymogen (Benjannet *et al.*, 1992*b*), hence furin is unlikely to be the proteinase responsible for the zymogen activation of these convertases.

In conclusion, the efficiency of the removal of the prosegment might represent a convenient regulatory step in the activation of PC1 and PC2, and hence in the processing of proproteins. The evidence suggests that such an activation could occur much faster for PC1 than for PC2, the latter proenzyme might then continue to be activated within the secretory granules. These data lead us to propose that cleavage of precursor substrates at pairs of basic residues by PC1 may occur earlier within the secretory pathway than processing by PC2.

Cleavage specificity of PC1, PC2 and furin

To investigate the cleavage specificity of PC1 and PC2 we used three different precursors as substrates: mouse POMC, human renin and rat prosomatostatin. Vaccinia virus recombinants of each enzyme were used to co-express these proteinases with the prosubstrates in a number of cell lines. After their purification, the intracellular and secreted peptide products were unambiguously characterized by microsequence.

Vaccinia virus recombinants of the mouse prohormone convertases PC1 and PC2 were co-expressed together with mouse POMC in constitutively secreting cells, such as the African green monkey kidney cell line BSC-40, and in the endocrine-derived cell lines PC12 and AtT-20 which exhibit regulated secretion. Fig. 5 summarizes the cleavage selectivity of PC1 and PC2 as deduced from the monitoring of POMC processing (Benjannet *et al.*, 1991; Seidah *et al.* 1992*d*). The data demonstrated the distinct cleavage specificities of PC1 and PC2, since in the cell lines analysed we found that: (i) PC1 cleaves POMC into adrenocorticotropic hormone (ACTH) and β-lipotropin; (ii) PC2 cleaves POMC into biologically active β-endorphin, an *N*-terminally-extended ACTH containing the joining peptide, and into either α-MSH or des-acetyl α-MSH; (iii) PC1 and PC2 cleave the *N*-terminal glycopeptide segment of porcine POMC (pPOMC) to yield pPOMC 1–107 which ends with the joining peptide sequence, and PC1 is much more efficient than PC2 in generating pPOMC 1–80 in cells contain-

Fig. 5 Composite model representing the end-products of POMC processing by PC1 and PC2 (Seidah *et al.*, 1992*d*)

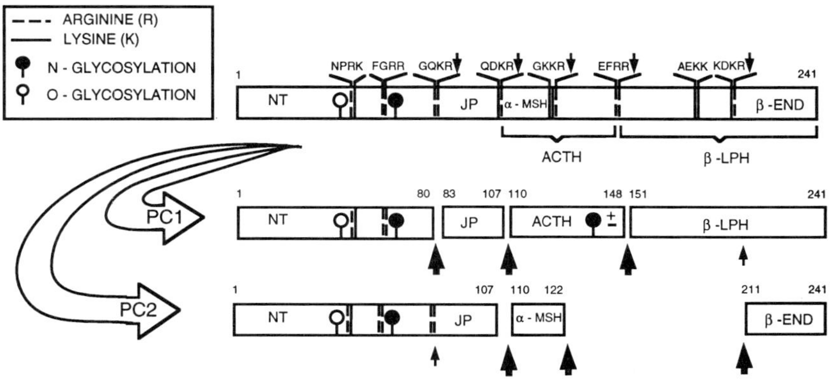

The major and minor cleavage sites following pairs of basic residues are represented by large and small arrows, respectively. The amino acid sequence at the cleavage sites are also shown. The numbers represent the start and end positions of the processed peptides based on the reported POMC sequence of pPOMC (Boileau et al., *1983). The* N- *and* O-*glycosylation (CHO) sites as well as the pairs of basic residues are emphasized.*

ing secretory granules; (iv) under the conditions and cell types used we could not detect a pPOMC 1–49 cleavage product with either enzyme. This cleavage, which is known to be inefficient in the pituitary pars intermedia (Seger and Bennett, 1986), probably occurs late in the intracellular transport route, and may therefore be affected by the vaccinia virus expression system we used, which does change somewhat the intracellular organization of the cell. Alternatively, the intracellular and intragranular environment conditions within the pituitary pars intermedia are not mimicked in either the BSC-40 or the PC12 cells used in this study, hence the cleavage conditions may not be optimal for such an inefficient cleavage reaction, which never exceeds 50% in the pars intermedia *in vivo* (Seger and Bennett, 1986). Finally, it is also possible that the Arg↓Lys cleavage needed for the production of pPOMC 1–49 and Lys-γ_3-MSH in the pituitary pars intermedia is effected by another enzyme not yet identified.

These data are consistent with the proposed hypothesis concerning the physiological role of PC1 and PC2 as distinct proprotein convertases acting alone or together to produce a set of tissue-specific maturation products in both the brain and peripheral tissues (Seidah *et al.*, 1990, 1991*a*). Interestingly, since some POMC processing can also occur in cells devoid of a regulated pathway of secretion, these results also show that POMC processing by either PC1 or PC2 is not absolutely dependent on the presence of secretory granules.

In contrast, recent data from our laboratory showed that neither PC1 nor PC2 can process human prorenin in constitutive cells such as BSC-40 or the Chinese hamster ovary cell line (CHO) (Benjannet *et al.*, 1992*a*). To test the hypothesis that, unlike POMC, human prorenin processing might require the presence of secretory granules, a somatomammotroph GH4 cell line that expresses a stable transfectant of human prorenin was infected with PC1 or PC2 vaccinia virus recombinants. The results of our analysis of the secreted renin activity demonstrated that only PC1 can activate human prorenin in this cell line, and that neither the endogenous furin nor the exogenous PC2 can activate this zymogen (Benjannet *et al.*, 1992*a*).

Pulse analysis of the prorenin maturation product confirmed that PC1 cleaved human prorenin after the Lys-Arg pair at amino acids 65 and 66 of the human pre-prorenin, the identical processing site reported for human kidney prorenin. These results point out that not all precursors can be processed efficiently in the absence of secretory granules, and that each case has to be carefully analysed before general conclusions can be reached. When cells expressing rat prorenin or human pre-prorenin mutated at either Lys_{65} (replaced by Ala, phR K/A-2) or Arg_{66} (replaced by Ala, phR R/A-1) at the native cleavage site were infected with a vaccinia virus recombinant of PC1, neither Ala-Arg_{66} nor Lys-Ala_{66} mutants were cleaved, suggesting that PC1 is not able to cleave at the monobasic Arg or Lys residues. Furthermore, the inability of PC1 to cleave rat prorenin implies that it cannot cleave the Lys-Lys_{64} bond in this precursor. Data from our work in AtT-20 tumour cells

(Ladenheim *et al.*, 1991) and those reported by Nagahama *et al.* (1991) in GH4 cells demonstrated that, like human prorenin, submaxillary gland mouse renin-2 is also cleaved by PC1.

A recent study involving the expression of rat prosomatostatin (rPSS) in COS-7, PC12, RIN5f and AtT-20 cells (Galanopoulo *et al.*, 1993) showed that these cells contain an endogenous convertase capable of cleaving rPSS at a single Arg residue generating SS-28. These data are in agreement with those of Devi (1991) which showed that the monobasic cleavage activities are widely distributed in many cells and tissues. However, the co-expression of rPSS with either PC1 or PC2, revealed that cleavage at the pair of basic residues Arg-Lys only occurred in constitutive cell lines expressing PC1 (not PC2). Furthermore, as in the case of POMC, although some cleavage did occur in cells devoid of secretory granules, the efficiency of cleavage of rPSS was enhanced in cells containing secretory granules, and PC2 was able to cleave rPSS in regulated cells.

In general, PC1 and PC2, which are localized within secretory granules (Steiner, 1991; Kirchmair *et al.*, 1992), were only detected in regulated cells containing secretory granules. They therefore represent major proprotein and prohormone convertases that are capable of cleaving at pairs of basic residue precursors negotiating the regulated pathway of secretion. In contrast, furin which is reported to be localized within the Golgi apparatus (Bresnahan *et al.*, 1990; Misumi *et al.*, 1991; Kirchmair *et al.*, 1992) is best able to process proproteins as they exit cells via the constitutive pathway of secretion, including pro-von Willebrand factor (Van de Ven *et al.*, 1990; Wise *et al.*, 1990), proalbumin, procomplement factor C3 (Misumi *et al.*, 1991) and pro-NGF (Bresnahan *et al.*, 1990).

The substrate specificity of furin was studied extensively, and it was shown that this enzyme prefers substrates with an Arg_{-4}-Xaa-Lys/Arg-Arg_{-1} motif at the site of cleavage (Hosaka *et al.*, 1991, 1992). The mutagenesis data (Watanabe *et al.*, 1992) also delineated the following sequence rules that govern constitutive precursor cleavage by furin. (a) A basic residue (Lys or Arg) at the fourth (P-4) or sixth (P-6) residue upstream from the cleavage site together with the basic residues at positions -1 and -2, is necessary. (b) At P-2, Lys is preferred to Arg. (c) At P-4, an Arg residue is preferable to Lys. (d) At position P'_1, i.e. the residue on the *C*-terminal side of the cleavage site, a hydrophobic aliphatic amino acid such as Val, Leu or Ile is not suitable.

It is important to note that these rules apply only to furin. In fact, rule (d) is not important for PC1 since in human prorenin the cleavage of the Pro-Met-Lys-Arg↓Leu bond by this convertase is quite efficient (Benjannet *et al.*, 1992*a*).

Cleavage of viral envelope glycoproteins

Aside from the cleavage of endogenous prohormones, convertases are also involved in the activation of other precursors such as certain cell surface receptors, e.g. the insulin receptor, adhesion molecules and viral envelope glycoproteins. In some of these cases it has been shown that the consensus sequence Arg_{-4}-Xaa-Lys/Arg-Arg_{-1} is critical for cleavage activation (Stieneke-Grober *et al.*, 1992). Evidence was presented that the glycoprotein prM of the yellow fever virus (Randolph *et al.*, 1990), the F-fusion protein of mumps (Yamada *et al.*, 1988), Newcastle disease virus F1 protein (Morrison *et al.*, 1985) and the haemagglutinin (HA) of fowl plague virus (Stieneke-Grober *et al.*, 1992) are activated in the Golgi apparatus, possibly within the *trans*-Golgi network. Furthermore, a Ca^{2+}-dependent proteinase that is similar, if not identical, to furin has been demonstrated to be the enzyme responsible for the activation of HA of fowl plague virus (Stieneke-Grober *et al.*, 1992). Since in most cases cleavage of the prosurface glycoprotein by host cells enzyme(s) is necessary for the virion to become infective, it is possible that furin or other convertases play an important role in the life cycle of various viruses and their interplay with host cells.

In Table 1 we compare the sequences surrounding the pair of basic residues which has to be cleaved to activate a number of viral membrane glycoprotein precursors. Clearly, an important clinical benefit will be gained once we can define exactly the convertase(s) responsible for the activation of each one of these viral membrane-fusion proteins, and whether the activation is specific only to the cells infected by the virus.

Transcriptional regulation of PC1, PC2 and furin

The expression levels of PC1 and PC2 were evaluated in the rat anterior and neuro-intermediate pituitary lobes, by *in situ* hybridization and Northern blot analysis, and after various pharmacological manipulations (Day *et al.*, 1992). All intermediate lobe melanotrophs expressed high levels of PC2 mRNA and lower levels of PC1 mRNA. PC1 mRNA was highly expressed throughout the anterior lobe; however, appreciable PC2 mRNA levels were also found. Based on co-localization studies, anterior lobe corticotrophs were found to express PC1 mRNA but very little PC2 mRNA. Neuro-intermediate lobe levels of PC1, PC2 and POMC mRNA increased 2–6-fold in rats treated with haloperidol, while they decreased to 10–25% of their control values after bromocriptine treatment. These results indicate that, in the intermediate lobe, dopamine is involved in the regulation of PC1 and PC2.

In the anterior lobe, haloperidol had a strong effect on PC2 mRNA, increasing its levels 8–12-fold compared with control, while PC1 mRNA was unaffected. Both PC1 and PC2 mRNA levels were increased 5–9-fold in animals made hypothyroid by treatment with 6-*n*-propyl-2-thiouracil. Adrenalectomy had no significant effect on anterior lobe PC1 mRNA levels. However, both PC1 and PC2 mRNA levels were shown to be responsive to dexamethasone treatment in the AtT-20 cell lines. These results indicate that

Table 1. Comparison of the cleavage sites of envelope glycoproteins found in some infectious diseases

Virus	Precursor	Bond cleaved
AIDS (HIV-1)	gp160	Arg-Val-Val-Gln-**Arg**-Glu-**Lys**-**Arg**↓Ala-Val-Gly-Ile-Gly-Ala
Newcastle disease	F_1 fusion protein	Ser-Gly-Gly-Arg-**Arg**-Gln-**Lys**-**Arg**↓Phe-Leu-Ile-Gly-Ala-Ile
Varicella-zoster	gpII	Thr-Arg-Asn-Thr-**Arg**-Ser-**Arg**-**Arg**↓Ser-Val-Pro-Val-Glu-Leu
Yellow fever	prM	Gly-Arg-Ser-Arg-**Arg**-Ser-**Arg**-**Arg**↓Ala-Ile-Asp-Leu-Pro-Thr
Mumps	F-fusion protein	Pro-Gly-Ser-Arg-**Arg**-His-**Lys**-**Arg**↓Phe-Ala-Gly-Ile-Ala-Ile
Measles	F_0-fusion protein	Ala-Ser-Ser-Arg-**Arg**-His-**Lys**-**Arg**↓Phe-Ala-Gly-Val-Val-Leu
Avian influenza virus	Haemagglutinin	Pro-Ser-Lys-Lys-**Arg**-Glu-**Lys**-**Arg**↓Gly-Leu-Phe-Gly-Ala-Ile

Cleavage at the Arg-Xaa-Lys/Arg-Arg site has been reported to be critical for infectious activity.

in the anterior lobe, dopamine, thyroid hormones and corticosteroids are involved in PC1 and/or PC2 gene expression. These results are indicative of the significant cellular heterogeneity of the anterior lobe and the extensive expression of PC1 and PC2 in this tissue.

When the mRNA levels of the ubiquitously expressed furin were examined in the intermediate lobe of the pituitary, after treatment with dopamine receptor agonists and antagonists which are known to regulate POMC, PC1 and PC2 levels (Day *et al.*, 1992), mRNA levels of furin were not affected (Birch *et al.*, 1991). These experiments support the notion that furin does not play an important role in the maturation of neuroendocrine precursors, but rather is involved in the processing of precursors negotiating the constitutive secretion pathway.

Chromosomal localization and gene structure

The chromosomal localization of the genes coding for PC1 (*Nec*-1), PC2 (*Nec*-2) and furin (*fur*) (Seidah *et al.*, 1991*a*, *c*) demonstrated that these three genes are not syntenic. The *Nec*-1, *Nec*-2 and *fur* loci were located on human chromosomes 5q15–21, 20p11.1–11.2, 15q25–26 and on mouse chromosomes 13 (C1–C3 band), 2 (F3–H2 region) and 7 (D1–E2 region), respectively. To further characterize the linkage relationship between *Nec*-1, *Nec*-2, *fur* and existing markers, the three loci were mapped in a mouse interspecific backcross (Copeland *et al.*, 1992). The results indicated that the *Nec*-1 locus is linked to the *Rasa*, *Dhrf* and *Ctla*-3 genes on the distal arm of mouse chromosome 13; *Nec*-2 mapped towards the distal region of mouse chromosome 2 between *Bmp*-2*a* and *Pax*-1.

Preliminary data on the analysis of the chromosomal localization of PC4 demonstrated that its gene is found in the central region of mouse chromosome 10, a region known to contain markers which are syntenic with human chromosome 21q22 (Seidah *et al.*, 1992*c*). In contrast to PC1, PC2, furin and PC4, which are all localized on different chromosomes, the gene coding for PACE4 has been found to localize on human chromosome 15 in close proximity to the *fur* gene. This observation suggests that furin and PACE4 genes probably evolved from a common ancestor by gene duplication (Kiefer *et al.*, 1991).

So far, these chromosomal localization data have not provided clues on the potential applications of these proteinases in any reported genetic disorder mapping close to the chromosomal loci of these convertases. To understand further the genetic organization of these convertases and their evolution, the gene structures of furin (Roebroek *et al.*, 1986; Van den Ouweland *et al.*, 1989; Barr *et al.*, 1991), PC1 (Ftouhi *et al.*, 1992, 1994), PC2 (Ohagi *et al.*, 1992; Sansson *et al.*, 1992; M.-L. Raffin-Sansson, M. Mbikay and N. G. Seidah, unpublished work) and PC4 (Mbikay *et al.*, 1992, 1994) were determined. The data showed a remarkable degree of conservation in the organization of the genes and of the exon/intron positions within the catalytic segment of the molecule, suggesting that all these enzymes arose from a common ancestral gene. The signal peptide, the four putative active site amino acids Asp, His, Asn and Ser and the RGD sequence were found on separate exons. Notably, the available data suggested that the length of the gene is inversely proportional to the size of the corresponding mRNA, with the furin gene being the smallest (11 kb) and PC2 the largest (>130 kb). Although the human *fur* gene contains putative TATA and CAAT boxes (Van den Ouweland *et al.*, 1989), the PC2 gene does not contain either a TATA or a CAAT box, but rather displays a G- and C-rich (average 80% over 250 nucleotides) 5′-end promoter sequence (Ohagi *et al.*, 1992). This is similar to the carboxypeptidase E gene which is TATA-less and CAAT-less and contains a GC box 50–60 nucleotides upstream of the major transcription initiation site (Jung *et al.*, 1991).

Conclusions

Since the discovery that pairs of basic amino acids represent a common motif found in processed prohormones and proproteins, and that this motif is recognized by specific processing enzymes, the amount of effort devoted to the definition of the processing enzymes involved in the final shaping of the active product has been monumental. The identification of the five processing enzymes (PC1, PC2, furin, PC4 and PACE4) has opened new avenues in our understanding of mammalian paired basic amino acid convertases and the phenomenon of tissue-specific processing. No doubt, in the future, new members of this growing family of convertases will be characterized and their cleavage selectivity, together with their cellular and tissue distribution, examined. In the years ahead, the genetic, biological and clinical implications of such findings are expected to be quite exciting.

The secretarial assistance of Mrs. S. Emond is appreciated. This work was supported by research grants from MRC Canada (PG-2 to N.G.S. and M.C.; MT-11268 to R.D.) and by J. A. DeSève Succession. Dr Robert Day is a scholar of the Fonds de la Recherche en Santé du Québec (FRSQ).

References

Ackland, J. F., Schwartz, N. B., Mayo, K. E. and Dodson, R. E. (1992) Nonsteroidal signals originating in the gonads. Physiol. Rev. **72**, 731–787

Bålöw, R. M., Tomkinsson, B., Ragnarsson, U. and Zetterqvist, Ö. (1986) Purification, substrate specificity, and classification of tripeptidyl peptidase II. J. Biol. Chem. **261**, 2409–2417

Barr, P. J., Mason, O. B., Landsberg, K. E., Wong, P. A., Kiefer, M. C. and Brake, A. J. (1991) cDNA and gene structure for a human subtilisin-like protease with cleavage specificity for paired basic amino acid residues. DNA Cell. Biol. **10**, 319–328

Bathurst, I. C., Brennan, S. O., Carrell, R. W., Cousens, L. S., Brake, A. J. and Barr, P. J. (1987) Yeast KEX2 protease has the properties of a human proalbumin converting enzyme. Science **235**, 348–350

Benjannet, S., Rondeau, N., Day, R., Chrétien, M. and Seidah, N. G. (1991) PC1 and PC2 are pro-protein convertases capable of cleaving POMC at distinct pairs of basic residues. Proc. Natl. Acad. Sci. U.S.A. **88**, 3564–3568

Benjannet, S., Reudelhuber, T., Mercure, C., Rondeau, N., Chrétien, M. and Seidah, N. G. (1992*a*) Pro-protein conversion is determined by a multiplicity of factors including convertase processing, substrate specificity and intracellular environment: cell type-specific processing of human prorenin by the convertase PC1. J. Biol. Chem. **267**, 11417–11423

Birch, N. P., Tracer, H. L., Hakes, D. J. and Loh, Y. P. (1991) Coordinate regulation of mRNA levels of pro-opiomelanocortin and the candidate processing enzymes PC2 and PC3, but not furin, in rat pituitary intermediate lobe. Biochem. Biophys. Res. Commun. **179**, 1311–1319

Bloomquist, B. T., Eipper, B. A. and Mains, R. E. (1991) Prohormone converting enzymes: regulation and evalution of function using antisense RNA. Mol. Endocrinol. **5**, 2014–2024

Boileau, G., Barbeau, C., Jeannotte, L., Chrétien, M. and Drouin, J. (1983) Complete structure of the porcine pro-opiomelanocortin mRNA derived from the nucleotide sequence of cloned cDNA. Nucleic Acids Res. **11**, 8063–8071

Bresnahan, P. A., Leduc, R., Thomas, L., Thorner, J., Gibson, H. L., Brake, A. J., Barr, P. J. and Thomas, G. (1990) Human *fur* gene encodes a yeast KEX2-like endoprotease that cleaves pro-β-NG *in vivo*. J. Cell Biol. **111**, 2851–2859

Bryan, P., Pantoliano, M. W., Quill, S. G., Hsiao, H.-Y. U. and Poulos, T. (1986) Site directed mutagenesis and the role of the oxyanion hole in subtilisin. Proc. Natl. Acad. Sci. U.S.A. **83**, 3743–3745

Burgess, T. L. and Kelly, R. B. (1987) Constitutive and regulated secretion of proteins. Annu. Rev. Cell Biol. **3**, 243–293

Chance, R. E., Ellis, R. M. and Bromer, W. W. (1968) Porcine proinsulin: characterization and amino acid sequence. Science **161**, 165–167

Chrétien, M. and Li, C. H. (1967) Isolation, purification and characterization of γlipotropic hormone from sheep pituitary glands. Can. J. Biochem. **45**, 1163–1174

Christie, D., Batchelor, D. C. and Palmer, D. J. (1991) Identification of kex2-related proteases in chromaffin granules by partial amino acid sequence analysis. J. Biol. Chem. **255**, 15679–15683

Copeland, N. G., Gilbert, D. J., Chrétien, M., Seidah, N. G. and Jenkins, N. A. (1992) Regional localization of three convertases, PC1 (*Nec*-1), PC2 (*Nec*-2), and furin (*Fur*) on mouse chromosomes. Genomics **13**, 1356–1358

Cullinan, W. E., Day, N. C., Schäfer, M. K.-H., Day, R., Seidah, N. G., Chrétien, M., Akil, H. and Watson, S. J. (1991) Neuroanatomical and functional studies of peptide precursors processing enzymes. Enzyme **45**, 285–300

Day, R., Schäffer, M. K. H., Watson, S. J., Chrétien, M. and Seidah, N. G. (1992) Distribution and regulation of the prohormone convertases PC1 and PC2 in the rat pituitary. Mol. Endocrinol. **6**, 485–497

Devi, L. (1991) Peptide processing at monobasic sites, in Peptide Biosynthesis and Processing (Fricker, L. D., ed.), pp. 175–198, CRC Press, Boca Raton

Foster, D. C., Holly, R. D., Sprecher, C. A., Walker, K. M. and Kumar, A. A. (1991) Endoproteolytic processing of the human protein C by the yeast Kex2 endopeptidase coexpressed in mammalian cells. Biochemistry **30**, 367–372

Fricker, L. D., Das, B. and Angeletti, R. H. (1990) Identification of the pH-dependent membrane anchor of carboxypeptidase E (EC 3.4.17.10). J. Biol. Chem. **265**, 2476–2482

Ftouhi, N., Day, R., Mbikay, M., Chretien, M. and Seidah, N. G. (1994) The gene organization of the mouse prohormone and proprotein convertase PC1. DNA Cell. Biol., in the press

Fuller, R. S., Sterne, R. E. and Thorner, J. (1988) Enzymes required for yeast prohormone processing. Annu. Rev. Biochem. **50**, 345–362

Fuller, R. S., Brake, A. and Thorner, J. (1989*a*) Yeast prohormone processing enzyme (KEX2 gene product) is a Ca^{2+}-dependent serine protease. Proc. Natl. Acad. Sci. U.S.A. **86**, 1434–1438

Fuller, R. S., Brake, A. J. and Thorner, J. (1989*b*) Intracellular targeting and structural conservation of a prohormone-processing protease. Science **246**, 482–486

Gainer, H., Russell, J. T. and Loh, Y. P. (1985) The enzymology and intracellular organization of peptide precursor processing: the secretory vesicle hypothesis. Neuroendocrinology **40**, 171–184

Galanopolou, A., Kent, G., Rabbani, S. N., Seidah, N. G. and Patel, Y. C. (1993) Heterologous processing of promatostatin in constitutive and regulated secretory pathways: putative role of the endoproteases furin, PC1 and PC2. J. Biol. Chem. **268**, 6041–6049

Hakes, D. J., Birch, N. P., Mezey, A. and Dixon, J. E. (1991) Isolation of two complementary deoxyribonucleic acid clones from a rat insulinoma cell line based on similarities to Kex2 and furin sequences and the specific localization of each transcript to endocrine and neuroendocrine tissues in rats. Endocrinology **129**, 3053–3063

Hatsuzawa, K., Hosaka, M., Nakagawa, T., Nagase, M., Shoda, A., Murakami, K. and Nakayama, K. (1991) Structure and expression of mouse furin, a yeast Kex2-related protease. J. Biol. Chem. **265**, 22075–22078

Hatsuzawa, K., Murakami, K. and Nakayama, K. (1992) Molecular and enzymatic properties of furin, a Kex2-like endoprotease involved in precursor cleavage at Arg-X-Lys/Arg-Arg sites. J. Biochem. (Tokyo) **111**, 296–301

Hosaka, M., Nagahama, M., Kim, W.-S., Watanabe, T., Hatsuzawa, K., Ikemizu, J., Murakami, K. and Nakayama, K. (1991) Arg-X-Lys/Arg-Arg motif as a signal for precursor cleavage catalyzed by furin within the constitutive secretory pathway. J. Biol. Chem. **266**, 12127–12130

Julius, D., Brake, A., Blair, L., Kunisawa, R. and Thorner, J. (1984) Isolation of the putative structural gene for the Lys-Arg- cleaving endopeptidase required for the processing of yeast prepro-α-factor. Cell **37**, 1075–1089

Jung, Y.-K., Kunczt, C. J., Pearson, R. K., Dixon, J. E. and Fricker, L. D. (1991) Structural characterization of the rat carboxypeptidase E gene. Mol. Endocrinol. **5**, 1257–1268

Kemmler, W., Peterson, J. D. and Steiner, D. (1971) Studies on the conversion of proinsulin to insulin. I. Conversion *in vitro* with trypsin and carboxypeptidase B. J. Biol. Chem. **246**, 6786–6791

Kiefer, M. C., Tucker, J. E., Joh, R., Landsberg, K. E., Saltman, D. and Barr, P. J. (1991) Identification of a second human subtilisin-like protease gene in the *fes/fps* region of chromosome 15. DNA Cell. Biol. **10**, 757–769

Kirchmair, R., Egger, C., Gee, P., Hogue-Angeletti, R., Fisher-Colbrie, R., Laslop, A. and Winkler, H. (1992) Differential subcellular distribution of PC1, PC2 and furin in bovine adrenal medulla and secretion of PC1 and PC2 from this tissue. Neurosci. Lett. **143**, 143–145

Korner, J., Chun, J., Harter, D. and Axel, R. (1990) Isolation and functional expression of a mammalian prohormone processing enzyme, murine prohormone convertase 1. Proc. Natl. Acad. Sci. U.S.A. **88**, 6834–6838

Ladenheim, R. G., Seidah, N. G. and Rougeon, F. (1991) *N*-linked glycosylation affects the processing of mouse submaxillary gland prorenin in transfected AtT20 cells. Eur. J. Biochem. **198**, 535–540

Lazure, C., Seidah, N. G., Pélaprat, D. and Chrétien, M. (1983) Proteases and post-translational processing of prohormones: a review. Can. J. Biochem. Cell Biol. **61**, 501–515

Leduc, R., Molloy S. S., Thomas, B. A. and Thomas, G. (1992) Activation of human furin precursor processing endoprotease occurs by an intramolecular autoproteolytic cleavage. J. Biol. Chem. **267**, 14304–14308

Mackin, R. B., Noe, B. D. and Spiess, J. (1991) Identification of a somatostatin-14-generating propeptide converting enzyme as a member of the kex2/furin/PC family. Endocrinology **129**, 2263–2265

Mains, R. E., Dickerson, I. M., May, V., Stoffers, D. A., Perkins, S. N., Ouafik, L., Husten, E. J. and Eipper, B. (1990) Cellular and molecular aspects of peptide hormone biosynthesis. Front. Neuroendocrinol. **11**, 52–89

Mbikay, M., Raffin-Sansson, M. L., Tadros, H., Sirois, F., Seidah, N. G. and Chretien, M. (1994) Structure of the gene for the testis-specific proprotein convertase 4 and of its alternate messenger RNA isoforms. Genomics, in the press

Misumi, Y., Oda, K., Fujiwara, T., Takami, N., Tashiro, K. and Ikehara, Y. J. (1991) Functional expression of furin demonstrating its intracellular localization and endoprotease activity for processing of proalbumin and complement Pro-C3*. J. Biol. Chem. **266**, 16954–16959

Mizuno, K., Nakamura, T., Oshima, T., Tanaka, S. and Matsuo, H. (1988) Yeast *KEX2* gene encodes an endopeptidase homologous to subtilisin-like serine proteases. Biochem. Biophys. Res. Commun. **156**, 246–254

Morrison, T., Ward, L. J. and Semerjian, A. (1985) Intracellular processing of the Newcastle disease virus fusion glycoprotein. J. Virol. **53**, 851–857

Nagahama, M., Nakayama, K. and Murakami, K. (1991) Sequence requirements for prohormone processing in mouse pituitary AtT-20 cells. Eur. J. Biochem. **197**, 135–140

Nakayama, K., Kim, W.-S., Torij, S., Hosaka, M., Nakagawa, X., Ikemizu, J., Baba, T. and Murakami, K. (1992) Identification of the fourth member of the mammalian endoprotease family homologous to the yeast Kex2 protease: its testis-specific expression. J. Biol. Chem. **267**, 5897–5900

Ohagi, S., LaMendola, J., LeBeau, M. M., Espinosa, R., Takeda, J., Smeekens, S. P., Chan, S. J. and Steiner, D. F. (1992) Identification and analysis of the gene encoding human PC2, a prohormone convertase expressed in neuroendocrine tissues. Proc. Natl. Acad. Sci. U.S.A. **89**, 4977–4981

Ohta, Y., Hojo, H., Aimoto, S., Kobayashi, T., Zhu, X., Jordan, F. and Inouye, M. (1991) Propeptide as an intramolecular chaperone: renaturation of denatured subtilisin E with a synthetic propeptide. Mol. Microbiol. **5**, 1507–1510

Randolph, V., Winkler, G. and Stollar, V. (1990) Acidotropic amines inhibit proteolytic processing of flavivirus prM protein. Virology **174**, 450–458

Roebroek, A. J. M., Schalken, J. A., Leunissen, J. A. M., Onnekink, C., Gloemers, H. P. J. and Van de Ven, W. J. M. (1986) Evolutionary conserved close linkage of the c-*fes/fps* proto-oncogene and genetic sequences encoding a receptor-like protein. EMBO J. **5**, 2197–2202

Roebroek, A. J. M., Pauli, I. G. L., Zhang, Y. and Van de Ven, W. J. M. (1991) cDNA sequence of a *Drosophila melanogaster* gene, D*fur*1, encoding a protein structurally related to the subtilisin-like proprotein processing enzyme furin. FEBS Lett. **289**, 133–137

Roebroek, A. J. M., Creemers, J. W. M., Pauli, I. G. L., Kurzik-Dumke, U., Rentrop, M., Gateff, E. A. F., Leunissen, J. A. M. and Van de Ven, W. J. M. (1992) Cloning and functional expression of Dfurin2, a subtilisin-like proprotein processing enzyme of *Drosophila melanogaster* with multiple repeats of a cysteine motif. J. Biol. Chem. **267**, 17208–17215

Ruoslahti, E. (1988) Fibronectin and its receptors. Annu. Rev. Biochem. **57**, 375–413

Sachs, H. and Takabatake, Y. (1964) Evidence for a precursor in vasopressin biosynthesis. Endocrinology **75**, 943–948

Schalken, J. A., Roebroek, A. J. M., Oomen, P. P. C. A., Wagenaar, S. S. C., Debruyne, F. M. J. and Van de Ven, W. J. M. (1987) *Fur* gene expression as a discriminating marker for small cell and nonsmall cell lung carcinomas. J. Clin. Invest. **80**, 1545–1549

Seger, M. A. and Bennett, H. P. J. (1986) Structure and bioactivity of the amino-terminal fragment of pro-opiomelanocortin. J. Steroid Biochem. **25**, 703–710

Seidah, N. G., Gaspar, L., Mion, P., Marcinkiewicz, M., Mbikay, M. and Chrétien, M. (1990) cDNA sequence of two distinct pituitary proteins homologous to Kex2 and furin gene products: tissue-specific mRNAs encoding candidates for pro-hormone processing proteinases. DNA **9**, 415–424

Seidah, N. G., Marcinkiewicz, M., Benjannet, S., Gaspar, L.,, Beaubien, G., Mattei, M. G., Lazure, C., Mbikay, M. and Chrétien, M. (1991*a*) Cloning and primary sequence of a mouse candidate prohormone convertase PC1 homologous to PC2, furin and Kex2: distinct chromosomal localization and mRNA distribution in brain and pituitary as compared to PC2. Mol. Endocrinol. **5**, 111–122

Seidah, N. G., Day, R., Marcinkiewicz, M., Benjannet, S. and Chrétien, M. (1991*b*) Mammalian neural and endocrine pro-protein and pro-hormone convertases belonging to the subtilisin family of serine proteinases. Enzyme **45**, 271–284

Seidah, N. G., Mattei, M. G., Gaspar, L., Benjannet, S., Mbikay, M. and Chrétien, M. (1991*c*) Chromosomal assignments of the genes of the pro-protein convertases PC1 (human 5q15-21), PC2 (human 20p11.1-11.2) and furin [mouse 7(D1–E2 region)]. Genomics 11, 103–107

Seidah, N. G., Hamelin, J., Gaspar, A., Day, R. and Chrétien, M. (1992*a*) The cDNA sequence of the human pro-hormone and pro-protein convertase PC1. DNA Cell. Biol. **11**, 283–289

Seidah, N. G. and Chrétien, M. (1992*b*) Pro-protein and pro-hormone convertase of the subtilisin family. Recent development and future perspectives. Trends Endocrinol. Metab. **3**, 133–140

Seidah, N. G., Day, R., Hamelin, J., Gaspar, A., Collard, M. W. and Chrétien, M. (1992*c*) Testicular expression of PC4 in the rat: Molecular diversity of a novel germ cell-specific Kex2/subtilisin-like pro-protein convertase. Mol. Endocrinol. **6**, 1559–1570

Seidah, N. G., Fournier, H., Boileau, G., Benjannet, S., Rondeau, N. and Chrétien, M. (1992*d*) The cDNA structure of the porcine pro-hormone convertase PC2 and the comparative processing by PC1 and PC2 of the N-terminal glycopeptide segment of porcine POMC. FEBS Lett. **310**, 235–239

Shields, D. (1992) Signals for intracellular protein sorting: implications for peptide hormone precursor processing and secretion, in Peptide Biosynthesis and Processing (Fricker, L. D., ed.), pp. 37–70, CRC Press, Boca Raton

Smeekens, S. P. and Steiner, D. F. (1990) Identification of a human insulinoma cDNA encoding a novel mammalian protein structurally related to the yeast dibasic processing protease Kex2. J. Biol. Chem. **265**, 2997–3000

Smeekens, S. P., Avruch, A. S., LaMendola, J., Chan, S. J. and Steiner, D. F. (1991) Identification of a cDNA encoding a second putative prohormone convertase related to PC2 in AtT20 cells and islets of Langerhans. Proc. Natl. Acad. Sci. U.S.A. **88**, 340–344

Steiner, D. (1991) Prohormone convertases revealed at last. Curr. Biol. **1**, 375–377

Steiner, D. F. and Oyer, P. E. (1967) The biosynthesis of insulin and a probable precursor of insulin by a human islet cell adenoma. Proc. Natl. Acad. Sci. U.S.A. **57**, 473–481

Steiner, D. F., Cunningham, D., Spiegelman, L. and Aten, B. (1967) Insulin biosynthesis: evidence for a precursor. Science **157**, 697–699

Steiner, D. F., Clark, J. L., Nolan, C., Rubenstein, A. H., Margoliash, E., Aten, B. and Oyer, P. E. (1969) Proinsulin and the biosynthesis of insulin. Recent Prog. Horm. Res. **25**, 207–282

Stieneke-Grober, A., Angliker, M. V. H., Shaw, E., Thomas, G., Roberts, C., Klenk, H.-D. and Garten, W. (1992) Influenza virus hemagglutinin with multibasic cleavage site is activated by furin, a subtilisin-like endoprotease. EMBO J. **11**, 2407–2414

Thomas, G., Thorne, B. A., Thomas, L., Allen, R. G., Hruby, D. E., Fuller, R. and Thorner, J. (1988) Yeast KEX2 endopeptidase correctly cleaves a neuroendocrine prohormone in mammalian cells. Science **241**, 226–230

Tomkinson, B. and Jonsson, A.-K. (1991) Characterization of cDNA for human tripeptidyl peptidase II: The N-terminal part of the enzyme is similar to subtilisin. Biochemistry **30**, 168–174

Van den Ouweland, A. M. W., Van Groningen, J. J. M., Roebroek, A. J. M., Onnekink, C. and Van de Ven, W. J. M. (1989) Nucleotide sequence analysis of the human *fur* gene. Nucleic Acids Res. **17**, 7101–7102

Van den Ouweland, A. M. W., Van Duijnhoven, H. L. P., Keizer, G. D., Dorssers, L. C. J. and Van de Ven, W. J. M. (1990) Structural homology between the human *fur* gene product and the subtilisin-like protease encoded by yeast KEX2. Nucleic Acids Res. **18**, 664

Van de Ven, W. J. M., Voorberg, J., Fontijn, R., Pannekoek, H., Van den Ouweland, A. M. W., Van Duijnhoven, H. L. P., Roebroek, A. J. M. and Siezen, R. J. (1990) Furin is a subtilisin-like proprotein processing enzyme in higher eukaryotes. Mol. Biol. Rep. **14**, 265–275

Watanabe, T., Nakagawa, T., Ikemizu, J., Nagahama, M., Murakami, K. and Nakayama, K. (1992) Sequence requirements for precursor cleavage within the constitutive pathway. J. Biol. Chem. **267**, 8270–8274

Wilcox, C. A. and Fuller, R. (1991) Post-translational processing of the prohormone-cleaving Kex2 protease in the *Saccharomyces cerevisiae* secretory pathway. J. Cell. Biol. **115**, 297–307

Wise, R. J., Barr, P. J., Wong, P. A., Kiefer, M. C., Brake, A. J. and Kaufman, R. J. (1990) Expression of a human proprotein processing enzyme: Correct cleavage of the von Willebrand factor precursor at a paired basic amino acid site. Proc. Natl. Acad. Sci. U.S.A. **87**, 9378–9382

Yamada, A., Takeuchi, K. and Hishiyama, M. (1988) Intracellular processing of mumps virus glycoproteins. Virology **165**, 268–273

Zhu, X., Ohta, Y., Jordan, F. and Inouye, M. (1989) Pro-sequence of subtilisin can guide the refolding of denatured subtilisin in an intermolecular process. Nature (London) **339**, 483–484

Enzymic α-amidation of neuropeptides and peptide hormones

A. F. Bradbury and D. G. Smyth

National Institute for Medical Research, The Ridgeway, Mill Hill, London NW7 1AA, UK

Introduction

The α-amide group is present at the *C*-terminus of a high proportion of peptide hormones and it is generally essential for biological activity. The α-amide group is generated as a post-translational processing event (Bradbury and Smyth, 1991), unlike the β- and γ-amide groups of asparagine and glutamine. The field of enzymic amidation has become the focus of much attention since the enzyme involved is of major importance in the biosynthesis of peptide hormones, transmitters and growth factors and was essentially the first processing enzyme to be identified (Bradbury and Smyth, 1991).

Detection and assay of amidating activity

Of the various substrates that have been employed for the detection and assay of enzymic amidating activity, virtually all are based on the use of synthetic peptides that are resistant to proteolytic degradation. This allows activity to be detected and measured in crude extracts which contain a variety of proteases. Since amidating activity can be determined with very small peptide substrates if they possess a *C*-terminal glycine residue, the synthetic tripeptide D-tyrosyl-valyl-glycine (D-Tyr-Val-Gly) has been widely employed (Bradbury *et al.*, 1982). The radiolabelled peptide ^{125}I-D-Tyr-Val-Gly is converted to the dipeptide amide ^{125}I-D-Tyr-Val which, lacking a carboxy-group, is easily separated by ion-exchange chromatography (Bradbury *et al.*, 1982; Landymore Lim *et al.*, 1983), t.l.c. (Miyazacki and Uemura, 1991) or by extraction into an organic solvent if the NH_2 group is blocked by acetylation (Mizuno *et al.*, 1986). A variation of this procedure has involved the use of dansyl-D-Tyr-Val-Gly; the final product is separated by t.l.c. which, while less accurate, may be more convenient when numerous assays are required (Bendig, 1986). The product D-Tyr-Val amide can also be separated by h.p.l.c. (Jones *et al.*, 1988). Similarly, a recent report has described a rapid, fluorimetric assay in which the substrate, intermediates and product are resolved by h.p.l.c. (Consalvo *et al.*, 1992). The sensitivity of these methods can be increased 4–5-fold by substituting phenylalanine for valine in the central position (Bradbury and Smyth, 1983*a*).

The reaction, which is usually carried out at neutral pH, can be optimized by the addition of cupric ion (Bradbury and Smyth, 1983*b*), together with ascorbate or dopamine and catalase, which appears to increase the stability of the enzyme (Eipper *et al.*, 1983; Bradbury and Smyth, 1985). The addition of BSA helps to maintain the copper concentration at an optimum level (Bradbury and Smyth, 1987*a*).

A tritiated precursor of thyrotropin-releasing hormone (TRH), ^{3}H-pGlu-His-Pro-Gly, has also been employed and the amidated product determined by t.l.c. (Hussain and Tate, 1983) or radioimmunoassay (Kizer *et al.*, 1986). It may be noted, however, that conversion of peptides terminating in prolyl-glycine is relatively slow (Bradbury and Smyth, 1983*a*) and the radioimmunoassay procedure is time consuming. The glycine-extended precursor of substance P has also been used as a substrate but only for assay of a purified enzyme (Jeng *et al.*, 1990).

Mechanism of amidation

A characteristic feature of enzymic amidation is the specificity exhibited by the amidating enzyme; with peptide substrates a *C*-terminal glycine residue is essential (Bradbury *et al.*, 1982; Bradbury and Smyth, 1983*a*). Peptides with other L-amino acids at the *C*-terminus are not amidated. Peptides with D-alanine at the *C*-terminus undergo slow amidation but peptides with other *C*-terminal D-amino acids do not react at all (Landymore Lim *et al.*, 1983).

Certain non-peptides are capable of acting as substrates, including a series of hydrazones and semi-carbazones of glyoxylic acid (Bradbury and Smyth, 1987*a*). With these non-peptides, the factors affecting binding to the enzyme were found to be similar to those that affect binding of peptide substrates: in particular a *C*-terminal carboxy group is an essential feature and the presence of an aromatic group favours the ability of the hydrazone or semi-carbazone to act as a substrate, while polar groups reduce the ability. It also appears that there has to be a minimum distance of three atoms between the *C*-terminal carboxy group and the aromatic group. When the peptide bond between the glycine residue and the penultimate amino acid is replaced by a methyleneoxy or ketomethylene group, the product retains weak affinity for the enzyme but the affinity is much reduced (Bradbury and Smyth, 1987*b*).

The nitrogen atom of the *O*-amide group is derived from the amino group of the *C*-terminal glycine residue. This was demonstrated by mass spectrometric identification of the amide produced from a synthetic substrate D-Tyr-Val-^{15}N-Gly; the D-Tyr-Val amide formed by the enzyme retained the ^{15}N isotope, showing that the peptide bond linking the *C*-terminal glycine is preserved in the formation of the amide. The simultaneous formation of glyoxylate, demonstrated by the use of ^{14}C-containing substrate, showed that the amidation reaction is oxidative (Bradbury *et al.*, 1982).

The initial step catalysed by the amidating enzyme involves the replacement of a hydrogen atom on the α-hydrogen of the glycine by an hydroxy group. This was shown using the phenylhydrazone of glyoxylic acid as a substrate. The hydroxylated intermediate rearranged to a stable product, oxalic acid phenylhydrazide, which differs from the initial hydrazone only in the replacement of a hydrogen atom by a hydroxy group (Bradbury and Smyth, 1987*a*) (Fig. 1). The hydroxylated intermediate formed during peptide amidation has more recently been identified (Suzuki *et al.*, 1990). Proof that the enzyme is a mono-oxygenase has been obtained by observing the ^{18}O-isotopic ^{13}C-n.m.r. shift produced in the n.m.r. spectrum of α-lactate generated from the pyruvate formed when the enzyme amidated dansyl Tyr-Val-α-^{13}C-D-Ala. It was shown that two one-electron reductions by ascorbate occurred for turnover of each molecule of α-amidating enzyme (Merkler *et al.*, 1992).

The amidating enzyme, like many enzymes that catalyse redox reactions, utilizes copper as an essential cofactor (Bradbury and Smyth, 1983*a*; Eipper *et al.*, 1983). In the case of the enzyme from porcine pituitary, the activity was abolished by dialysis against EDTA but it was restored by addition of cupric ion (Bradbury and Smyth, 1985). No reactivation occurs on addition of other metal ions. The activity of a partially purified amidating enzyme from *Xenopus* appeared to be stimulated by ferric ion (Mollay *et al.*, 1986) but the highly purified enzyme has a specific requirement for copper (Mizuno *et al.*, 1986).

The optimal copper concentration for maximum activity varies with the purity of the enzyme. Its removal by EDTA and the restoration of activity when copper is re-introduced suggests that the metal is only loosely attached to the enzyme and in the presence of other proteins there may be competition for copper binding. Addition of cupric ion often increases the activity of partially pure enzyme preparations but further addition leads to a reduction in activity.

The amidating enzyme from porcine pituitary can function without addition of exogenous cofactors, but its activity is significantly increased by the addition of ascorbate or quinols such as dopamine and catechol and, as

Fig. 1. **Hydroxylation of glyoxylic acid phenylhydrazone to oxalic acid phenylhydrazide catalysed by the peptide-amidating enzyme**

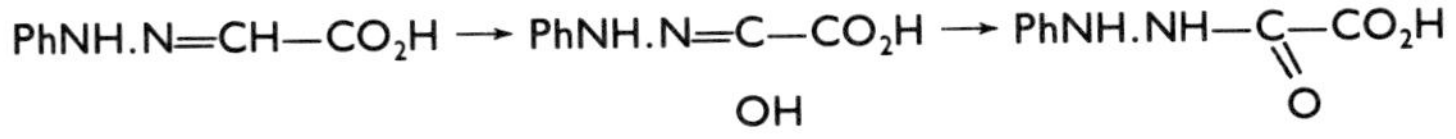

with the addition of copper, it has been observed that there is an optimal concentration beyond which activity declines (Bradbury and Smyth, 1985). The identity of the natural cofactors of the amidating enzyme has not been finally established. It has been reported that severe depletion of ascorbate in the guinea-pig, which is unable to synthesize this reducing agent, leads to a decrease in the degree of amidation of pituitary α-MSH (melanocyte-stimulating hormone) (Fenger and Hilsted, 1988). Ascorbate occurs naturally in most cells and is capable of functioning as a cofactor. In agreement with the view that ascorbate is the natural cofactor, it has been shown in experiments with cultured pituitary cells that addition of ascorbate to the medium is necessary to maintain the continuing synthesis of α-MSH (Glembotski, 1986). Furthermore, intracellular inhibition of amidating activity by a mechanism-based inhibitor has been found to be increased by elevated levels of ascorbate in the medium (Bradbury *et al.*, 1990).

The hydroxylation reaction, which is the first stage of amidation, is catalysed by an enzyme that has been called peptidylglycine hydroxylase (Bradbury and Smyth, 1987*a*), or peptidylglycine α-hydroxylating monooxygenase (Martinez *et al.*, 1993). This reaction is then followed by rearrangement to peptide amide and glyoxylate (Fig. 2). While this rearrangement will occur spontaneously at either high or low pH, in biological systems it is catalysed by another enzyme activity which has been called peptidylhydroxyglycine N–C lyase (Merkler *et al.*, 1992) or peptidyl α-hydroxyglycine *O*-amidating lyase (Martinez *et al.*, 1993). The two complementary activities occur in adjacent regions of a single polypeptide chain; the amidating enzyme is thus a bifunctional enzyme. The overall activity has been referred to previously as peptidylglycine *O*-amidating monooxygenase (Eipper *et al.*, 1983) but the simpler title *O*-amidating enzyme is preferred.

The finding that the amidating enzyme contained two complementary activities emerged from a series of studies showing that a 41 kDa protein, which co-localized with the amidating activity, possessed the ability to lower the pH optimum of the amidation reaction (Noguchi *et al.*, 1989).

Fig. 2. Mechanism of peptide amidation

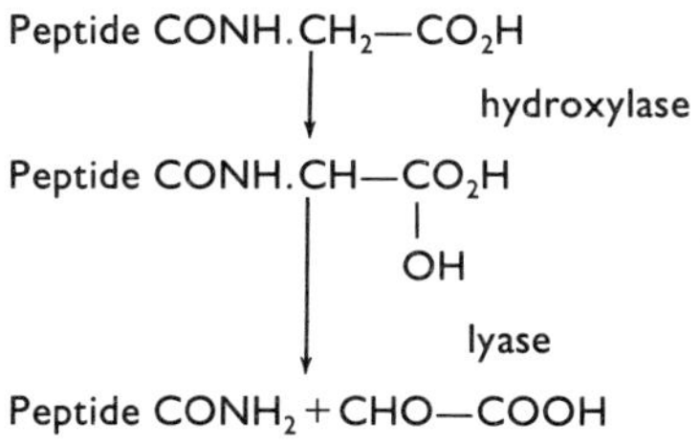

Subsequently the 41 kDa protein was purified and evidence was produced that it was actually an enzyme which catalysed the conversion of the hydroxylated intermediate to the peptide amide (Tajima *et al.*, 1990; Takahashi *et al.*, 1990). In another study, this second enzyme was isolated by using an assay based on the conversion of *O*-hydroxybenzoylglycine to benzamide (Katopodis *et al.*, 1990), thus showing that the amidation reaction proceeds by a two-step reaction. Final confirmation that the hydroxylation and lyase activities reside in a common precursor was demonstrated by the finding that the two enzymes are encoded by a single messenger RNA (Perkins *et al.*, 1990).

Multiple forms of amidating enzyme

A variety of cDNAs coding for the *O*-amidating enzyme have been cloned from frog skin (Mizuno *et al.*, 1987; Ohsuye *et al.*, 1988), mammalian pituitary pars intermedia (Eipper *et al.*, 1988), rat atrium (Stoffers *et al.*, 1989) and rat medullary thyroid carcinoma (Bertelsen *et al.*, 1990). In the rat, the various forms appear to have been formed by alternative splicing from a single gene but in the frog the two mRNAs arise by transcription from different genes. The two cDNAs identified in frog skin (AE-I and AE-II) display a high degree of homology in the *N*-terminal region of the proteins they produce but they differ greatly in molecular size: AE-I codes for a protein of 45 kDa whereas AE-II codes for a protein of 97 kDa. The larger frog DNA is homologous to one of the rat cDNAs; the smaller frog cDNA does not appear to have a mammalian counterpart.

In the cDNAs from ox, and in two of the three cDNAs identified from rat as well as in the larger frog cDNA, the coding is for an approximately 100 kDa protein. A single putative transmembrane domain can be discerned and each protein contains an *N*-terminal signal sequence and a short propeptide. The catalytic domain of the hydroxylating enzyme must be present in the *N*-terminal third of the precursor molecule since the main form of the enzyme found in the tissues has a molecular mass in the region of 40 kDa (Murthy *et al.*, 1986; Bradbury and Smyth, 1988*a*) and corresponds to the *N*-terminal region of the 100 kDa protein deduced by cDNA analysis. In addition, the smaller frog enzyme, encoded by AE-I, contains only the *N*-terminal third of AE-II but it possesses the full hydroxylating activity (Mizuno *et al.*, 1987).

The *N*-terminal section leads into a transmembrane region. This is followed by a region containing the α-amidating lyase activity (Eipper *et al.*, 1991), a putative membrane-spanning domain, and a putative cytoplasmic tail which contains several potential phosphorylation sites. It is of particular interest that the sequence of the amidating enzyme contains a number of pairs of basic amino acid residues which seem likely to serve as sites for endoproteolytic cleavage and, indeed, it has been known for some time that the forms of the enzyme that occur in pituitary are markedly heterogeneous. In contrast, comparison of the sequences of the amidating enzyme in the three species reveals that all of the 14 cysteine residues in the catalytic and intragranular regions are conserved, along with the three clusters of histidine residues which are probably involved in the binding of copper. This conservation extends to the clusters of methionine residues and the potential phosphorylation and some of the *N*-glycosylation sites.

Fractionation of crude enzyme preparations by gel exclusion chromatography of pituitary extracts indicates the presence of several forms of the amidating enzyme which differ in molecular size; it is not clear which of the different forms arise from a common precursor and which from different precursors. This variety of forms could be generated, as mentioned previously, by post-translational processing or it might arise from the expression of different mRNAs formed by alternative splicing, or by both mechanisms. In *Xenopus*, more than one gene is expressed: not only is a 97 kDa form encoded but there is also a 40 kDa molecule. In the rat, 41 kDa and 75 kDa enzymes have been described (Mehta *et al.*, 1988; Beaudry and Bertelsen, 1989). In the ox, where so far only a single gene has been reported, a series of enzymes with overlapping sequences occurs which appears to arise either by alternative gene splicing or by post-translational processing, and the presence of several pairs of basic residue sites offers the probable points of cleavage (Eipper *et al.*, 1988). In rat thyroid carcinoma CA77 cells, two proteins of 104 and 94 kDa are synthesized; the larger is processed to 41 kDa and 43 kDa enzymes while the 94 kDa precursor is processed to a 75 kDa enzyme. The 41, 43 and 75 kDa molecules are secreted as active enzymes (Gilligan *et al.*, 1989). Interestingly, studies on amidating enzyme mRNA in rat tissues have revealed the presence of three major size classes; the heart and central nervous system expressed mRNA of 4.2 and 3.8 kb size classes whereas other tissues expressed predominantly 3.8 and 3.6 kb classes (Braas *et al.*, 1989). It is apparent that *O*-amidating activity is displayed by a highly heterogeneous collection of molecules.

It is not known whether the various forms of the amidating enzyme have different functional roles. The 40 kDa core region of one of the *Xenopus* enzymes which accounts for the whole of one of the gene sequences must contain the active site of the hydroxylating enzyme and in this context it is relevant that the smallest size of active enzyme that has been isolated has a molecular mass in the region of 40 kDa. This enzyme lacks both the carbohydrate-containing sites and the putative membrane-attachment sites that are present in the larger forms. It is reasonable to conclude, therefore, that neither of those structural features is essential for the hydroxylating activity of the enzyme.

The role of the *C*-terminal domain of the amidating enzyme has been clarified in a study of the expression of natural and mutant forms of the enzyme in a stably transfected cell line that lacked secretory granules. It was found that deletion of the *C*-terminal domain eliminated the ability of these cells to internalize membrane-bound enzyme after its secretion at the surface of the cell, leading to accumulation at that site. This suggests that the *C*-terminal region performs a role as a 'routing domain' (Tausk *et al.*, 1992).

It is of interest that 20% of the amidating activity in porcine pituitary is retained by a column of concanavalin A–Sepharose (Bradbury and Smyth, 1988*b*) which shows that only some of the enzyme molecules are glycosylated. In addition, it is notable that the degree of glycosylation varies according to the tissue of origin. The carbohydrate does not appear to be necessary for targeting into the secretory granule or for the process of secretion since the 40 kDa enzyme which lacks carbohydrate was secreted readily by *Spodoptera frugiperda* sf9 cells after transfection of its gene (Merkler *et al.*, 1992). Similar uncertainty exists concerning the specific requirement for a membrane-spanning domain, which is absent in some forms of the enzyme. The larger enzyme molecules, as mentioned above, contain paired basic residue sites where cleavage may take place during post-translational processing and this is likely to account in part for the heterogeneity in molecular size that is exhibited by the enzyme in many tissues.

An expression vector containing the cDNA for the 40 kDA *Xenopus* enzyme has been transfected into an insect cell line using a baculovirus promoter. The cells secreted the 40 kDa enzyme and it was shown that this enzyme was capable only of catalysing hydroxylation which is the first stage of the amidation reaction. Previous studies in which *Xenopus* AE-I cDNA was expressed in *Escherichia coli* produced a protein of the same molecular mass as the native enzyme but with relatively low activity and this was attributed to impaired folding of the protein in the bacillus (Mizuno *et al.*, 1987).

Expression vectors for bovine cDNAs encoding the 108 kDa precursor protein transfected into AtT-20 cells produced considerably increased levels of mRNA and enzyme activity compared with wild-type cells (Mains *et al.*, 1990). The increase in amidating activity was observed in both the soluble and membrane-containing fractions; this suggests that the 110 kDa precursor molecule, which includes a putative membrane-spanning domain, can give rise to soluble activity. A similar result was obtained on transfection of a cDNA encoding a 40 kDa protein (corresponding to the *N*-terminal end of the longer precursor molecule), confirming the previous reports that the full length precursor is not necessary for production of active enzyme (Bradbury and Smyth, 1988*b*). The enzyme produced in these cells exhibited a similar substrate specificity to the purified bovine enzyme.

Recent studies have given information on the nature of the gene(s) involved in biosynthesis of amidating enzymes. It has been shown that rat α-amidating enzyme is encoded by a single copy gene containing >160 kb. The hydroxylase is encoded from 12 exons of >70 kb and these range in size from 49 to 185 bp. Alternative splicing gives rise either to an inactive or to a monofunctional enzyme. The lyase is encoded from eight exons of >19 kb and ranging in size from 54 to 209 bp and there is no evidence, in this case, for alternative splicing. The two catalytic domains are separated by a single exon which codes for a sequence containing a paired basic residue site where cleavage can take place to separate the two catalytic domains. The ensuing exon codes for the transmembrane domain and finally the *C*-terminal domain is encoded by two exons. Evidence was also obtained that the human gene is situated on the long arm of chromosome 5 (Ouafik *et al.*, 1992).

Tissue distribution and localization

Amidating activity is widely distributed in a variety of mammalian tissues, including pituitary (Beaudry and Bertelsen, 1989), heart (Kojima *et al.*, 1989), thyroid (Bradbury and Smyth, 1983*c*; Sakata *et al.*, 1986), hypothalamus (Gomez *et al.*, 1984), cerebrospinal fluid (Wand *et al.*, 1985; Vaeroy *et al.*, 1987), submandibular gland (Mains *et al.*, 1985), parotid gland (von-Zastrow *et al.*, 1986), pancreas (Sakata *et al.*, 1986; Ouafik *et al.*, 1987), adrenal medulla (Bastiaensen and De-Potter, 1989), intestine (Noguchi *et al.*, 1988), bone (Mains and Eipper, 1984), seminiferous tubules (Braas *et al.*, 1992) and plasma (Eipper *et al.*, 1985). The levels of amidating activity in the circulation are hardly affected by hypophysectomy, adrenalectomy, sialectomy or castration but they have been reported to increase after thyroidectomy (Mains *et al.*, 1985). This would suggest that the enzymes in blood originate from a number of different sources. In this context it was shown that an amidated peptide (*O*-MSH) is secreted together with the intracellular enzyme that catalyses its formation (Mains and Eipper, 1984) and it is probable therefore that the amidating activity in serum is released as a byproduct of processing events that took place prior to secretion of the hormone. Much higher levels of amidating activity occur in seminal fluid, however, and in this case it is possible that amidation reactions may take place outside the confines of the cell.

Amidating activity and amidated peptides have been observed in species ranging from Angler fish (Mackin *et al.*, 1987) to moths (Raina *et al.*, 1989) and it seems likely that the amide group is formed by the same mechanism in each case. Recently it has been demonstrated that the enzyme in Angler fish has similar properties to mammalian amidating enzymes (Noe *et al.*, 1991). Indeed, even in the bee the amidated venom peptide mellitin originates from a precursor that contains the *C*-terminal glycine (Suchanek and Kreil, 1977), which is essential to the formation of the amide group in mammals. That this same feature recurs implies that the amidating enzyme in the bee catalyses the same two reactions of hydroxylation and cleavage.

Immunocytochemical and *in situ* hybridization studies have been performed which determined the distribution of the amidating enzyme(s) in different tissues. For example, in heart very high levels were found in all of the atrial cardiomyocytes whereas much lower levels were found in ventricular cells (Braas *et al.*, 1992). A detailed investigation of amidating activity in brain has shown a wide distribution, with the highest levels in the hypothalamus and very low levels in the cerebellum and pineal gland (Lew *et al.*, 1992). At the subcellular level, the enzyme has been shown in AtT-20 cells to be located both in the perinuclear region adjacent to the Golgi apparatus and in the secretory granules (Milgram *et al.*, 1992).

Regulation

Regulatory control of processing reactions might take place at a number of different levels, from enzyme synthesis to post-translational processing and availability of cofactors. Furthermore, control of the access of a processing enzyme to its substrate offers another possible level for regulation. In the case of amidation, effective regulation can be regarded as a possibility only if the relative amounts of the inactive glycine-extended precursor are significant in comparison with the amidated product. In human and rat anterior pituitary, adrenocorticotropic hormone (ACTH) 1–13 (des-acetyl-*O*-MSH), for example, exists almost entirely in the amidated form, which suggests that the amidation process is not regulated in this tissue. In contrast, the levels of a glycine-extended TRH-immunoreactive peptide in rat prostate have been reported to be much greater than the levels of the corresponding peptide amide (Pekary *et al.*, 1987) and in this case *O*-amidation could be an important regulatory step in the generation of active hormone. In the antrum, significant amounts of both gastrin amide and gastrinyl glycine occur and major changes in their proportions take place during development (Hilsted *et al.*, 1988).

Some understanding of regulation has been gained by the use of a specific inhibitor of the amidating enzyme, 4-phenyl 3-butenoic acid, which acts as a mechanism-based inhibitor (Bradbury *et al.*, 1990). In experiments carried out with CA77 rat thyroid carcinoma cells this inhibitor was shown to reduce the intracellular amidating activity with a maximum effect at 15 h after its addition to the medium. The activity then recovered to the control levels even though the intracellular levels of the inhibitor remained the same. This observation points to the existence of a homeostatic mechanism possessed by the cells for maintaining the intracellular levels of amidating activity. It is of particular interest that such a mechanism should operate within an isolated cell but this would be consistent with a report that release of growth hormone (an amidated peptide) and amidating activity from rat pituitary cells plated at high density in culture is much greater than when secreted from cells at low density (Sugimoto *et al.*, 1991), suggesting that this regulatory mechanism may involve a paracrine interaction.

Numerous reports have appeared that intracellular levels of amidating enzyme mRNA and amidating activity rise and fall in parallel with the intracellular levels of amidated peptide. In developmental studies in rat pancreas it has been found that the levels of TRH and amidating enzyme increase immediately after birth and then decline to the levels present in the adult (Ouafik *et al.*, 1987). The maximal levels of both amidating enzyme mRNA and TRH occur at parturition (Maltese *et al.*, 1989). Again, expression of *O*-amidating activity and TRH levels increase rapidly and in unison in mouse hypothalamus during the first two weeks after birth (Faivre-Bauman *et al.*, 1988). In all these cases, however, it is not certain whether the regulation of amidating activity is simply a part of an overall process in which the primary regulated event is the demand for more amidated peptide or whether external messages might act directly to control the levels of enzyme activity.

On occasion, amidating activity can be present in a tissue not known to produce an amidated peptide or in a tissue where peptides occur but have not been identified with a specific function. For example, rat atrium has the highest concentration of amidating activity of any mammalian tissue studied but it is not known to contain amidated peptides. In both the atrium and the ventricle, a 4.2 kb form of mRNA predominates on the day before birth while a 3.8 kb mRNA predominates on the day when birth occurs (Ouafik *et al.*, 1989). There are also changes during development in the proportions of soluble versus membrane-associated activity. In the prostate, a number of amidated peptides structurally related to TRH have been identified (Cockle *et al.*, 1989; Kahn *et al.*, 1992) but no biological function has so far been identified. Similarly in the anterior pituitary large amounts of amidating activity are found but the only amidated peptides known to occur there, are des-acetyl-*O*-MSH and the 'hinge peptide' (joining peptide, which links the *N*-terminal region of pro-opiomelanocortin to the *N*-terminus of ACTH). The physiological role of these amidated peptides is not yet known.

It has been reported that the ability of certain cofactors to support amidation is specific to cell type. Tetrahydrobiopterin and catecholamines can stimulate amidation of the hinge peptide in AtT-20 cells but not in primary melanotrophs, where ascorbate is required (Mains *et al.*, 1990). It was concluded that the cofactors involved in regulation may differ in different tissues.

Amidating activity is generally confined to cells of endocrine or neuroendocrine origin, which would be consistent with the view that production of peptide amides is the primary role. The ability of the hydroxylating enzyme to accept a variety of non-peptides as substrates, however, raises the interesting possibility that this enzyme may have more diverse functions.

Biotechnological aspects of amidation

With the increasing interest that has been shown in the production of hormone amides by biotechnology, production of the precursor peptides has been achieved but difficulty is experienced in the introduction of the *C*-terminal amide group. Overproduction of the precursor peptide in bacteria leads to the formation of products that do not possess the α-amide group since the bacteria lack the necessary procesing enzymes. In principle, however, the amide could be produced by treating the glycine-extended precursor *in vitro* with amidating enzyme. Isolation of this enzyme from pituitary for commercial purposes has proved too time consuming and expensive to be of practical use, but the availability of recombinant enzyme offers an attractive alternative. As mentioned earlier, a recent report has described the expression of a 39 kDa enzyme from *Xenopus* in insect cell lines using a baculovirus expression system (Merkler *et al.*, 1992). This enzyme contained the hydroxylating activity but not the lyase necessary for conversion of the intermediate to the amide. More recently, a 75 kDa α-amidating enzyme produced in recombinant C127 cells (Beaudry *et al.*, 1990) has been shown to possess both hydroxylating and lyase activities and consequently should have the potential for use on a commercial scale (Merkler and Young, 1991).

A report has appeared on the oxidation *in vitro* of a synthetic glycine-extended precursor of growth hormone releasing factor (GRF) using a recombinant α-amidating enzyme (Bongers *et al.*, 1992). This approach gave a good yield of the bioactive form of GRF. Another report (Bongers *et al.*, 1991) described essentially quantitative conversion of the glycine-extended GRF to the amide using the amidating enzyme whereas attempts to achieve the same result using trypsin or carboxypeptidase gave lower yields. These experiments indicate that the production of peptide amides by biotechnology will become feasible in the near future.

References

Bastiaensen, E. and De-Potter, ?. (1989) A peptidyl alpha-amidation activity in chromaffin granules of bovine adrenal medulla. FEBS Lett. **244**, 477

Beaudry, G. A. and Bertelsen, A. H. (1989) Secreted alpha-amidating enzymes are generated by specific posttranslational processing of precursors containing trans-membrane domains. Biochem. Biophys. Res. Commun. **163**, 959

Beaudry, G. A., Mehta, N. M., Ray, M. L. and Bertelsen, A. H. (1990) Purification and characterisation of functional alpha-amidating enzyme secreted from mammalian cells. J. Biol. Chem. **265**, 17694

Bendig, M. M. (1986) Posttranslational processing in *Xenopus* oocytes includes carboxyl-terminal amidation. J. Biol. Chem. **261**, 11935

Bertelsen, A. H., Beaudry, G. A., Galella, A. E. Jones, B. N., Ray, M. L. and Mehta, N. M. (1990) Cloning and characterization of two alternatively spliced rat alpha-amidating enzyme cDNAs from rat medullary thyroid carcinoma. Arch. Biochem. Biophys. **279**, 87

Bongers, J., Offord, R. E., Felix, A. M., Lambros, T., Liu, W., Ahmad, M., Campbell, R. M. and Heimer, E. P. (1991) Comparison of enzymatic semisynthesis of peptide amides: human growth hormone releasing factor and analogues. Biomed. Biochim. Acta **50**, S157

Bongers, J., Felix, A. M., Campbell, R. M., Merkler, D. J. and Heimer, E. P. (1992) Semisynthesis of human growth hormone-releasing factors by alpha-amidating enzyme catalysed oxidation of glycine extended precursors. Peptide Res. **5**, 183

Braas, K. M., Stoffers, D. A., Eipper, B. A. and May, V. (1989) Tissue specific expression of rat peptidylglycine alpha-amidating monooxygenase activity and mRNA. Mol. Endocrinol. **3**, 1387

Braas, K. M., Harakall, S. A., Ouafik, L., Eipper, B. A. and May, V. (1992) Expression of alpha-amidating monooxygenase: an *in situ* hybridisation and immunocytochemical study. Endocrinology **130**, 2778

Bradbury, A. F. and Smyth, D. G. (1983*a*) Substrate specificity of an amidating enzyme in porcine pituitary. Biochem. Biophys. Res. Commun. **112**, 372

Bradbury, A. F. and Smyth, D. G. (1983*b*) Amidation of synthetic peptides by a pituitary enzyme: specificity and mechanism of the reaction, in Peptides (Blaha, K. and Malon, P., eds.), pp. 381–386, Walter de Gruyter, Berlin

Bradbury, A. F. and Smyth, D. G. (1983*c*) Amide forming enzymes in thyroid and pituitary, in Peptides, Structure and Function (Hruby, V. J. and Rich, D. H., eds.), p. 249, Pierce Chemical Co.

Bradbury, A. F. and Smyth, D. G. (1985) C-terminal amide formation in peptide hormones, in Biogenetics of Neurohormonal Peptides (Hakanson, R. and Thorell, S., eds.), pp. 171–186, Academic Press, New York

Bradbury, A. F. and Smyth, D. G. (1987*a*) Enzyme catalysed peptide amidation: isolation of a stable intermediate formed by reaction of the amidating enzyme with an amino acid. Eur. J. Biochem. **169**, 579

Bradbury, A. F. and Smyth, D. G. (1987*b*) Preparation of substrate inhibitors of the amidating enzyme in porcine pituitary, in Peptides (Theodoropoulos, D., ed.), p. 591, Walter de Gruyter, Berlin

Bradbury, A. F. and Smyth, D. G. (1988*a*) *O*-Amidation of peptide hormones: multiple amidating enzymes, in Progress in Endocrinology (Imura, H., Shizume, K. and Yoshida, S., eds.), p. 335, Elsevier, Amsterdam

Bradbury, A. F. and Smyth, D. G. (1988*b*) Peptide amidation: evidence for multiple molecular forms of the amidating enzyme. Biochem. Biophys. Res. Commun. **154**, 1293

Bradbury, A. F. and Smyth, D. G. (1991) Peptide amidation. Trends Biochem. Sci. **16**, 112

Bradbury, A. F., Finnie, M. D. A. and Smyth, D. G. (1982) Mechanism of C-terminal amide formation by pituitary enzymes. Nature (London) **298**, 686

Bradbury, A. F., Mistry, J., Roos, B. A. and Smyth, D. G. (1990) 4-Phenyl-3-butenoic acid, an *in vivo* inhibitor of peptidylglycine hydroxylase (peptide amidating enzyme). Eur. J. Biochem. **189**, 363

Cockle, S. M., Aitken, A., Beg, F. and Smyth, D. G. (1989) A novel peptide pyroglutamylglutaminyl-proline amide, in the rabbit prostate complex, structurally related to thyrotropin releasing hormone. J. Biol. Chem. **264**, 7788

Consalvo, A. P., Young, S. D. and Merkler, D. J. (1992) Rapid fluorimetric assay for the detection of the peptidyl alpha-amidating enzyme intermediate using high-performance liquid chromatography. J. Chromatogr. **607**, 25

Eipper, B. A., Mains, R. E. and Glembotski, C. C. (1983) Identification in pituitary tissue of a peptide *O*-amidating activity that acts on glycine-extended peptides and requires molecular oxygen, copper and ascorbic acid. Proc. Natl. Acad. Sci. U.S.A. **80**, 5144

Eipper, B. A., Myers, A. C. and Mains, R. E. (1985) Peptidyl-glycine alpha-amidation activity in tissues and serum of the adult rat. Endocrinology **116**, 2497

Eipper, B. A., Park, L. P., Dickerson, I. M., Keutmann, H. T., Thiele, E. A., Rodriguez, H., Schofield, P. R. and Mains, R. E. (1988) Structure of the precursor to an enzyme mediating COOH-terminal amidation in peptide biosynthesis. Mol. Endocrinol. **1**, 777

Eipper, B. A., Perkins, S. N., Husten, J. E., Johnson, R. C., Keutman, H. T. and Mains, R. E. (1991) Peptidyl-α-hydroxy-α-amidating lyase. J. Biol. Chem. **266**, 1827

Faivre-Bauman, A., Loudes, C., Barrett, A., Patte, C. and Tixier-Vidal, A. (1988) Outogenesis of peptidylglycine *O*-amidating activity in the mouse hypothalamus *in vivo* and in serum free medium cultures. Dev. Brain Res. **40**, 261

Fenger, M. and Hilsted, L. (1988) Influence of ascorbic acid on *in vivo* amidation of alpha-melanocyte stimulating hormone in guinea pig pituitary. Acta Endocrinol. **118**, 119

Gilligan, J. P., Lovat, S. J., Mehta, N. M., Bertelsen, A. H., Jeng, A. Y. and Taburini, P. P. (1989) Multiple forms of peptidyl alpha-amidating enzyme: purification from rat medullary thyroid carcinoma CA77 cell-conditioned medium. Endocrinology **124**, 2729

Glembotski, C. C. (1986) The alpha-amidation of alpha-melanocyte stimulating hormone in intermediate pituitary requires ascorbic acid. J. Biol. Chem. **259**, 13041

Gomez, S., di-Bello, C., Lam, T. H., Genet, R., Morgan, J. L., Fromageot, P. and Cohen, P. (1984) C-Terminal amidation of neuropeptides. Gly-Lys-Arg extension an efficient precursor of C-terminal amides. FEBS Lett. **167**, 160

Hilsted, L., Bardram, L. and Rehfeld, J. F. (1988) Progastrin maturation during outogenesis. Biochem. J. **255**, 397

Hussain, I. and Tate, S. S. (1983) Formation of the COOH-terminal amide group of thyrotropin-releasing factor. FEBS Lett. **152**, 277

Jeng, A. Y., Wong, M., Lovato, S. J., Erion, M. D. and Gilligan, J. P. (1990) A radioimmunoassay for measuring alpha-amidating activity. Ann. Biochem. **185**, 213

Jones, B. M., Tamburini, P. P., Consalvo, A. P., Young, S. D., Lovato, S. J., Gilligan, J. P., Jeng, A. Y. and Wennogle, L. P. (1988) A fluorimetric assay for peptidyl alpha-amidating activity using high performance liquid chromatography. Ann. Biochem. **168**, 272

Kahn, Z., Aitken, A., del Rio Garcia, J. and Smyth, D. G. (1992) Isolation and identification of two neutral thyrotropin releasing hormone like peptides, pyroglutamyl-phenylalanyl-proline amide and pyroglutamyl-glutaminyl-proline amide from human seminal fluid. J. Biol. Chem. **267**, 7464

Katopodis, A. G., Ping, D. and May, S. W. (1990) A novel enzyme from bovine neurointermediate pituitary catalyses dealkylation of *O*-hydroxyglycine derivatives, thereby functioning sequentially with peptidylglycine *O*-amidating monooxygenase in peptide amidation. Biochemistry **29**, 6116

Kizer, J. S., Bateman, R. C., Miller, C. R., Humm, J., Busby, W. H. and Youngblood, W. W. (1986) Purification and characterization of a peptidylglycine monooxygenase from porcine pituitary. Endocrinology **118**, 2262

Kojima, M., Mizuno, K., Kangawa , K. and Matsuo, H. (1989) Purification and characterization of a peptide C-terminal alpha-amidating enzyme from porcine atrium. J. Biochem. (Tokyo) **105**, 440

Landymore Lim, A., Bradbury, A. F. and Smyth, D. G. (1983) Amidating enzyme in pituitary will accept a peptide with D-alanine as substrate. Biochem. Biophys. Res. Commun. **117**, 289

Lew, R. A., Clarke, I. J. and Smith, A. I. (1992) Distribution and characterisation of peptidylglycine alpha-amidating monooxygenase activity in the ovine brain and hypothalamo-pituitary axis. Endocrinology **130**, 994

Mackin, R. B., Flacker, J. M., Mackin, J. A. and Noe, B. D. (1987) Peptidylglycine alpha-amidating activity is present in islet secretory granules of the anglerfish, ***Lophius americanus***. Gen. Comp. Endocrinol. **67**, 263

Mains, R. E. and Eipper, B. A. (1984) Secretion and regulation of two biosynthetic enzyme activities, peptidyl-glycine monooxygenase and a carboxypeptidase, by mouse pituitary corticotropic tumour cells. Endocrinology **115**, 1683

Mains, R. E., Myers, A. C. and Eipper, B. A. (1985) Alpha-amidating monooxygenase activity in various tissues of the adult male rat. Endocrinology **116**, 2505

Mains, R. E., Dickerson, I. M., May, V., Stoffers, D. A., Perkins, S. N., Ouafik, L., Husten, E. J. and Eipper, B. A. (1990) Cellular and molecular aspects of peptide hormone biosynthesis. Front. Neuroendocrinol. **11**, 52

Maltese, J. Y., Giraud, P., Kowalski, C., Ouafik, L., Salers, P., Pelen, F. and Oliver, C. (1989) Outogenic expression of PAM mRNA in the rat pancreas. Biochem. Biophys. Res. Commun. **158**, 244

Martinez, A., Montuenga, L. M., Springall, D. R., Treston, A., Cuttitta, F. and Polak, J. M. (1993) Immunocytochemical localisation of peptidylglycine alpha-amidating monooxygenase enzymes (PAM) in human endocrine pancreas. J. Histochem. Cytochem. **41**, 375

Mehta, N. M., Gilligan, J. P., Jones, B. N., Bertelsen, A. H., Roos, B. A. and Birnbaum, R. S. (1988) Purification of a peptidylglycine alpha-amidating enzyme from transplantable rat medullary thyroid carcinoma. Arch. Biochem. Biophys. **261**, 44

Merkler, D. J. and Young, S. D. (1991) Recombinant type A rat 75 kDa alpha-amidating enzyme catalyses the conversion of glycine extended peptides to peptide amides via an alpha-hydroxyglycine intermediate. Arch. Biochem. Biophys. **289**, 192

Merkler, D. J., Kulathila, R., Consalvo, A. P., Young, D. S. and Ash, D. E. (1992) ^{18}O isotopic ^{13}C NMR shift as proof that bifunctional peptidylglycine alpha-amidating enzyme is a monooxygenase. Biochemistry **31**, 7282

Milgram, S. L., Johnson, R. C. and Mains, R. E. (1992) Expression of individual forms of peptidylglycine alpha-amidating monooxygenase in AtT-20 cells: endoproteolytic processing and routing to secretory granules. J. Cell Biol. **117**, 717

Miyazacki, N. and Uemura, T. (1991) Determination of peptidylglycine alpha-amidating activity in human serum by thin-layer chromatography. Anal. Biochem. **197**, 108

Mizuno, K., Sakata, J., Kojima, M., Kangawa, K. and Matsuo, H. (1986) Peptide C-terminal alpha-amidating enzyme purified to homogeneity from *Xenopus laevis* skin. Biochem. Biophys. Res. Commun. **137**, 984

Mizuno, K., Ohsuye, K., Wada, Y., Fuchimura, K., Tanaka, S. and Matsuo, H. (1987) Cloning and sequence of cDNA encoding a peptide C-terminal *O*-amidating enzyme from *Xenopus laevis*. Biochem. Biophys. Res. Commun. **148**, 546

Mollay, C., Wichita, J. and Kreil, G. (1986) Detection and partial characterisation of an amidating enzyme in skin secretion of *Xenopus laevis*. FEBS Lett. **202**, 251

Murthy, A. S. N., Mains, R. E. and Eipper, B. A. (1986) Purification and characterization of peptidylglycine O-amidating monooxygenase from bovine neurointermediate pituitary. J. Biol. Chem. **261**, 1815

Noe, B. D., Katopodis, A. G. and May, S. W. (1991) Kinetic analyses of peptidylglycine alpha-amidating monooxygenase from pancreatic islets. Gen. Comp. Endocrinol. **83**, 183

Noguchi, M., Takahashi, K. and Okamoto, H. (1988) Characterization of peptidylglycine alpha-amidating activities in rat pituitary, brain and small intestine using glycine-extended C-terminal analogues of vasoactive intestinal polypeptide as substrate. Tohoku J. Exp. Med. **156**, 191

Noguchi, M., Takahashi, K. and Okamoto, H. (1989) Rat peptidylglycine alpha-amidating enzyme: the relation between activities at neutral and alkaline pH values. Arch. Biochem. Biophys. **275**, 505

Ohsuye, K., Kitano, K., Wada, Y., Fuchimura, K., Tanaka, S., Mizuno, K. and Matsuo, H. (1988) Cloning of cDNA encoding a new peptide C-terminal alpha-amidating enzyme having a putative membrane-spanning domain from *Xenopus laevis* skin. Biochem. Biophys. Res. Commun. **150**, 1275

Ouafik, L., Giraud, P., Salers, P., Dutour, A., Castanas, E., Boudouresque, F. and Oliver, C. (1987) Evidence for high peptide alpha-amidating activity in the pancreas from neonatal rats. Proc. Natl. Acad. Sci. U.S.A. **84**, 261

Ouafik, L., May, V., Keutmann, H. T. and Eipper, B. A. (1989) Developmental regulation of peptidylglycine O-amidating monooxygenase (PAM) in rat heart, atrium and ventricle. J. Biol. Chem. **264**, 5839

Ouafik, L. H., Stoffers, D. A., Campbell, T. A., Johnson, R. C., Bloomquist, B. T., Mains, R. E. and Eipper, B. A. (1992) The multifunctional peptidylglycine alpha-amidating monooxygenase gene: exon/intron organisation of catalytic, processing, and routing domains. Mol. Endocrinol. **6**, 1571

Pekary, A. E., Bhasin, S., Smith, V., Sugawara, M., Swerdloff, R. S. and Herschman, J. N. (1987) Thyroid hormone modulation of TRH and TRH Gly levels in the rat reproductive system. J. Endocrinol. **114**, 271

Perkins, S. N., Husten, E. J. and Eipper, B. A. (1990) The 108 kDa peptidylglycine O-amidating monooxygenase precursor contains two separable enzymatic activities involved in peptide amidation. Biochem. Biophys. Res. Commun. **171**, 926

Raina, A. K., Jaffe, H., Kempe, T. G., Keim, P., Blatcher, I. W., Fales, H. M., Riley, C. T., Klun, J. A., Ridgway, R. L. and Hayes, D. K. (1987) Identification of a neuropeptide hormone that regulates sex pheromone production in female moths. Science **244**, 798

Sakata, J., Mizuno, K. and Matsuo, H. (1986) Tissue distribution and characterization of peptide C-terminal alpha-amidating activity in rat. Biochim. Biophys. Res. Commun. **140**, 230

Stoffers, D. A., Green, C. B.-R. and Eipper, B. A. (1989) Alternative mRNA splicing generates multiple forms of peptidyl-glycine O-amidating monooxygenase in rat atrium. Proc. Natl. Acad. Sci. U.S.A. **86**, 735

Suchanek, G. and Kreil, G. (1977) Translation of melittin messenger RNA *in vitro* yields a product terminating with glutaminylglycine rather than with glutaminoamide. Proc. Natl. Acad. Sci. U.S.A. **74**, 975

Sugimoto, H., Suzuki, M., Takeuchi, T. and Ishikawa, K. J. (1991) Effect of cell density on growth hormone release from, cyclic AMP levels in, and peptide alpha-amidation activity of primary cultured rat anterior pituitary cells. Endocrinology **131**, 237

Suzuki, J., Shimoi, H., Iwasak, Y., Kawahara, T., Matsuura, Y. and Nishikawa, Y. (1990) Peptidylglycine O-amidating monooxygenase, expressed in insect cell culture is a peptidylglycine hydroxylase. EMBO J. **9**, 4259

Tajima, M., Iida, T., Yoshida, S., Komatsu, K., Namba, R., Yanagi, M., Noguchi, M. and Okamoto, H. (1990) The reaction product of a peptidylglycine O-amidating enzyme is a hydroxyl derivative of O-carbon of the carboxyl-terminal glycine. J. Biol. Chem. **265**, 9602

Takahashi, K., Okamoto, H., Seino, H. and Noguchi, M. (1990) Peptidylglycine O-amidating reaction: evidence for a two-step mechanism involving a stable intermediate at neutral pH. Biochem. Biophys. Res. Commun. **169**, 524

Tausk, F. A., Milgram, S. L., Mains, R. E. and Eipper, B. A. (1992) Expression of a peptide processing enzyme in cultured cells: truncation mutants reveal a routing domain. Mol. Endocrinol. **6**, 2185

Vaeroy, H., Nyberg, F., Franzen, H. and Terenius, L. (1987) Characterization of a substance P-Gly 12 amidating enzyme in human cerebrospinal fluid. Biochem. Biophys. Res. Commun. **148**, 24

von-Zastrow, M., Tritton, T. R. and Castle, J. D. (1986) Exocrine granules contain peptide amidation activity. Proc. Natl. Acad. Sci. U.S.A. **83**, 3297

Wand, G. S., Ney, R. L., Mains, R. E. and Eipper, B. A. (1985) Characterization of peptide alpha-amidating activity in cerebrospinal fluid and central nervous system tissue. Neuroendocrinology **41**, 482

9

Angiotensin I-converting enzyme: biochemical and structural properties of a duplicated enzyme

Florent Soubrier, Lei Wei, Christine Hubert, Eric Clauser, François Alhenc-Gelas* and Pierre Corvol†

INSERM U 36, Collège de France, 3 rue d'Ulm; and *INSERM U 367, 17 rue du Fer-à-Moulin, 75005 Paris, France

Distribution and enzymic characteristics of angiotensin I-converting enzyme (ACE)

ACE (kininase II, dipeptidylcarboxypeptidase I; EC 3.4.15.1) acts primarily as a dipeptidyl carboxypeptidase and is involved in the metabolism of two major vasoactive peptides, angiotensin II (A11) and bradykinin. ACE generates the potent vasopressor AII by cleaving the *C*-terminal dipeptide from angiotensin I (AI), and inactivates the vasodepressor bradykinin (BK), by the sequential removal of two *C*-terminal dipeptides (Erdös, 1990). The most favourable ACE substrate is BK ($K_m = 0.2\ \mu M$) for which the K_m is approximately 80 times lower than for AI ($K_m = 16\ \mu M$).

ACE can hydrolyse a wide range of oligopeptides *in vitro* via its dipeptidyl carboxypeptidase action. These substrates include neurotensin (Skidgel *et al.*, 1984), the enkephalins and *C*-terminal-extended pro-enkephalins (Skidgel and Erdös, 1987); however, the K_m values of ACE for these substrates are substantially higher than for BK or AI. ACE also displays some endopeptidase activity in the cleavage of certain substrates which are blocked at the *C*-terminus. For example, ACE can release the *C*-terminal tripeptide amide from substance P (Yokosawa *et al.*, 1983; Skidgel *et al.*, 1984) and from luteinizing hormone-releasing hormone (LH-RH; Skidgel and Erdös, 1985). In addition, and perhaps more surprisingly, ACE can cleave the tripeptide from the blocked *N*-terminus of LH-RH (Skidgel and Erdös, 1985). However, the tripeptide amide is not invariably released from *C*-terminal-blocked substrates; cholecystokinin-8 and various gastrin

†To whom correspondence should be addressed.

analogues are hydrolysed by ACE to release the amidated *C*-terminal dipeptides as the initial and major products (Dubreuil *et al.*, 1989). Although these enzymic activities on natural peptides have been observed *in vitro*, there is no convincing evidence that they bear any physiological relevance, except perhaps for substance P in the brain and the upper respiratory tract (Skidgel and Erdös, 1967; Subissi *et al.*, 1990). To establish a role for ACE in the metabolism of a given peptide, the co-localization of enzyme and substrate must be demonstrated, the presence of other competing enzymes should be considered, and pharmacological experiments using a specific ACE inhibitor should clearly result in substrate accumulation and/or potentiation of action.

Monovalent anions, especially chloride, enhance the activity of ACE towards all known substrates although the presence of chloride is not always essential (Bünning and Riordan, 1983); for example, BK is hydrolysed in the absence of chloride. This activation is thought to be a consequence of a change in protein conformation occurring on chloride binding to a putative lysine residue located in the region of the active site (Shapiro *et al.*, 1983).

ACE is a monomeric zinc metallopeptidase which exists predominantly as a single molecular form, with a molecular mass of 170 kDa, in man. It is an extensively glycosylated protein: the pulmonary enzyme contains a higher proportion of sialic acid than kidney ACE (Weare *et al.*, 1982).

ACE is a membrane-bound enzyme and is orientated such that the catalytic sites are exposed at the extracellular surface of the cell. Owing to the association of ACE with the plasma membrane of vascular endothelial cells, the enzyme displays a ubiquitous tissue distribution. In addition to an endothelial location, high levels of ACE are also found in the brush borders of absorptive epithelia, such as the apical microvilli of the small intestine and the kidney proximal convoluted tubule (Bruneval *et al.*, 1986). Other epithelial locations of ACE include the choroid plexus, where the enzyme is found at a high concentration and is probably the source of ACE in the cerebrospinal fluid (Arregui and Iversen, 1978; Schweisfurth and Schioberg-Schiegnitz, 1984), the prostate and epididymis of the male genital tract (Cushman and Cheung, 1971; Yokohama *et al.*, 1982). ACE is also found in mononuclear cells, such as monocytes (after macrophage differentiation) and T-lymphocytes, and in fibroblasts (Friedland *et al.*, 1978; Weinberg *et al.*, 1982; Costerousse *et al.*, 1993).

Autoradiography, employing radiolabelled specific ACE inhibitors, and immunohistochemical studies have mapped the locations of ACE in brain *in vitro*. ACE was found primarily in the choroid plexus, ependyma, subfornical organ, basal ganglia (caudate putamen and globus pallidus) but, most notably, the enzyme was found to be highly concentrated in the striatonigral neuronal pathway (Defendini *et al.*, 1983; Barnes *et al.*, 1988; Strittmatter *et al.*, 1989). This suggests a role for ACE in the metabolism of a neuropeptide present in the same pathway. Substance P co-localizes with ACE to the same neuronal pathway, however, neutral endopeptidase (endopeptidase-24.11), also highly abundant in this area, is more likely to be involved in the degradation of substance P *in vivo* (Matsas *et al.*, 1984; Oblin *et al.*, 1988). Neuronal ACE is slightly smaller in size than the endothelial enzyme and this observation has been attributed to differences in the extent of *N*-glycosylation (Hooper and Turner, 1987). However, both forms of the enzyme display the same neuropeptide specificities and inhibition profiles and, therefore, do not represent distinct isoforms (Williams *et al.*, 1991).

A soluble form of ACE also exists and has been detected in biological fluids such as plasma (Das *et al.*, 1977), semen (El-Dorry *et al.*, 1983), amniotic fluid (Yasui *et al.*, 1984) and cerebrospinal fluid (Schweisfurth and Schioberg-Schiegnitz, 1984). Soluble ACE appears to originate predominantly from endothelial cells on the basis of similar glycosylation content and the secretion of ACE by endothelial cell cultures.

In the testis, in addition to the endothelial form of ACE, a distinct isoform is present in germinal cells. This germinal form of ACE is smaller, with a molecular mass of 100 kDa in humans, and its expression is specifically regulated. The presence of germinal ACE is dependent on the maturation of germinal cells occurring at the onset of puberty, since ACE is only present in spermatids and not in immature forms of germinal cells (Cushman and Cheung, 1971; Brentjens *et al.*, 1986).

The molecular cloning of the endothelial and germinal ACE cDNAs and of the ACE gene from different species allowed the elucidation of the primary structure of the two ACE isoforms. The determination of the primary structure of the enzyme displayed an unexpected organization of the molecule with two active sites and raised new questions concerning ACE function. Polymorphic genetic markers of the ACE gene were developed and used for investigating the effect of the ACE gene on the control of ACE expression.

ACE gene structure results from the duplication of an ancestral gene

Molecular cloning of the endothelial ACE cDNA revealed that the enzyme is composed of two highly homologous domains, thus indicating a gene duplication event in the course of evolution. This hypothesis was further substantiated by the elucidation of the structure of the germinal form of ACE; this form is encoded by an mRNA which corresponds to the putative ancestral form of the ACE gene and contains only one of the two homologous domains. The discovery of two initiation transcription sites and the determination of the ACE gene structure allowed the elucidation of the mechanism of transcription of the two ACE mRNAs.

Somatic ACE

Affinity chromatography of ACE, employing the specific ACE inhibitor lisinopril as the affinity ligand together with other appropriate steps for solubilization and selection of the inhibitor-bound form of ACE, enabled the purification of the enzyme to homogeneity (Bull *et al.*, 1985; Hooper *et al.*, 1987; Soubrier *et al.*, 1988). Oligonucleotide probes designed on the basis of the *N*-terminal sequence and of several short internal peptides of the enzyme were used to screen cDNA libraries. Complete ACE cDNAs were isolated from human endothelial and mouse kidney cDNA libraries. The human endothelial ACE mRNA is 4.3 kb long although there appear to be some species differences: two kidney ACE mRNAs were detected in mice (4.9 and 4.15 kb) in contrast to the rabbit where the pulmonary ACE mRNA is 5.0 kb long (Bernstein *et al.*, 1989; Thekkumkara *et al.*, 1992). Human endothelial ACE comprises 1306 amino acids, possesses 17 potential *N*-glycosylation sites and 14 cysteinyl residues (Fig. 1). Comparison of the *N*-terminal end of the human ACE precursor, as deduced from the cDNA sequence, with the *N*-terminal sequence of the purified human enzyme enabled us to propose a model for the intracellular processing of ACE and its mode of membrane

Fig. 1. Schematic representation of the testicular and endothelial ACE enzymes

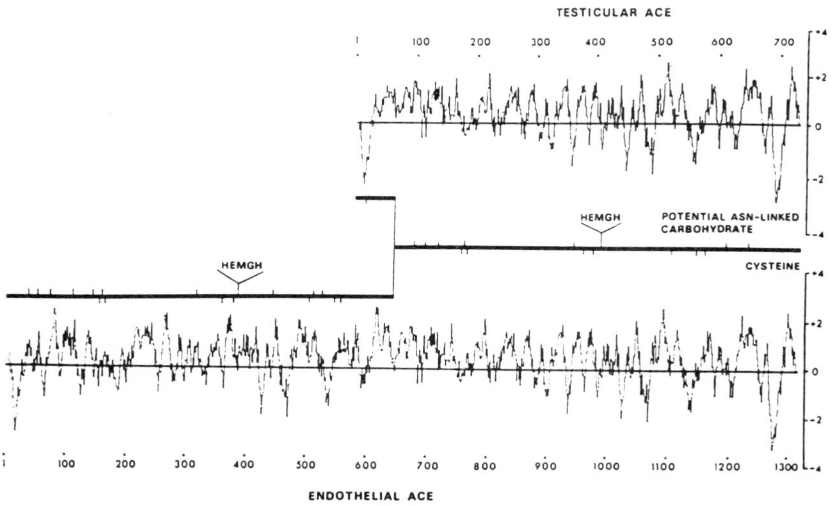

(Middle) Diagram showing the cysteine positions, and potential asparagine-linked glycosylation sites, and the positions of the putative residues of the active site of the two enzymes. Beyond the point of divergence, the testicular enzyme is given on the upper line and the endothelial enzyme on the lower line. (Top and bottom) Hydropathy plots of the predicted testicular (top) and endothelial (bottom) amino acid sequences. Amino acid numbering is presented above and under the hydropathy plots. From Lattion et al. *(1989), with permission.*

anchorage (Soubrier *et al.*, 1988). The sequence of a 29 amino acid hydrophobic peptide at the *N*-terminus was absent in the mature enzyme, and therefore constitutes the cleaved signal peptide. A second sequence of 17 hydrophobic amino acids, is located near the *C*-terminal end of the molecule. Computerized prediction of the secondary structure revealed that this *C*-terminal hydrophobic sequence corresponds to a membrane-associated α-helix, and is, therefore, the candidate region for anchoring the enzyme to the cell membrane. *In vitro* expression of wild-type and mutated recombinant ACE allowed us to define the mechanism of anchorage, as described below.

One surprising feature of the enzyme, disclosed by molecular cloning of the cDNA, is the presence of two large, highly homologous domains, referred to here as *N*- and *C*-domains. This suggests that the ancestral gene was duplicated during evolution. This duplication accounts for the higher molecular mass of ACE, when compared with other zinc metallopeptidases, such as endopeptidase-24.11 (Devault *et al.*, 1987). The overall sequence similarity between the two domains of human ACE is more than 60% at the nucleotide level and at the amino acid level. However, in the latter case, this homology increases to 89% in a region spanning 40 amino acids and containing essential residues of the active site (Soubrier *et al.*, 1988). There is no sequence similarity between the *C*- and *N*-terminal extremities of the molecule, and the central region which separates the two similar domains is also unique. Each domain contains a pentapeptide sequence which complies with the consensus sequence His-Glu-Xaa-Xaa-His for the active site of zinc metallopeptidases (Vallee and Auld, 1990). The two histidine residues provide two of the three zinc-co-ordinating ligands and the glutamic acid acts as a general base in the catalytic reactions, as suggested from crystallographic studies of another zinc metallopeptidase, thermolysin (Kester and Matthews, 1977) and from amino acid comparisons with this and other metallopeptidases (Fig. 2). The presence of two putative active sites in the ACE molecule was unexpected since previous studies had

Fig. 2. Consensus sequence of the active site of zinc metallopeptidases

hACE	T	V	H	H	E	M	G	H	365
hACE	V	A	H	H	E	M	G	H	963
THERM	V	V	A	H	E	L	T	H	146
rNEP	V	I	G	H	E	I	T	H	588

hACE: first and second domain of human ACE (Soubrier et al., *1988). Therm:* Bacillus thermoproteolyticus thermolysin *(Titani* et al., *1972). rNEP: rabbit neutral endopeptidase (Devault* et al., *1987). Numbers refer to the last amino acid position in each protein segment. Identical residues are boxed.*

suggested that ACE contained a single zinc atom per molecule (Bünning and Riordan, 1985) and a single high-affinity inhibitor binding site per molecule of ACE (Cumin *et al.*, 1989). However, by *in vitro* mutagenesis studies, we have demonstrated that ACE in fact contains two functional catalytic sites (see below).

Other residues have a key role in catalysis by zinc metallopeptidase such as His-231 of thermolysin, which stabilizes the transition state of the substrate, and Glu-166, which provides the third zinc-co-ordinating ligand (Kester and Matthews, 1977). Although we proposed His-404 and -1002, and Glu-389 and -987 as candidate residues for these functions, respectively, their role has not been confirmed. A tyrosine and a lysine residue have also been identified as essential residues in rabbit lung ACE by inactivation of the purified enzyme with 1-fluoro-2,4-dinitrobenzene (Dnp-F) (Bünning *et al.*, 1990), confirming previous findings (Bünning *et al.*, 1978). By peptide mapping and sequencing, these amino acids have been identified as Lys-694 and Tyr-776 in the *C*-domain. Similarly Dnp-F reacts with Lys-122 and Tyr-204 of rabbit testicular ACE, which correspond to Lys-118 and Tyr-200 of human testicular ACE (Chen and Riordan, 1990). A mutant human testicular ACE in which Tyr-200 was replaced by a phenylalanine, was expressed in Chinese hamster ovary (CHO) cells (Chen *et al.*, 1992). The mutant ACE exhibited a 15-fold decrease in k_{cat} for the hydrolysis of the synthetic substrate furanacryloyl-Phe-Gly-Gly and a 7-fold decrease in AI hydrolysis. The K_m values of the mutant enzyme for these substrates were slightly increased, but the specific ACE inhibitor, lisinopril, was 100-fold less-tightly bound to the mutant enzyme compared with the wild-type enzyme. These results may suggest that the Tyr-200 residue is involved in stabilizing the transition state. The two Dnp-F reactive residues, which are absent in the corresponding sequence of the *N*-domain of ACE, could be responsible, at least in part, for the increased catalytic activity of the *C*-domain, as described below.

A high degree of sequence identity was found between the different mammalian ACE cDNA sequences cloned to date. Human (Soubrier *et al.*, 1988) and mouse (Bernstein *et al.*, 1989) ACE coding sequences are 83% homologous at both the nucleotide and amino acid levels, and the identity is even higher between human and rabbit (Kumar *et al.*, 1989) ACE sequences. The most conserved sequences are in the regions of the homologous domains. The bacterial dipeptidyl carboxypeptidase shares common synthetic and natural substrates with ACE and is also inhibited by captopril (Deutch and Soffer, 1978), with an IC_{50} of 40 nM as compared with 18 nM for the pig endothelial enzyme (Williams *et al.*, 1991). The *Salmonella typhimurium* gene, which codes for this enzyme, does not display any sequence similarity with mammalian ACE except for the consensus zinc-binding motif (Hamilton and Miller, 1992). This is in contrast to the clear homology between the *Escherichia coli* and human genes coding for another zinc metallopeptidase, aminopeptidase N (Olsen *et al.*, 1988).

Germinal ACE

Early experiments, using *in vitro* cell-free translation of rabbit testis mRNA followed by immunoprecipitation of ^{35}S-labelled proteins with an ACE antiserum showed that the testicular ACE isoform was translated as a shorter precursor (90 kDa) compared with the lung ACE precursor (140 kDa) (El-Dorry *et al.*, 1982*b*), thus implying that the smaller molecular mass did not result from a post-translational modification. Immunohistochemical studies showed that testicular ACE is recognized by antibodies raised against the somatic enzyme, demonstrating the presence of common epitopes (El-Dorry, 1982*a*).

Using the endothelial ACE cDNA as a probe in Northern blot experiments, a single 3 kb hybridizing mRNA species was detected in the human testis, in contrast to the 4.3 kb mRNA detected in endothelial cells (Soubrier *et al.*, 1988).

The primary structure of testicular ACE was deduced from the sequences of cDNA clones covering the entire coding sequence (Ehlers *et al.*, 1989; Lattion *et al.*, 1989). As expected from the size of the mRNA, testicular ACE contains only 732 amino acids, compared with 1306 amino acids present in endothelial ACE. A schematic representation of endothelial and testicular ACE is shown in Fig. 2.

The testicular ACE sequence corresponds to the *C*-domain of endothelial ACE, and therefore contains only one of the two putative catalytic sites identified in endothelial ACE. The absolute sequence identity downstream from residue 1944 of endothelial ACE suggests transcription of both mRNAs from a single gene. The 5′ end of the testicular ACE mRNA, however, contains a 227 bp sequence, which is absent in endothelial ACE, suggesting differential splicing of a common primary RNA transcript or the use of an alternative promoter. This *N*-terminal sequence of 67 residues encodes the signal peptide and a serine- and threonine-rich region presumed to be *O*-glycosylated (Lattion *et al.*, 1989). By site-directed mutagenesis, it was established that a segment of 36 amino acid residues encoded by the testicular ACE-specific 5′ sequence contained more than 90% of all *O*-linked sugars and was not required for enzyme stability and activity (Ehlers *et al.*, 1992).

The complete sequence similarity includes, at the 3′ end of testicular ACE, the hydrophobic segment which represents the anchoring domain (see below), suggesting a similar mechanism of membrane anchorage for the germinal and somatic enzymes. The rabbit pulmonary and testis ACE mRNAs use different 3′ polyadenylation sites located within the same exon (Thekkumkara *et al.*, 1992). The pulmonary mRNA polyadenylation site is 628 bp from the testis polyadenylation site on the 3′ side, as demonstrated by hybridization with specific probes. The presence of an mRNA destabilization sequence (ATTTA) between the two sites might indicate a possible functional role of the alternative use of one of the two polyadenylation sites.

Structure of the ACE gene

By Southern blot experiments and gene cloning, it was demonstrated that the somatic and germinal ACE mRNAs are transcribed from a unique gene (Hubert *et al.*, 1991). The somatic and germinal ACE mRNAs could be generated either by differential splicing of a common primary RNA transcript or by transcription from alternative initiation sites under the control of separate promoters. Several studies have demonstrated the presence of two functional promoters in the ACE gene. The somatic promoter is located on the 5′ side of the first exon of the gene. Fusions of various size fragments of the somatic ACE promoter to a reporter gene in transfection experiments showed the presence of positive regulatory elements inside the 132 bp region upstream of the transcription start site and also suggested the presence of negative regulatory elements between −132 and −343 and between positions −472 and −754 bp (Testut *et al.*, 1993). The germinal promoter is located on the 5′ side of the specific 5′ end of germinal ACE mRNA. Primer extension and RNase protection assays on testicular RNAs were performed in mice, rabbits and humans (Hubert *et al.*, 1991; Kumar *et al.*, 1991). In all species, transcription of the germinal ACE mRNA was initiated inside the gene. Therefore, intron 12, corresponding to the genomic sequence flanking the 5′ region of the testicular-specific exon 13, as deduced from complete analysis of the ACE gene in humans, was proposed as the putative germinal ACE promoter (Hubert *et al.*, 1991).

The promoter function of this sequence was firmly established by using intron 12 to drive the transcription of a reporter gene in a germinal-specific fashion (Langford *et al.*, 1991). A mouse genomic fragment, 689 bp long, containing intron 12, exon 12 and a part of intron 11 of the mouse ACE gene was fused to the β-galactosidase coding sequence and micro-injected into pro-nuclei of one-cell stage embryos. A histochemical analysis of the transgenic mice revealed that β-galactosidase was only expressed, together with ACE, in elongating spermatozoa. In another series of transgenic mice, a 91 bp fragment of intron 12 of the ACE gene was used as promoter and was able to confer to the transgene a germinal cell-restricted pattern of transcription (Howard *et al.*, 1993). Further mapping of elements controlling transcription by DNAse footprint experiments and gel mobility shift assays showed that a sequence between positions −42 and −62 specifically binds to nuclear factors from testicular extracts and contains a consensus cyclic AMP responsive element (Howard *et al.*, 1993). The two alternative promoters of the ACE gene exhibit highly contrasting cell specificities, as the somatic promoter is active in several cell types, whereas the germinal promoter is only active in a stage-specific manner in male germinal cells (Nadaud *et al.*, 1992). To our knowledge, this is the only example of an intragenic, alternative promoter, present inside a duplicated gene and driving transcription of the ancestral non-duplicated form of the gene.

The complete intron–exon structure of the human ACE gene was determined by restriction mapping of genomic clones and by sequencing of the intron–exon boundaries (Hubert *et al.*, 1991). The human ACE gene contains 26 exons. The somatic ACE mRNA is transcribed from exons 1 to 26, but exon 13 is spliced during maturation of the somatic ACE transcript. The germinal mRNA is transcribed from exons 13 to 26.

The structure of the human ACE gene provides further support for the duplication of an ancestral ACE gene (Fig. 3). Exons 4–11 and 17–24, encoding the two homologous domains of the ACE molecule are highly similar both in size and in sequence (Hubert *et al.*, 1991). In contrast, intron size separating homologous exons is not conserved. The ACE gene duplication appears to have occurred early in evolution. In all mammalian species where the ACE gene has been cloned, that is, rabbits, mice and humans, the ACE gene appears to be duplicated. A dipeptidyl carboxypeptidase, affinity-purified from the electric organ of *Torpedo marmorata*, was recognized by a polyclonal antiserum raised against pig kidney ACE and was activated by chloride (Turner *et al.*, 1987). The molecular mass of *Torpedo* ACE is 190 kDa, and therefore this enzyme would also appear to be transcribed from a duplicated ACE gene. If this is the case, the duplication of the ACE gene must have occurred more than 600 million years ago. Conservation in remote species of the two transcription units inside the ACE duplicated gene reflects their physiological significance, although a local and specific substrate of testicular ACE, which would definitively signify its function, is not yet known. A gene duplication has also been postulated for another zinc ectopeptidase, aminopeptidase N, although there is a single active site. A duplicated structure with two active sites has been found in other epithelial ectoenzymes, such as lactase-phlorizin hydrolase (Mantei *et al.*, 1988) and sucrase-isomaltase (Hunziker *et al.*, 1986). However, in contrast to ACE, these two brush-border hydrolases contain two active sites with less homology and a clear difference in substrate specificity.

An ACE-like enzyme has been characterized in the housefly, *Musca domestica* (Lamango and Isaac, 1993). This enzyme is able to cleave the ACE

Fig. 3. Organization of the human ACE gene

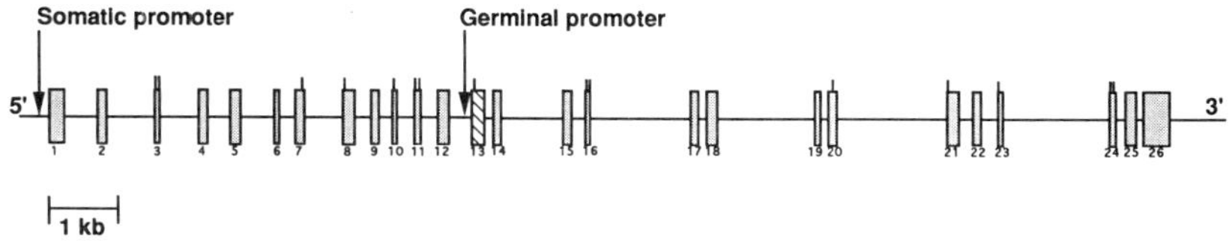

Location of the 26 numbered exons (vertical bars). Exon 13 (hatched bar) is specific to the testicular ACE mRNA. The two promoters are indicated by vertical arrows. Vertical bars above the exon boxes indicate the location of the cysteine residues; adapted from Hubert et al. *(1991).*

substrate Hip-His-Leu and is inhibited by captopril with an IC_{50} of 0.4 μM. A cDNA coding for an enzyme with high sequence similarity to ACE has also been cloned in *Drosophila* (Cornell *et al.*, 1993). The amino acid identity reaches 65% in the region of the zinc-binding motif. Interestingly, the *D. melanogaster* ACE-like enzyme gene does not seem to be duplicated as the full length cDNA is 2.1 kb, an observation which is in agreement with the 87 kDa found for the ACE-like enzyme from *M. domestica.*

Function and substrate specificity of the two active sites

The primary structure of human endothelial ACE indicated the presence of two putative active sites. A series of studies was performed to establish whether both domains of ACE were indeed enzymically active and whether both active sites had similar or different kinetic properties and substrate specificities.

N- and *C*-domains of human ACE are both catalytically active

The sequence identity between the *N*- and *C*-domains is more than 60% over a stretch of 357 amino acids. Each of the two homologous domains contains the consensus sequence found at the active sites of thermolysin, endopeptidase-24.11 and other zinc peptidases (Soubrier *et al.*, 1988). A similar structure, with two domains, has also been determined for the mouse (Bernstein *et al.*, 1989), rabbit (Thekkumkara *et al.*, 1992) and bovine (Shai *et al.*, 1992) somatic enzymes.

To establish whether both putative active sites of the somatic enzyme were functional, a series of ACE mutants, each containing only one intact domain, was constructed by deletion or point mutations of putative critical residues of the other domain (Wei *et al.*, 1991*a*). Wild-type ACE and the different mutants were expressed in stable heterologous CHO cell lines. As negative controls, ACE constructs were designed in which either the zinc-binding histidines or the catalytic glutamic acid of the two domains were mutated. Fig. 4 shows a schematic representation of these different constructions.

The recombinant wild-type ACE was structurally, immunologically and enzymically identical to affinity-purified human kidney ACE (Wei *et al.*, 1991*b*). Conversely, the two mutants which possessed critical mutations in both *N*- and *C*-domains were devoid of enzymic activity. The mutants possessing only an intact *N*-domain were able to hydrolyse a synthetic substrate which constituted the *C*-terminus of AI (Hip-His-Leu). For these mutants, the K_m values for both substrates, Hip-His-Leu and AI, were similar to the wild-type enzyme but the $k_{cat.}$ values were ten and four times lower, respectively. Similar results were obtained with the truncated mutants containing only the *C*-domain or the full-length mutants containing mutations which inactivated the *N*-domain although, in this case, the $k_{cat.}$

Fig. 4. Diagram of ACE constructions

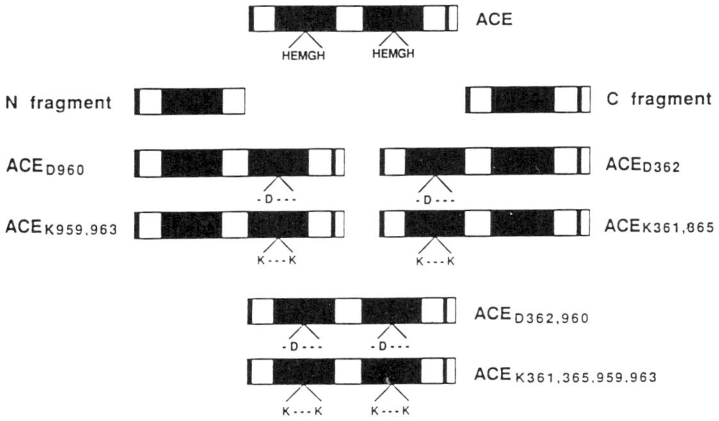

The wild-type ACE (ACE) contains the signal peptide (left black line), the two large homologous domains (black box), and the transmembrane domain (right black line) near the C-terminus. The sequence containing the two putative zinc-binding histidines and the putative catalytic glutamic acid is shown under each domain. The truncated mutant, N fragment, contains the signal peptide and the N-domain. The truncated mutant, C fragment, contains the signal peptide, the C-domain and the transmembrane domain. All full-length mutants contain the mutated sequence domain. Unchanged amino acids are represented by dashes. The position(s) of mutated amino acids is indicated in the mutant name. From Wei et al. *1991a with permission. HEMGH indicates the amino acid residues of the active site.*

value for AI was three times higher than for the *N*-domain active site. Both the *N*- and *C*-domains had an absolute zinc requirement for activity (Table 1). In all these experiments, both domains appeared to function independently since the activity of the wild-type enzyme was equal to the sum of the activities of the *N*- and *C*-domains (Wei *et al.*, 1991). These studies established that ACE possessed two functional catalytic sites, which were both dependent on a zinc cofactor. In agreement with these findings, the zinc content of somatic ACE has been re-investigated and found to be two atoms of zinc per molecule (Ehlers *et al.*, 1991; Williams *et al.*, 1992).

It is well established that chloride has a marked influence on ACE activity (Shapiro *et al.*, 1983) and it is interesting to note that the catalytic activities of the two domains were affected differently by chloride (Wei *et al.*, 1991). For AI hydrolysis, the *N*-active site exhibited low levels of activity in the absence of chloride and optimal activity at a concentration of 10 mM chloride. In contrast, the *C*-active site was essentially inactive in the absence of chloride and required a higher chloride concentration for optimal activity

Table 1. Kinetic parameters and chloride dependence for hydrolysis of Hip-His-Leu and AI

Enzyme	Hip-His-Leu*			Angiotensin I†		
	K_m (μM)	$k_{cat.}$ (s^{-1})	Optimal [Cl^-]‡ (mM)	K_m (μM)	$k_{cat.}$ (s^{-1})	Optimal [Cl^-]‡ (mM)
Wild-type	1540	408	800	16	40	30
N-domain	2000	40	10	16	12	10
C-domain	2000	360	800	16	35	30

Assay conditions for hydrolysis of Hip-His-Leu were: 100 mM potassium phosphate, pH 8.3; 300 mM NaCl; 10 μM $ZnSO_4$, 37°C. †Assay conditions for AI hydrolysis were: 50 mM Hepes, pH 7.5; 50 mM NaCl; 1 μM $ZnSO_4$, 37°C. ‡Optimal [Cl^-]: optimal chloride concentration for hydrolysis of Hip-His-Leu and AI. Adapted from Wei et al., 1991a

($\geq$ 30 mM) (Table 2). BK was hydrolysed efficiently by both active sites with similar $k_{cat.}$ values in the presence of NaCl but again different chloride activation profiles. Therefore, the two domains both act as dipeptidyl carboxypeptidases towards AI and BK but display different catalytic constants for these substrates and different chloride activation profiles.

The interaction of the two active sites of ACE with competitive ACE inhibitors has recently been investigated (Wei *et al.*, 1992). Both the *N*- and *C*-domains contain a high-affinity binding site for [^{3}H]trandolaprilat, a potent ACE inhibitor. Chloride stabilizes the enzyme–inhibitor complex and slows the dissociation rate, an effect which is more marked for the *C*-domain than for the *N*-domain. In addition, various ACE inhibitors show a difference in their potency towards the *N*- and the *C*-domains when chloride concentration varies thus providing further evidence for the existence of structural differences between the two active sites of ACE (Table 2).

The binding of different ACE inhibitors was studied, using purified lung and testis rat ACE (Perich *et al.*, 1992). Binding parameters for the two binding sites of lung ACE were deduced from displacement of the ACE inhibitor, ^{125}I-Ro 31-8472 with either Ro 31-8472 or 351A, a *p*-hydroxybenzamidine analogue of lisinopril. Whereas analogous K_d values were found for the two binding sites of lung ACE with Ro 31-8472, contrasting K_d values were found for 351A, the K_d for the *N*-domain being more than 100-times higher than for the *C*-domain. As the main structural difference between these two inhibitors concerns the length of their side chains, it was suggested that the side chain may encounter steric hindrance at a deeply recessed active site in the *N*-terminal domain of ACE. Thus, further studies with various ACE inhibitors may help to elucidate the structure/

Table 2. Comparison of the inhibitor binding properties of the *N*- and *C*-domains of trandolaprilat in the presence of 300 mM NaCl

Enzyme	K_D* ($\times 10^{10}$ M)	K_{+1}† ($\times 10^5$ M^{-1} s^{-1})	K_{-1}† ($\times 10^5$ s^{-1})	K_i‡ ($\times 10^{10}$ M)	Relative inhibitory potency§
N-domain	2.9	8.6	24	3.1	T>C>E>L
C-domain	0.3	4.3	1.4	0.3	T>L>E>C

**K_D was determined by equilibrium dialysis in 100 mM Hepes, pH 7. †H_{+1} and K_{-1} are the rat constants of [^{3}H]trandolaprilat binding. ‡K_i is the inhibitory constant using Hip-His-Leu as a substrate, as in Table 1. §Relative inhibitory potency of T (trandolapril), C (captopril), E (enalapril) and L (lisinopril) on the binding properties of [^{3}H]trandolaprilat to various ACE molecules. Adapted from Wei et al., 1992.*

function of these two active sites and possibly lead to the design of specific inhibitors for each active site.

Substrate specificity of each active site

ACE is the only zinc metallopeptidase known to date to possess two active sites. Some other enzymes contain two active sites and are truly bifunctional since they exhibit distinct functions, for example, the brush-border hydrolases sucrase–isomaltase (Hunziker *et al.*, 1986) and lactase–phlorizin hydrolase (Mantei *et al.*, 1988). In these cases, a duplication of an ancestral gene occurred during evolution forming two active sites. Subsequently a divergent evolution resulted in a different catalytic activity for each domain. In the case of ACE, the two active sites display highly similar kinetic parameters for the most well-known natural substrates. However, studies with natural and synthetic peptide substrates, chloride and competitive inhibitors, as discussed above, suggest structural and functional differences between subsites raising the possibility of different substrate specificities of the *N*- and *C*-active sites.

The demonstration of another natural substrate with more favourable kinetic parameters for the *N*-domain compared with the *C*-domain would allow us to define ACE as a bifunctional enzyme. Ehlers *et al.* (1991) reported that the somatic enzyme hydrolysed LH-RH and cleaved the Trp_3–Ser_4 bond of this peptide more efficiently compared with the testicular enzyme, suggesting that the *N*-domain may be preferentially involved in this cleavage. This would be a strong argument for a distinct specificity of the *N*-domain, although it is not firmly established that ACE inactivates this peptide in the pituitary. Our own studies also indicate that LH-RH is hydrolysed by both domains, releasing a *C*-terminal tripeptide. However, this primary cleavage between Trp_3 and Ser_4 is performed more efficiently by the *N*-domain (Jaspard *et al.*, 1993). These results indicate the occurrence of functional and structural differences between the two active sites, despite the high degree of sequence identity. However, a truly specific substrate for each active site has not yet been identified.

In addition to the possible bifunctionality of ACE, alternative hypotheses can be proposed to account for the specific properties of each active site: first, the *N*-domain of ACE could be more efficient under particular physico-chemical conditions, for example, in the intracellular medium where the chloride concentration is low; secondly, a lower selection pressure on the *N*-domain may have been responsible for mutational changes of key residues involved in the catalytic mechanism.

Mechanisms of anchorage and solubilization

Conventional studies on the purification of ACE revealed that the enzyme was bound to the membrane of the cell and that a solubilization step, either by proteolytic cleavage with trypsin or by detergent, was required to purify the enzyme (Hooper *et al.*, 1987). Immunohistochemical studies localized the enzyme to the surface of the cell on various epithelia and on endothelium (Ryan *et al.*, 1975; Defendini *et al.*, 1983). In addition to the membrane-bound form of the enzyme, ACE is also found as a soluble form in plasma, semen, cerebrospinal fluid and other body fluids. The relationship between the membrane-bound and the soluble forms of the enzyme is not clear, although their molecular sizes are similar, indicating that the solubilization process does not greatly affect the size of the enzyme.

Molecular mechanism of ACE anchorage

The mode of anchorage of ACE to the plasma membrane and the mechanism of its solubilization was first studied by Hooper *et al.* (1987) who demonstrated that ACE was inserted via its *C*-terminus, as the trypsin-solubilized (hydrophilic) and the detergent-solubilized (amphipathic) enzymes had identical *N*-terminal sequences.

The study by Hooper *et al.* was further supported by the determination of the primary structure of ACE which revealed the presence of a highly hydrophobic sequence located at the *C*-terminal extremity of the molecule (residues 1231–1247). This sequence almost certainly constitutes the membrane anchor since this is the only region, with the exception of the cleaved *N*-terminal signal sequence, of high hydrophobicity capable of spanning the lipid bilayer.

Further evidence for *C*-terminal anchorage was obtained from site-directed mutagenesis studies on human endothelial ACE (Wei *et al.*, 1991*b*) and human testicular ACE (Ehlers *et al.*, 1991). Transfection of CHO cells with the wild-type recombinant ACE cDNA resulted in 95% of newly synthesized enzyme bound to the cell membrane whereas 5% was secreted into the cell supernatant. However, when a stop codon was introduced into the cDNA on the 5′ side immediately before the sequence encoding the putative membrane anchor, ACE was only recovered in the cell supernatant. Among zinc ectopeptidases, ACE is the only enzyme to be inserted via its *C*-terminus in contrast to endopeptidase-24.11 and aminopeptidase N which retain the membrane-inserted signal peptide.

Mechanism and physiological significance of ACE secretion

The relationship between the membrane-bound form of ACE and its circulating form has been elucidated. The plasma enzyme is probably derived from vascular endothelial cells and lacks the *C*-terminus of ACE, as it is not recognized by a specific antibody directed against this part of the enzyme (Wei *et al.*, 1991*a*). Therefore, it seems that the secreted, circulating form of ACE is derived from the membrane-bound form by a post-translational event. Such a mechanism has been described for the solubilization of several ectoproteins. A post-translational processing of the membrane-bound form of ACE into a soluble form, probably similar to that occurring in endothelial cells, has been observed during the culture of CHO cells which express recombinant ACE (Wei *et al.*, 1991*b*; Ehlers *et al.*, 1991). The enzymic cleavage, which generates soluble ACE from the membrane-bound form, occurs intracellularly or at the plasma membrane. An endogenous EDTA-sensitive protease, identified in lung and kidney, was able to release a soluble form of the enzyme from pig kidney membranes (Hooper *et al.*, 1987). The cellular localization, the identification of this putative proteolytic enzyme, and the exact site(s) of ACE cleavage are unknown.

The functional significance of endothelial ACE versus circulating plasma ACE was first investigated by Ng and Vane (1967) who showed that most of the conversion of AI into AII occurred in the lung. It is interesting to note that many investigators failed to correlate the antihypertensive effect of ACE inhibitors with the degree of plasma ACE inhibition (Waeber *et al.*, 1989). Alterations in the tissue levels of ACE, as opposed to plasma ACE, may be of importance for vasoactive peptide metabolism.

Conclusion

The molecular cloning of the somatic and germinal forms of ACE, and structure–function studies of the protein have radically modified our knowledge of the structure, biosynthesis and expression of ACE. However, important questions still remain. Among these questions are the possible substrate specificity of each active site, the nature and the localization of the enzyme involved in the solubilization of ACE. The discovery of a non-duplicated ACE gene in *Drosophila* emphasizes the interest in finding out whether a physiological function is fulfilled by the appearance of the gene duplication. The precise regulatory mechanisms allowing a cell- and stage-specific transcription of the germinal ACE as well as its function are also unknown.

It is clear that ACE is not only an important enzyme for blood pressure regulation and neuropeptide metabolism but also represents a fascinating model from a biochemical and evolutionary point of view.

References

Arregui, A. and Iversen, L. L. (1978) Angiotensin-converting enzyme: presence of high activity in choroid plexus of mammalian brain. Eur. J. Pharmacol. **52**, 147–150

Barnes, K., Matsas, R., Hooper, N. M., Turner, A. J. and Kenny, A. J. (1988) Endopeptidase 24.11 is striosomally ordered in pig brain and, in contrast to aminopeptidase N and dipeptidyl dipeptidase A, angiotensin converting enzyme, is a marker for a set of striatal efferent fibers. Neuroscience **27**, 799–817

Bernstein, K. E., Martin, B. M., Edwards, A. S. and Bernstein, E. A. (1989) Mouse angiotensin I-converting enzyme is a protein composed of two homologous domains. J. Biol. Chem. **264**, 11945–11951

Brentjens, J. R., Matsuo, S., Andres, G. A., Caldwell, P. R. B. and Zamboni, L. (1986) Gametes contain angiotensin converting enzyme (kininase II). Experientia **42**, 399–402

Bruneval, P., Hinglais, N., Albenc-Gelas, F., Tricottet, V., Corvol, P., Ménard, J., Camilleri, J. P. and Bariety, J. (1986) Angiotensin I converting enzyme in human intestine and kidney. Ultrastructural immunohistochemical localization. Histochemistry **85**, 73–80

Bull, H. G., Thornberry, N. A. and Cordes, E. S. (1985) Purification of angiotensin-converting enzyme from rabbit lung and human plasma by affinity chromatography. J. Biol. Chem. **260**, 2963–2972

Bünning, P. and Riordan, J. F. (1983) Activation of angiotensin converting enzyme by monovalent anions. Biochemistry **22**, 110–116

Bünning, P. and Riordan, J. F. (1985) The functional role of zinc in angiotensin converting enzyme. Implication for the enzyme mechanism. J. Inorg. Biochem. **24**, 183–198

Bünning, P., Holmquist, B. and Riordan, J. F. (1978) Functional residues of the active site of angiotensin-converting enzyme. Biochem. Biophys. Res. Commun. **83**, 1442–1449

Bünning, P., Kleemann, S. G. and Riordan, J. F. (1990) Essential residues in angiotensin converting enzyme: modification with 1-fluoro-2,4-dinitrobenzene. Biochemistry **29**, 10488–10492

Chen, Y. N.-P. and Riordan, J. F. (1990) Identification of essential tyrosine and lysine residues in angiotensin converting enzyme: Evidence for a single active site. Biochemistry **29**, 10493–10498

Chen, Y. N.-P., Ehlers, M. R. W. and Riordan, J. F. (1992) The functional role of tyrosine-200 in human testis angiotensin-converting enzyme. Biochem. Biophys. Res. Commun. **184**, 306–309

Cornell, M. J., Coates, D. and Isaac, R. E. (1993) Characteristics of putative *Drosophila* angiotensin converting enzyme cDNA clones. Biochem. Soc. Trans. **21**, 243S

Costerousse, O., Allegrini, J., Lopez, M. and Alhenc-Gelas, F. (1993) Angiotensin I-converting enzyme in human circulating mononuclear cells. Genetic polymorphism of expression in T-lymphocytes. Biochem. J. **290**, 33–40

Cumin, F., Vallaud, V., Corvol, P. and Alhenc-Gelas, F. (1989) Evidence for a single active site in the human angiotensin I-converting enzyme from inhibitor binding studies with [^{3}H]RU44403: Role of chloride. Biochem. Biophys. Res. Commun. **163**, 718–725

Cushman, D. W. and Cheung, H. S. (1971) Concentrations of angiotensin-converting enzyme in tissues of rat. Biochim. Biophys. Acta **250**, 261–265

Das, M. and Soffer, R. L. (1975) Pulmonary angiotensin-converting enzyme: structural and catalytic properties. J. Biol. Chem. **250**, 6762–6768

Das, M., Hartley, J. L. and Soffer, R. L. (1977) Serum angiotensin-converting enzyme. Isolation and relationship to the pulmonary enzyme. J. Biol. Chem. **252**, 1316–1319

Defendini, R., Zimmerman, E. A., Weare, J. A., Alhenc-Gelas, F. and Erdös, E. G. (1983) Angiotensin-converting enzyme in epithelial and neuroepithelial cells. Neuroendocrinology **37**, 32–40

Deutch, C. E. and Soffer, R. L. (1978) *Escherichia coli* mutants defective in dipeptidyl carboxypeptidase. Proc. Natl. Acad. Sci. U.S.A. **75**, 5998–6001

Devault, A., Lazure, C., Nault, C., Le Moual, H., Seidah, N. G., Chrétien, M., Kahn, P., Powell, J., Mallet, J., Beaumont, A., *et al.* (1987) Amino acid sequence of rabbit kidney neutral endopeptidase 24.11 (enkephalinase) deduced from a complementary DNA. EMBO J. **6**, 1317–1322

Dubreuil, P., Fulcrand, P., Rodriguez, M., Fulcrand, H., Laur, J. and Martinez, J. (1989) Novel activity of angiotensin-converting enzyme hydrolysis of cholecystokinin and gastrin analogues with release of the amidated carboxy-terminal dipeptide. Biochem. J. **262**, 125–130

Ehlers, M. R. W. and Riordan, J. F. (1991) Angiotensin-converting enzyme: Zinc- and inhibitor-binding stoichiometries of the somatic and testis isozymes. Biochemistry **30**, 7118–7126

Ehlers, M. R. W., Fox, E. A., Strydom, D. J. and Riordan, J. F. (1989) Molecular cloning of human testicular angiotensin-converting enzyme: The testis isozyme is identical to the C-terminal half of endothelial angiotensin-converting enzyme. Proc. Natl. Acad. Sci. U.S.A. **86**, 7741–7745

Ehlers, M. R. W., Chen, Y.-N. P. and Riordan, J. F. (1991) Spontaneous solubilization of membrane-bound human testis angiotensin-converting enzyme expressed in chinese hamster ovary cells. Proc. Natl. Acad. Sci. U.S.A. **88**, 1009–1013

Ehlers, M. R. W., Chen, Y.-N. P. and Riordan, J. F. (1992) The unique *N*-terminal sequence of testis angiotensin-converting enzyme is heavily *O*-glycosylated and unessential for activity or stability. Biochem. Biophys. Res. Commun. **183**, 199–205

El-Dorry, H. A., Bull, H. G., Iwata, K., Thornberg, N. A., Cordes, E. H. and Soffer, R. L. (1982*a*) Molecular and catalytic properties of rabbit testicular dipeptidyl carboxypeptidase. J. Biol. Chem. **257**, 14128–14133

El-Dorry, H. A., Pickett, C. B., McGregor, J. S. and Soffer, R. L. (1982*b*) Tissue-specific expression of mRNAs for dipeptidyl carboxypeptidase isoenzymes. Proc. Natl. Acad. Sci. U.S.A. **79**, 4295–4297

El-Dorry, H. A., MacGregor, J. S. and Soffer, R. L. (1983) Dipeptidyl carboxypeptidase from seminal fluid resembles the pulmonary rather than the testicular isoenzyme. Biochem. Biophys. Res. Commun. **115**, 1096–1100

Erdös, E. G. (1990) Angiotensin I converting enzyme and the changes in our concepts through the years; Lewis K. Dahl memorial lecture. Hypertension **16**, 363–370

Friedland, J., Setton, C. and Silverstein, E. (1978) Induction of angiotensin-converting enzyme in human monocytes in culture. Biochem. Biophys. Res. Commun. **83**, 843–849

Hamilton, S. and Miller, C. G. (1992) Cloning and nucleotide sequence of the *Salmonella typhimurium* dcp gene encoding dipeptidyl carboxypeptidase. J. Bacteriol. **174**, 1626–1630

Hooper, N. M. and Turner, A. J. (1987) Isolation of two differentially glycosylated forms of peptidyl dipeptidase A (angiotensin converting enzyme) from pig brain: a re-evaluation of their role in neuropeptide metabolism. Biochem. J. **241**, 625–633

Hooper, N. M., Keen, J., Pappin, D. J. C. and Turner, A. J. (1987) Pig kidney angiotensin-converting enzyme. Purification and characterization of amphipathic and hydrophilic forms of the enzyme establishes C-terminal anchorage to the plasma membrane. Biochem. J. **247**, 85–93

Howard, T. E., Shai, S.-Y., Langford, K. G., Martin, B. M. and Bernstein, K. E. (1990) Transcription of testicular angiotensin-converting enzyme (ACE) is initiated within the 12th intron of the somatic ACE gene. Mol. Cell. Biol. **10**, 4294–4302

Howard, T., Balogh, R., Overbeek, P. and Bernstein, K. E. (1993) Sperm-specific expression of angiotensin-converting enzyme (ACE) is mediated by a 91-base-pair promoter containing a CRE-like element. Mol. Cell. Biol. **13**, 1–27

Hubert, C., Houot, A.-M., Corvol, P. and Soubrier, F. (1991) Structure of the angiotensin I-converting enzyme gene. Two alternate promoters correspond to evolutionary steps of a duplicated gene. J. Biol. Chem. **266**, 15377–15383

Hunziker, W., Spiess, M., Semenza, G. and Lodish, H. F. (1986) The sucrase isomaltase complex: Primary structure, membrane orientation and evolution of stalked, intrinsic brush border protein. Cell **46**, 227–234

Jaspard, E., Wei, L. and Alhenc-Gelas, F. (1993) Differences in properties and enzymatic specificities between the two active sites of angiotensin I-converting enzyme. Studies with bradykinin and other natural peptides. J. Biol. Chem., **268**, 9496–9503

Kester, W. R. and Matthews, B. W. (1977) Crystallographic study of the binding of dipeptide inhibitors to thermolysin: Implications for the mechanism of catalysis. Biochemistry **16**, 2506–2516

Kumar, R. S., Kusari, J., Roy, S. N., Soffer, R. L. and Sen, G. C. (1989) Structure of testicular angiotensin-converting enzyme. A segmental mosaic isozyme. J. Biol. Chem. **264**, 16754–16758

Lamango, N. and Isaac, R. E. (1993) Identification of an ACE-like peptidyl dipeptidase activity in the housefly, *Musca domestica*. Biochem. Soc. Trans. **21**, 245S

Langford, K. G., Shai, S.-Y., Howard, T. E., Kovac, M. J., Overbeek, P. A. and Bernstein, K. E. (1991) Transgenic mice demonstrate a testis-specific promoter for angiotensin-converting enzyme. J. Biol. Chem. **266**, 15559–15562

Lattion, A.-L., Soubrier, F., Allegrini, J., Hubert, C., Corvol, P. and Alhenc-Gelas, F. (1989) The testicular transcript of the angiotensin I-converting enzyme encodes for the ancestral, non-duplicated form of the enzyme. FEBS Lett. **252**, 99–104

Mantei, N., Villa, M., Enzler, T., Wacker, H., Boll, W., James, P., Hunziker, W. and Semenza, G. (1988) Complete primary structure of human and rabbit lactase–phlorizin hydrolase: implication for biosynthesis, membrane anchoring and evolution of the enzyme. EMBO J. **7**, 2705–2713

Matsas, R., Kenny, A. J. and Turner, A. J. (1984) The metabolism of neuropeptides. The hydrolysis of peptides including enkephalins, tachykinins and their analogues by endopeptidase 24.11 (E.C. 3.4.24.11). Biochem. J. **223**, 433–440

Nadaud, S., Houot, A. M., Hubert, C., Corvol, P. and Soubrier, F. (1992) Functional study of the germinal angiotensin I-converting enzyme promoter. Biochem. Biophys. Res. Commun. **189**, 134–140

Ng, K. K. F. and Vane, J. R. (1967) Conversion of angiotensin I to an angiotensin II. Nature (London) **216**, 762–766

Oblin, A., Danse, M. J. and Zivkovic, B. (1988) Degradation of substance P by membrane peptidases in the rat substantia nigra: effect of selective inhibitors. Neurosci. Lett. **84**, 91–96

Olsen, J., Cowell, G. M., Konigshofer, E., Danielsen, E. M., Moller, J., Laustsen, L., Hansen, O. C., Welinder, K. G., Engberg, J., Hunziker, W., *et al.* (1988) Complete amino acid sequence of human intestinal aminopeptidase N as deduced from cloned cDNA. FEBS Lett. **238**, 307

Perich, R. B., Jackson, B., Rogerson, F., Mendelsohm, F. A. O., Paxton, D. and Johnston, C. I. (1992) Two ligand sites on angiotensin-converting enzyme: Evidence from radioligand binding studies. Mol. Pharmacol. **42**, 286–293

Ryan, J. W., Ryan, U. S., Schultz, D. R., Whitaker, C., Shung, A. and Dorer, F. E. (1975) Subcellular localization of pulmonary angiotensin converting enzyme (kininase II). Biochem. J. **146**, 497–499

Schweisfurth, H. and Schioberg-Schiegnitz, S. (1984) Assay and biochemical characterization of angiotensin I-converting enzyme in cerebrospinal fluid. Enzyme **32**, 12–19

Shai, S.-Y., Fishel, R. S., Martin, B. M., Berk, B. C. and Bernstein, K. B. (1992) Bovine angiotensin converting enzyme cDNA cloning and regulation. Increased expression during endothelial cell growth arrest. Circ. Res. **70**, 1274–1281

Shapiro, R., Holmquist, B. and Riordan, J. F. (1983) Anion activation of angiotensin-converting enzyme: dependence on nature of substrate. Biochemistry **22**, 3850–3857

Skidgel, R. A. and Erdös, E. G. (1987) Angiotensin I converting enzyme and its role in neuropeptide metabolism, in Neuropeptides and Their Peptidases (Turner, A. J., ed.), pp. 165–182, Ellis Horwood, Chichester

Skidgel, R. A. and Erdös, E. G. (1985) Novel activity of human angiotensin I converting enzyme: release of the NH_2- and COOH-terminal tripeptides from the luteinizing hormone-releasing hormone. Proc. Natl. Acad. Sci. U.S.A. **82**, 1025–1029

Skidgel, R. A., Engelbrecht, S., Johnson, A. R. and Erdös, E. G. (1984) Hydrolysis of substance P and neurotensin by converting enzyme and neutral endopeptidase. Peptides **5**, 769–776

Soubrier, F., Alhenc-Gelas, F., Hubert, C., Allegrini, J., John, M., Tregear, G. and Corvol, P. (1988) Two putative active centers in human angiotensin I-converting enzyme revealed by molecular cloning. Proc. Natl. Acad. Sci. U.S.A. **85**, 9386–9390

Strittmatter, S. M., Lo, M. M. S., Javitch, J. A. and Snyder, S. H. (1984) Autoradiographic visualization of angiotensin-converting enzyme in rat brain with [^{3}H]captopril: localization to a striatonigral pathway. Proc. Natl. Acad. Sci. U.S.A. **81**, 1599–1603

Subissi, A., Guelfi, M. and Criscuoli, M. (1990) Angiotensin converting enzyme inhibitors potentiate the bronchoconstriction induced by substance P in the guinea-pig. Br. J. Pharmacol. **100**, 502–506

Testut, P., Soubrier, F., Corvol, P. and Hubert, C. (1993) Functional analysis of the somatic angiotensin I converting enzyme promoter. Biochem. J. **293**, 843–848

Thekkumkara, T. J., Livingstone, W., III, Kumar, R. S. and Sen, G. C. (1992) Use of alternative polyadenylation sites for tissue-specific transcription of two angiotensin-converting enzyme mRNAs. Nucleic Acids Res. **20**, 683–687

Turner, A. J., Hryszko, J., Hooper, N. M. and Dowdall, M. J. (1987) Purification and characterization of a peptidyl dipeptidase resembling angiotensin converting enzyme from the electric organ of *Torpedo marmorata*. J. Neurochem. **48**, 910–916

Vallee, B. L. and Auld, D. S. (1990) Zinc coordination, function, and structure of zinc enzymes and other proteins. Biochemistry **29**, 5647–5659

Waeber, B., Nussberger, J., Juillerat, L. and Brunner, H. R. (1989) Angiotensin converting enzyme inhibitor: discrepancy between antihypertensive effect and suppression of enzyme activity. J. Cardiovasc. Pharmacol. **14**, S53

Weare, J. A., Gafford, J. T., Lu, H. S. and Erdös, E. G. (1982) Purification of human kidney angiotensin I-converting enzyme using reverse-immunoadsorption chromatography. Anal. Biochem. **123**, 310–319

Wei, L., Alhenc-Gelas, F., Corvol, P. and Clauser, E. (1991*a*) The two homologous domains of the human angiotensin I-converting enzyme are both catalytically active. J. Biol. Chem. **266**, 9002–9008

Wei, L., Alhenc-Gelas, F., Soubrier, F., Michaud, A., Corvol, P. and Clauser, E. (1991*b*) Expression and characterization of recombinant human angiotensin I-converting enzyme. Evidence for a C-terminal transmembrane anchor and for a proteolytic processing of the secreted recombinant and plasma enzymes. J. Biol. Chem. **266**, 5540–5546

Wei, L., Clauser, E., Alhenc-Gelas, F. and Corvol, P. (1992) The two homologous domains of human angiotensin I-converting enzyme interact differently with competitive inhibitors. J. Biol. Chem. **267**, 13398–13405

Weinberg, K. S., Douglas, W. H. J., MacNamee, D. R., Lanzillo, J. J. and Fanburg, B. L. (1982) Angiotensin I-converting enzyme localization on cultured fibroblasts by immunofluorescence. *In Vitro* **18**, 400–406

Williams, T. A., Hooper, N. M. and Turner, A. J. (1991) Characterization of neuronal and endothelial forms of angiotensin converting enzyme in pig brain. J. Neurochem. **57**, 193–199

Williams, T. A., Barnes, K., Kenny, A. J., Turner, A. J. and Hooper, N. M. (1992) A comparison of the zinc content and substrate specificity of the endothelial and testicular forms of porcine angiotensin converting enzyme, and the isolation of isoenzyme specific antisera. Biochem. J. **288**, 878–881

Yasui, T., Alhenc-Gelas, F., Corvol, P. and Ménard, J. (1984) Angiotensin I-converting enzyme in amniotic fluid. J. Lab. Clin. Med. **104**, 741–751

Yokosawa, H., Endo, S., Ogura, Y. and Ishii, S. (1983) A new feature of angiotensin-converting enzyme in the brain: hydrolysis of substance P. Biochem. Biophys. Res. Commun. **116**, 735–742

Yokoyama, M., Takada, Y., Iwata, H., Ochi, K., Takeuchi, M., Hiwada, K. and Kokubu, T. (1982) Correlation between angiotensin-converting enzyme activity and histologic patterns in benign prostatic hypertrophy tissue. J. Urol. **127**, 368–370

Structure and expression of G-protein-linked neuropeptide receptors

Craig R. Hurrell, Dan Donnelly, Nirmala Bhogal, Alan Cox and John B. C. Findlay

The Department of Biochemistry and Molecular Biology, University of Leeds, Leeds LS2 9JT, U.K.

Introduction

Multicellular organisms have evolved an elaborate cell-to-cell communication network based on an enormous variety of signalling molecules. All hormones and neurotransmitters exert their influence on cells by interacting with target receptors that specifically recognize a particular chemical signal. Hydrophobic effectors such as steroid and thyroid hormones can pass through the plasma membrane to interact with intracellular proteins, although the mechanism of transport is still unclear. The majority of signalling molecules, however, exert their action from the external surface of the cell. These molecules elicit intracellular changes via specific membrane-spanning receptors. Binding of the correct ligand to a receptor stimulates a conformational change in the protein which initiates an intracellular pathway. Receptors involved in transmembrane signalling fall into three functional categories: (1) receptors with intrinsic enzyme activity (e.g. the insulin receptor, a tyrosine kinase); (2) ligand-gated ion channels (e.g. the nicotinic acetylcholine receptor); and (3) receptors that couple to G-proteins (e.g. the muscarinic acetylcholine receptor). Most neuropeptide receptors belong to the family of G-protein-coupled receptors (GPCRs).

GPCRs are a diverse super-family of integral membrane proteins that are believed to share many structural (Attwood and Findlay, 1993*a*) and functional properties (reviewed in Savarese and Fraser, 1992). They are involved in diverse processes ranging from visual and olfactory perception to neurotransmission, and have been identified in organisms as evolutionary divergent as yeast and man. Similarly, they bind a variety of unrelated ligands, ranging from small, charged molecules (e.g. acetylcholine) to whole proteins (e.g. parathyroid hormone-like protein). On binding a signal molecule, the receptor transmits the information across the bilayer through conformational changes that activate particular G-proteins. The latter proteins trigger a specific cascade of intracellular events that lead to the amplified activation of enzymes or transport systems. Thus, there are a number of points at which specificity can be introduced.

Cloning of receptors

Neuropeptide receptors

Most of our current knowledge regarding neuropeptide receptors is fairly recent, but the characterization of the first neuropeptide began in 1931 when von Euler and Gaddum (1931) described the biological activity of a novel component in a crude tissue extract. They proposed that the activity was due to a peptide and named the active ingredient 'substance P'. Attention was focused on substance P in the 1950s when it was suggested that the molecule was a neurotransmitter of primary sensory neurones, but this undecapeptide was not isolated and sequenced until 1971 (Chang *et al.*, 1971). Cloning revealed that the peptide was part of a longer gene product that also contained the gene for substance K (Nawa *et al.*, 1983). Finally, in 1989, a cDNA clone for the rat substance P receptor was reported (Yokota *et al.*, 1989).

There are, at present, over 40 published sequences for more than 20 different types of neuropeptide receptor (reviewed in Burbach and Meijer, 1992; Wank *et al.*, 1992*a*, *b*; Christians *et al.*, 1993). The receptors cloned to date fall loosely into three categories. Most are of the 'neurokinin-type'; they are GPCRs and bind small peptide ligands (e.g. angiotensin II, bradykinin and substance P). The secretin, calcitonin, vasoactive intestinal peptide and parathyroid hormone receptors share little sequence similarity with neurokinin-type receptors, but share a high degree of similarity with one another (Attwood and Findlay, 1993*b*). These all bind large peptide ligands and are classed as 'secretin-type' receptors. The third category comprises the *mas*, *mrg* and rat thoracin aorta gene products. These all share significant sequence similarity with one another, but vary considerably from other peptide receptors. It has been proposed that the *mas* oncogene encodes an angiotensin receptor (Jackson *et al.*, 1988), but this group of proteins still remains something of a mystery.

Receptor subtypes

The characterization of neuropeptide receptors has been complicated by the existence of multiple receptor subtypes, as manifested by the angiotensin, bombesin, bradykinin, cholecystokinin, endothelin, tachykinin, somatostatin and neuropeptide Y receptors. Receptor subtypes are traditionally distinguished pharmacologically but, more recently, cloning approaches have confirmed the pharmacological evidence (e.g. the gastrin-releasing peptide and neuromedin B bombesin-like receptors in man; Corjay *et al.*, 1991). The situation appears most complicated for the mammalian tachykinin receptors. Substance P, substance K (neurokinin A) and neuromedin K (neurokinin B) are the preferred natural ligands for the neurokinin type 1, 2 and 3 receptors, respectively, but each ligand can activate all three receptor types if present in a high enough concentration. Furthermore, isoforms of each receptor may exist. Two versions of the human neurokinin-1 receptor have been cloned and expressed (Fong *et al.*, 1992*a*). They differ only in the length of their C-termini, which may reflect coupling to different transduction systems. There is also some evidence for heterogeneity amongst neurokinin-2 receptors. Selective ligands can show markedly different affinities for receptors from different tissues or different species (reviewed in Buck *et al.*, 1991) and there has been a preliminary report describing neurokinin-2 receptor subtypes in rat bladder (Carstairs *et al.*, 1993). However, two putative neurokinin-2 receptor subtypes could only be distinguished by Van Giersbergen *et al.* (1991) using antagonists, not agonists. Only one neurokinin-2 receptor has been cloned to date and it is possible that different behaviour towards ligands can arise from the environment of the receptor. Cell-type specific receptor glycosylation may be another mechanism by which superficial subtypes of the same protein may be generated.

The cloning advantage

Human receptors are of particular interest to the pharmaceuticals industry since species variability, especially to synthetic agents, is common. The problems associated with developing therapeutic agents for humans have pushed research towards molecular cloning and expression. Cultured cells can often accurately mimic the human cell in producing a correctly folded and modified protein. Moreover, the use of cloned receptors offers additional advantages; expression levels can be many times greater than in nature with no background from endogenous receptors and the manipulation of receptor genes *in vitro* permits novel experimental approaches. Cloning a receptor also provides valuable sequence data which can be used to generate three-dimensional (3-D) models of the protein, with a view to designing experiments to examine its structural and functional properties.

Cloning strategies

Protein purification

There exist a number of different strategies to clone membrane receptors. Protein purification is the traditional starting point. The success of this approach relies upon having a rich source of receptor and means to purify and monitor the purification process, e.g. a high-affinity ligand. Unfortunately, neuropeptide receptors are only found in small numbers embedded in the plasma membrane of certain tissues, so isolating sufficient material for protein sequencing and antibody production is difficult.

The purification of membrane proteins uses the same basic techniques as employed for water-soluble proteins, i.e. gel filtration, ion-exchange chromatography, reversed-phase chromatography and affinity chromatography. However, the amphipathic nature of membrane proteins requires the presence of detergents and often makes the use of these methods problematic. Solubilization with detergents can dramatically reduce ligand binding, so hindering the most effective purification method, affinity chromatography.

Fortunately, a small amount of protein sequence is often sufficient to design oligonucleotides suitable for screening a gene library by the polymerase chain reaction (PCR) or hybridization. This approach has been successful for several peptide receptors, including those for cholecystokinin type A (Wank *et al.*, 1992*a*), endothelin type B (Nakamuta *et al.*, 1991; Saito *et al.*, 1991) and gastrin-releasing peptide (Battey *et al.*, 1991).

Functional expression in *Xenopus* oocytes

When the protein is difficult to purify, is present in very small amounts or is from an unobtainable human source, expression cloning is an alternative approach. Oocytes from the frog *Xenopus laevis* are large cells (approx. 1 mm in diameter when fully grown) that can efficiently translate mRNAs injected into their cytoplasm or transcribe a variety of exogenous DNAs introduced into the nucleus. They are capable of expressing a wide variety of functional receptors and channels which makes them a powerful assay system for isolating functional cDNA clones.

Masu *et al.* (1987) first exploited the oocyte system to clone the bovine substance K receptor — the first receptor for a neural peptide to be cloned. They constructed a bovine stomach cDNA library in a phage lambda vector in which the cDNA was flanked by the promoter for SP6 RNA polymerase. Poly(A^+) mRNA synthesized from the library *in vitro* was injected into the cytoplasm of *Xenopus* oocytes and the electrophysiological response resulting from exposure to substance K was measured using conventional two-electrode voltage-clamping. This electrophysiological response is thought to be due to the opening of calcium-gated chloride channels in the plasma membrane of the oocyte. The inositol (IP_3) second messenger system leads to mobilization of this calcium from intracellular stores, and so it is assumed that the neurokinin-2 receptor functions through this pathway. cDNA mixtures that produced response-evoking mRNA were repeatedly fractionated and re-tested until a single clone was obtained. This method isolates functional clones only and is extremely sensitive; a single clone in a mixture of approx. 3×10^5 cDNA clones produced a clear response to substance K.

Several neuropeptide receptors activate the IP_3 second messenger system, thereby leading to elevated intracellular calcium levels and hence activation of the chloride channel. This has allowed the bradkykinin type 2 (McEachern *et al.*, 1991), endothelin type A (Arai *et al.*, 1990), gastrin-releasing peptide (Spindel *et al.*, 1990), neuromedin K (Shigemoto *et al.*, 1990), neurotensin (Tanaka *et al.*, 1990), substance P (Yokota *et al.*, 1989) and thyrotropin-releasing hormone (Straub *et al.*, 1990) receptors to be cloned. Many other receptors have been expressed in oocytes. These have been shown to be functionally active by either standard electrophysiology, ligand-induced production of radioactive inositol phosphates in oocytes loaded with tritiated myo-inositol (McIntosh and Catt, 1987), measurement of $^{45}Ca^{2+}$ following ligand-challenge (Williams *et al.*, 1988), or the detection of light emission from oocytes containing aequorin, a jellyfish protein that is luminescent in the presence of calcium ions (Sandberg *et al.*, 1988).

Functional expression in COS cells

Expression cloning in COS (CV-1, Origin of SV40) cells has been used to isolate the angiotensin type 1 (Murphy *et al.*, 1991; Sasaki *et al.*, 1991), calcitonin (Lin *et al.*, 1991 *a*) and endothelin (Lin *et al.*, 1991 *b*; Sakurai *et al.*, 1990) receptors. In a typical procedure, pools of clones in a suitable expression vector are transfected into COS cells, which are then incubated with radiolabelled ligand. Cells expressing the receptor can be detected by autoradiography or ligand-binding assay. Pools of clones giving a positive signal can be subdivided and screened until a single clone is isolated. A high-affinity ligand is essential to the success of this procedure, which can be lengthy and laborious. However, it has the advantage of detecting receptors that do not mobilize intracellular calcium.

PCR

The use of PCR is a powerful cloning strategy. The PCR approach is effective when looking for known receptors or variants of known receptors, but it is most powerful when used to identify putative new receptors. PCR can exploit conserved sequences in the transmembrane segments of GPCRs. For example, degenerate oligonucleotide primers, designed towards consensus sequences of the third and sixth transmembrane segments, have been used to isolate four new GPCRs (Libert *et al.*, 1989). However, the ligands for these receptors cannot be deduced from the sequence alone and it can be far more difficult to assign a function to a newly cloned receptor than to clone it initially.

Screening a library by PCR usually produces products corresponding to only part of a gene. These fragments may then be used to isolate full-length clones from a gene library by hybridization, but this is not the only approach available. Using one primer directed towards the gene, and the other primer towards the cloning vector, it is possible to amplify both ends of a cDNA. DNA sequences on either side of the coding region can then be used to isolate a full-length clone in a single PCR, or it may be possible to splice together overlapping 5′- and 3′-terminal regions to produce the full-length cDNA. In the case of a previously cloned receptor, PCR can be used to generate a full-length clone with relative ease in a very short time.

The PCR approach is a rapidly developing technique, but has suffered from several shortcomings. The *Taq* polymerase is susceptible to producing DNA errors, is limited to amplifying DNA fragments no longer than a few thousand bases and can often produce false positives. This enzyme can now be substituted for one of many known thermostable polymerases, the most notable of which is *Pfu*, which can replicate DNA with high fidelity but does not overcome the other problems. PCR should be very useful in isolating rare neuropeptide receptors, especially those that do not couple to the endogenous IP_3 second messenger system in *Xenopus* oocytes.

Gene libraries

There are two basic types of gene library, each of which offers particular advantages: (1) genomic libraries represent the entire chromosome, including any introns that are present in genes. (2) cDNA libraries represent the various populations of poly(A^+) mRNA found in the tissue from which they were constructed and do not contain introns or 'junk DNA'. Many neuropeptide receptors are translated from rare mRNAs. The genes for these receptors will be found more frequently in a genomic library than in a cDNA library. However, although most GPCR genes have no introns, genomic clones of the substance K (Gerard *et al.*, 1990), substance P (Gerard *et al.*, 1991) and neuromedin K (Buell *et al.*, 1992) receptors were found to contain four introns and were therefore not suitable for use in heterologous expression. It is interesting to note that all three tachykinin receptors have introns in identical positions, but these positions differ from those in bovine rhodopsin (Nathans and Hogness, 1983). All four introns occur at the boundary region between a loop and a transmembrane segment, the first of which occurs in the third transmembrane segment in the Glu-Arg-Tyr sequence (Asp-Arg-His in substance P receptors), one of the most highly conserved sequences in GPCRs (Fig. 1).

Overview

At the latest count, there were nearly 400 GPCRs in the database. A large part of this rapid progress is based on the conservation of primary structure within particular regions of the transmembrane segments and, as in the case of neuropeptide receptors, on the development of expression cloning.

The discovery of new receptors and receptor subtypes highlights the variety of receptors that are still to be discovered. The development of new techniques is likely to aid the cloning of species that have proved difficult to identify using present technology. One technique under development makes use of activated cell sorting. Transfected cells are loaded with indo-1-acetoxy-methylester, which is hydrolysed in the cell to yield a membrane-impermeable dye. Exposure of transfected cells to ligand mobilizes intracellular calcium, which is bound by the dye. The chelated dye has an altered fluorescence emission maxima which is utilized by an activated cell sorter. Further cycles of this procedure will lead to the isolation of clonal cell lines expressing receptor-encoding DNA. Like the oocyte system, this strategy relies on the receptor coupling to an endogenous IP_3 second messenger system. There is some evidence, however, that GPCRs couple to G-proteins with a certain degree of promiscuity and there is the hope that this approach will be generally applicable.

The structure of GPCRs

Seven-helical model of GPCRs

Virtually all GPCRs bear detectable sequence similarity to one another (Attwood *et al.*, 1991) and are thought to possess a common structural prototype. They all contain seven predominantly hydrophobic segments, which may each take up a largely α-helical conformation to collectively form a seven-helix bundle. The 3-D structure of one or more of these receptors would be of great value for the design of novel ligands and for a full understanding of receptor function. Unfortunately, only a very low resolution appreciation (Corless *et al.*, 1982) and a projection map (Schertler *et al.*, 1993) are so far available for any member of the super-family and a high resolution 3-D structure is likely to be many years away. We must therefore rely at present upon the cruder science of protein modelling, and the results of protein chemistry and recombinant DNA technology, to provide us with useful structural information.

The seven-helical model of GPCRs was derived from protein chemistry studies on the vision receptor rhodopsin. Protein sequencing was used to determine residues in the protein subjected to biological or chemical modification (reviewed in Findlay and Pappin, 1986). Antipeptide antibodies were later used with the β-adrenergic receptor to produce a similar topology (Wang *et al.*, 1989). A number of biophysical methods (reviewed in Findlay and Pappin, 1986) suggested that the overall structure and intramembranous cross-sectional area of rhodopsin is similar to that of bacteriorhodopsin — a light-driven proton pump for which a medium resolution 3-D model is available (Henderson and Unwin, 1975; Henderson *et al.*, 1990). It may be possible, therefore, that all GPCRs have a similar chain-fold to bacteriorhodopsin despite the fact that the two proteins share no obvious sequence homology. This assumption has largely been substantiated by the publication of a 9 Å projection map for bovine rhodopsin (Schertler *et al.*, 1993).

Modelling GPCRs

How do we get from a seven-helical topological representation of a GPCR to a 3-D representation based on bacteriorhodopsin? The side-chains that are orientated towards the inside of the helical bundle need to be discriminated from those that face the fatty acid portion of the bilayer. In addition, the precise point at which the segments enter and leave the bilayer needs to be determined so that the residues in these regions can be positioned at the correct 'depth' in the bilayer and hence with respect to one another.

Fig. 1. Protein sequence alignment for the transmembrane segments of GPCRs that bind peptide ligands

	Helix 1	Helix 2
ET1R_BOVIN	AFKYINTVISCTIFIVGMVGNATLLRIIYQNKCM	NALIASLALGDLIYVVIDLPINVFKLLAG
ET1R_HUMAN	AFKYINTVISCTIFIVGMVGNATLLRIIYQNKCM	NALIASLALGDLIYVVIDLPINVFKLLAG
ET1R_RAT	AFKYINTVISCTIFIVGMVGNATLLRIIYQNKCM	NALIASLALGDLIYVVIDLPINVFKLLAG
ETBR_RAT	TFKYINTIVSCLVFVLGIIGNSTLLRIIYKNKCM	NILIASLALGDLLHIIIDIPINAYKLLAG
ET1B_BOVIN	TFKYINTVVSCLVFVLGIIGNSTLLRIIYKNKCM	NILIASLALGDLLHIIIDIPINTYKLLAK
ETBR_HUMAN	TFKYINTVVSCLVFVLGIIGNSTLLRIIYKNKCM	NILIASLALGDLLHIVIDIPINVYKLLAE
GRPR_MOUSE	GFIYVIPAVYGLIIVIGLIGNITLIKIFCTVKSM	NLFISSLALGDLLLLVTCAPVDASKYLAD
A41007	GILYVIPAVYGVIILIGLIGNITLIKIFCTVKSM	NLFISSLALGDLLLLITCAPVDASRYLAD
NMBR_RAT	VIRCVIPSLYLLIITVGLLGNIMLVKIFITNSAM	NIFISNLAAGDLLLLLTCVPVDASRYFFD
B41007	VIRCVIPSLYLLIITVGLLGNIMLVKIFITNSAM	NIFISNLAAGDLLLLLTCVPVDASRYFFD
NTR_RAT	YSKVLVTAIYLALFVVGTVGNSVTAFTLARKKSL	HYHLGSLALSDLLILLLAMPVELYNFIWV
C5AR_HUMAN	VPDILALVIFAVVFLVGVLGNALVVWVTAFEAKR	AIWFLNLAVADFLSCLALPILFTSIVQHH
FMLR_HUMAN	FLDIITYLVFAVTFVLGVLGNGLVIWVAGFRMTH	TISYLNLAVADFCFTSTLPFFMVRKAMGG
FML1_HUMAN	VLWIFSLLVHGVTFVFGVLGNGLVIWVAGFRMTR	TICYLNLALADFSFSAILPFRMVSVAMRE
A36309	FLDIITYLVFAVTFVLGVLGNGLVIWVAGFRMTH	TISYLNLAVADFCFTSTLPFFMVRKAMGG
FML2_HUMAN	VLRILPLVVLGVTFVLGVLGNGLVIWVAGFRMTR	TICYLNLALADFSFTATLPFLIVSMAMGE
IL8B_HUMAN	SLEINKYFVVIYALVFLLSGNSLVMLVILYSRVG	DVYLLNLALADLLFALTLPIWAASKVNGW
IL8A_HUMAN	LNKYVVIIAYALVFLLSLLGNSLVMLVILYSRVG	DVYLLNLALADLLFALTLPIWAASKVNGW
IL8A_RABIT	LNKYVVVVIYALVFLLSLLGNSLVMLVILYSRSN	DVYLLNLAMADLLFALTMPIWAVSKEKGW
NK1R_HUMAN	WQIVLWAAAYTVIVVTSVVGNVVVMWIILAHKRM	NYFLVNLAFAEASMAAFNTVVNFTYAVHN
NK1R_RAT	WQIVLWAAAYTVIVVTSVVGNVVVIWIILAHKRM	NYFLVNLAFAEACMAAFNTVVNFTYAVHN
NK2R_BOVIN	WQLALWTAAYLALVLVAVMGNATVIWIILAHQRM	NYFIVNLALADLCMAAFNAAFNFVYASHN
NK2R_RAT	WQLALWATAYLALVLVAVTGNATVIWIILAHERM	NYFIINLALADLCMAAFNATFNFIYASHN
NK2R_HUMAN	WQLALWAPAYLALVLVAVTGNAIVIWIILAHRRM	NYFIVNLALADLCMAAFNAAFNFVYASHN
NK3R_RAT	WRIALWSLAYGLVVAVAVFGNLIVIWIILAHKRM	NYFLVNLAFSDASVAAFNTLINFIYGLHS
S20435	WRIALWSLAYGVVVAVAVLGNLIVIWIILAHKRM	NYFLVNLAFSDASMAAFNTLVNFIYALHS
NKR1_DROME	WRQVLWSILFGGMVIVATGGNLIVVWIVMTTKRM	NYFIVNLSIADAMVSSLNVTFNYYYLMDS
NKR2_DROME	EQKTIWAIIFGLMMFVAIAGNGIVLWIVTGHRSM	NYFLLNLSIADLLMSSLNCVFNFIFMLNS
NYR_DROME	YFKIIVYMLYIPIFIFALIGNGTVCYIVYSTPRM	NYFIASLAIGDILMSFFCEPSSFISLFIL
NY1R_HUMAN	AMIFTLALAYGAVIILGVSGNLALIIIILKQKEM	NILIVNLSFSDLLVAIMCLPFTFVYTLMD
NY1R_RAT	AVIFTLALAYGAVIILGVSGNLALIIIILKQKEM	NILIVNLSFSDLLVAVMCLPFTFVYTLMD
NY3R_BOVIN	MEGIFLPTVYSIIFLTGIVGNGLVILVMGYQKKL	DKYRLHLSVADLLFVLTLPFWAVDAVANW
AG2R_RAT	YIFVMIPTLYSIIFVVGIFGNSLVVIVIYFYMKL	SVFLLNLALADLCFLLTLPLWAVYTAMEY
AG2R_BOVIN	YIFIMIPTLYSIIFVVGIFGNSLVVIVIYFYMKL	SVFLLNLALADLCFLLTLPLWAVYTAMEY
AG2R_HUMAN	YIFVMIPTLYSIIFVVGIFGNSLVVIVIYFYMKL	SVFLLNLALADLCFLLTLPLWAVYTAMEY
VIP_HUMAN	VLLYTLSFIYIFIFVIGMIANSVVVWVNIQAKTT	HCYILNLAIADLWVVLTIPVWVVSLVQHN
S74767	GSAILISFIYSVVCLVGLCGNSMVIYVILRYAKM	NIYILNLAIADELLMLSVPFLVTSTLLRH
S74769	TSNAVLTFIYFVVCIIGLCGNTLVIYVILRYAKM	NIYILNLAIADELFMLGLPFLAMQVALVH
MUSSRI2A	TSNAVLTFIYFVVCVVGLCGNTLVIYVILRYAKM	NIYILNLAIADELFMLGLPFLAMQVALVH
MUSSRI1A	GSAILISFIYSVVCLVGLCGNSMVIYVILRYAKM	NIYILNLAIADELLMLSVPFLVTSTLLRH
S102315	TSNAVLTFIYFVVCVVGLCGNTLVIYVILRYAKM	NIYILNLAIADELFMLGLPFLAMQVALVH
BRB2_RAT	WLNAIQAPFLWVLFLLAALENIFVLSVFCLHKTN	EIYLGNLAGADLILACGLPFWAITIANNF
TRFR_MOUSE	EYQVVTILLVVIICGLGIVGNIMVVLVVMRTKHM	NCYLVSLAVADLMVLVAAGLPNITDSIYG

	Helix 3	Helix 4
ET1R_BOVIN	FLCKLFPFLQKSSVGITVLNLCALSVDRYRAV	IPLVTAIEIVSIWILSFILAIPEAIG
ET1R_HUMAN	FLCKLFPFLQKSSVGITVLNLCALSVDRYRAV	IPLVTAIEIVSIWILSFILAIPEAIG
ET1R_RAT	FLCKLFPFLQKSSVGITVLNLCALSVDRYRAV	IPLITAIEIVSIWILSFILAIPEAIG
ETBR_RAT	EMCKLVPFIQKASVGITVLSLCALSIDRYRAV	VPKWTAVEIVLIWVVSVVLAVPEAIG
ET1B_BOVIN	EMCKLVPFIQKASVGITVLSLCALSIDRYRAV	VPKWTAVEIVLIWVVSVVLAVPEAVG
ETBR_HUMAN	EMCKLVPFIQKASVGITVLSLCALSIDRYRAV	VPKWTAVEIVLIWVVSVVLAVPEAIG
GRPR_MOUSE	IGCKLIPFIQLTSVGVSVFTLTALSADRYKAI	ALMKICLKAALIWIVSMLLAIPEAVF
A41007	IGCKLIPFIQLTSVGVSVFTLTALSADRYKAI	ALMKICLKAAFIWIISMLLAIPEAVF
NMBR_RAT	VGCKLIPVIQLTSVGVSVFTLTALSADRYRAI	ALLRTCVKAMGIWVVSVLLAVPEAVF
B41007	VGCKLIPVIQLTSVGVSVFTLTALSADRYRAI	ALLRTCVKAMGIWVVSVLLAVPEAVF
NTR_RAT	AGCRGYYFLRDACTYATALNVASLSVERYLAI	SRSRTKKFISAIWLASALLAIPMLFT
C5AR_HUMAN	AACSILPSLILLNMYASILLLATISADRFLLV	GAGLAWIACAVAWGLALLLTIPSFLY
FMLR_HUMAN	FLCKFLFTIVDINLFGSVFLIALIALDRCVCV	TVSLAKKVIIGPWVMALLLTLPVIIR
FML1_HUMAN	FLCKLVHVMIDINLFVSVYLITIIALDRCICV	TMSLAKRVMTGLWIFTIVLTLPNFIF
A36309	FLCKFVFTIVDINLFGSVFLIALIALDRCVCV	TVSLAKKVIIGPWVMALLLTLPVIIR
FML2_HUMAN	FLCKLIHIVVDINLFGSVFLIGFIALDRCICV	TVSLAMKVIVGPWILALVLTLPVFLF
IL8B_HUMAN	FLCKVVSLLKEVNFYSGILLLACISVDRYLAI	KRYLVKFICLSIWGLSLLLALPVLLF
IL8A_HUMAN	FLCKVVSLLKEVNFYSGILLLACISVDRYLAI	KRHLVKFVCLGCWGLSMNLSLPFFLF
IL8A_RABIT	PLCKVVSLVKEVNFYSGILLLACISVDRYLAI	KRHLVKFICLGIWALSLILSLPFFLF
NK1R_HUMAN	FYCKFHNFFPIAAVFASIYSMTAVAFDRYMAI	SATATKVVICVIWVLALLLAFPQGYY
NK1R_RAT	FYCKFHNFFPIAALFASIYSMTAVAFDRYMAI	SATATKVVIFVIWVLALLLAFPQGYY
NK2R_BOVIN	AFCYFQNLFPITAMFVSIYSMTAIAADRYMAI	SAPGTRAVIAGIWLVALALAFPQCFY
NK2R_RAT	AFCYFQNLFPITAMFVSIYSMTAIAADRYMAI	SAPSTKAIIAGIWLVALALASPQCFY
NK2R_HUMAN	AFCYFQNLFPITAMFVSIYSMTAIAADRYMAI	SAPSTKAVIAGIWLVALALASPQCFY
NK3R_RAT	NYCRFQNFFPITAVFASIYSMTAIAVDRYMAI	SATATKIVIGSIWILAFLLAFPQCLY
S20435	NYCRFQNFFPITAVFASIYSMTAIAVDRYMAI	SATATKIVIGSIWILAFLLAFPQCLY
NKR1_DROME	FYCKLSQFIAMLSICASVFTLMAISIDRYVAI	SKRCNLAIAAVIWLASTLISCPMMII
NKR2_DROME	IYTCINNFVANVTVSTSVFTLWAISFDRYIAI	SRRKVRIILVLIWALSCVLSAPCLLY
NYR_DROME	ALCHFVNYSQAVSVLVSAYTLVAISIDRYIAI	TKRYATFIIAGVWFIALATALPIPIV
NY1R_HUMAN	AMCKLNPFVQCVSITVSIFSLVLIAVERHQLI	NNRHAYVGIAVIWVLAVASSLPFLIY
NY1R_RAT	TMCKLNPFVQCVSITVSIFSLVLIAVERHQLI	NNRHAYIGITVIWVLAVASSLPFVIY
NY3R_BOVIN	FLCKAVHVIYTVNLYSSVLILAFISLDRYLAI	KLLAEKVVYVGVWLPAVLLTIPDLIF
AG2R_RAT	HLCKIASASVSFNLYASVFLLTCLSIDRYLAI	TMLVAKVTCIIIWLMAGLASLPAVIH
AG2R_BOVIN	YLCKIASASVSFNLYASVFLLTCLSIDRYLAI	TMLVAKVTCIIIWLLAGLASLPTIIH
AG2R_HUMAN	HLCKIASASVSFNLYASVFLLTCLSIDRYLAI	TMLVAKVTCIIIWLMAGLASLPAVIH
VIP_HUMAN	LTCKVTHLIFSINLFSGIFFLTCMSVDRYLSI	KKMVRRVVCILVWLLAFCVSLPDTYY
S74767	LLCRLVLSVDAVNMFTSIYCLTVLSVDRYVAV	RPTVAKVVNLGVWVLSLLVILPIVVF
S74769	AICRVVMTVDGINQFTSIFCLTVMSIDRYLAV	RPRTAKMITMAVWGVSLLVILPIMIY
MUSSRI2A	AICRVVMTVDGINQFTSIFCLTVMSIDRYLAV	RPRTAKMINVAVWCVSLLVILPIMIY
MUSSRI1A	LLCRLVLSVDAVNMFTSIYCLTVLSVDRYVAV	RPTVAKVVNLGVWVLSLLVILPIMIY
S102315	AICRVVMTVDGINQFTSIFCLTVMSIDRYLAV	RPRTAKMINVAVWGVSLLVILPIMIY
BRB2_RAT	VLCRVVNTMIYMNLYSSICFLMLVSIDRYLAL	GVRWAKLYSLVIWSCTLLLSSPMLVF
TRFR_MOUSE	VGCLCITYLQYLGINASSCSITAFTIERYIAI	TFSRAKKIIIFVWAFTSIYCMLWFFL

	Helix 5	Helix 6
ET1R_BOVIN	QDVKDWWLFGFYFCMPLVCTAIFYTLMTCEMLN	QRREVAKTVFCLVVIFALCWFPLHLSRILKKTV
ET1R_HUMAN	QDVKDWWLFGFYFCMPLVCTAIFYTLMTCEMLN	QRREVAKTVFCLVVIFALCWFPLHLSRILKKTV
ET1R_RAT	QDVKDWWLFGFYFCMPLVCTAIFYTLMTCEMLN	QRREVAKTVFCLVVIFALCWFPLHLSRILKKTV
ETBR_RAT	KTAKDWWLFSFYFCLPLAITAIFYTLMTCEMLR	QRREVAKTVFCLVLVFALCWLPLHLSRILKLTL
ET1B_BOVIN	KTAKDWWLFSFYFCLPLAITALFYTLMTCEMLR	QRREVAKTVFCLVLVFALCWLPLHLSRILKLTL
ETBR_HUMAN	KTAKDWWLFSFYFCLPLAITAFFYTLMTCEMLR	QRREVAKTVFCLVLVFALCWLPLHLSRILKLTL
GRPR_MOUSE	PKIHSMASFLVFYVIPLAIISVYYYFIARNLIQ	SRKRLAKTVLVFVGLFAFCWLPNHVIYLYRSYH
A41007	PKIHSMASFLVFYVIPLSIISVYYYFIAKNLIQ	SRKRLAKTVLVFVGLFAFCWLPNHVIYLYRSYH
NMBR_RAT	PKIHSVLIFLVYFLIPLAIISIYYYHIAKTLIK	TRKRLAKIVLVFVGCFIFCWFPNHILYMYRSFN
B41007	PKIHSVLIFLVYFLIPLAIISIYYYHIAKTLIK	TRKRLAKIVLVFVGCFIFCWFPNHILYMYRSFN
NTR_RAT	VKVVIQVNTFMSFLFPMLVISILNTVIANKLTV	ALRHGVLVLRAVVIAFVVCWLPYHVRRLMFCYI
C5AR_HUMAN	ERAVAIVRLVLGFLWPLLTLTICYTFILLRTWS	RSTKTLKVVVAVVASFFIFWLPYQVTGIMMSFL
FMLR_HUMAN	LTVRGIIRFIIGFSAPMSIVAVSYGLIATKIHK	KSSRPLRVLSFVAAAFFLCWSPYQVVALIATVR
FML1_HUMAN	AKVFLILHFIIGFTVPMSIITVCYGIIAAKIHR	KSSRPLRVFAAVVASFFICWFPYELIGILMAVW
A36309	LTVRGIIRFIIGFSAPMSIVAVSYGLIATKIHK	KSSRPLRVLSFVAAAFFLCWSPYQVVALIATVR
FML2_HUMAN	LTARGIIRFVIGFSLPMSIVAICYGLIAAKIHK	KSSRPLRVLTAVVASFFICWFPFQLVALLGTVW
IL8B_HUMAN	RMLLRILPQSFGFIVPLLIMLFCYGFTLRTLFK	QKHRAMRVIFAVVLIFLLCWLPYNLVLLADTLM
IL8A_HUMAN	RMVLRILPHTFGFIVPLFVMLFCYGFTLRTLFK	QKHRAMRVIFAVVLIFLLCWLPYNLVLLADTLM
IL8A_RABIT	RMVLRILPHTFGFILPLLVMLFCYGFTLRTLFQ	QKHRAMRVIFAVVLIFLLCWLPYNLVLLADTLM
NK1R_HUMAN	EKVYHICVTVLIYFLPLLVIGYAYTVVGITLWA	AKRKVVKMMIVVVCTFAICWLPFHIFFLLPYIN
NK1R_RAT	EKAYHICVTVLIYFLPLLVIGYAYTVVGITLWA	AKRKVVKMMIVVVCTFAICWLPFHVFFLLPYIN
NK2R_BOVIN	LLLYHLIVIALIYFLPLVVMFVAYSVIGLTLWR	AKKKFVKTMVLVVVTFAICWLPYHLYFILGTFQ
NK2R_RAT	LLLYHLVVFVLIYFLPLLVMFGAYSVIGLTLWK	AKKKFVKAMVLVVLTFAICWLPYHLYFILGTFQ
NK2R_HUMAN	LLLYHLVVIALIYFLPLAVMFVAYSVIGLTLWR	AKKKFVKTMVLVVLTFAICWLPYHLYFILGSFQ
NK3R_RAT	HFTYHIIVIILVYCFPLLIMGVTYTIVGITLWG	AKRKVVKMMIIVVVTFAICWLPYHVYFILTAIY
S20435	HFTYHIIVIILVYCFPLLIMGITYTIVGITLWG	AKRKVVKMMIIVVMTFAICWLPYHIYFILTAIY
NKR1_DROME	ESLYNILIIILTYFLPIVSMTVTYSRVGIELWG	SKRRVVKMMIVVVLIFAICWLPFHSYFIITSCY
NKR2_DROME	DYAYNLIILVLTTGIPMIVMLICYSLMGRVPGG	SKRKVVRMFIAIVSIFAICWLPYHLFFIYAYHN
NYR_DROME	EYYYTLSLFALQFVVPLGVLIFTYARITIRVWA	SKRKMVKMMLTVVIVFTCCWLPFNILQLLLNDE
NY1R_HUMAN	RLSYTTLLLVLQYFGPLCFIFICYFKIYIRLKR	ETKRINIMLLSIVVAFAVCWLPLTIFNTVFDWN
NY1R_RAT	RLSYTTLLLVLQYFGPLCFIFICYFKIYIRLKR	ETKRINVMLLSIVVAFAVCWLPLTIFNTVFDWN
NY3R_BOVIN	LVVFQFQHIVVGLLLPGIVILSCYCIIISKLSH	QKRKALKTTVILILTFFACWLPYYIGISIDSFI
AG2R_RAT	PIGLGLTKNILGFLFPFLIILTSYTLIWKALKK	RNDDIFRIIMAIVLFFFFSWVPHQIFTFLDVLI
AG2R_BOVIN	PVGLGLTKNILGFLFPFLIILTSYTLIWKTLKK	RKDDIFKIILAIVLFFFFSWVPHQIFTFMDVLI
AG2R_HUMAN	PIGLGLTKNILGFLFPFLIILTSYTLIWKALKK	RNDDIFRIIMAIVLFFFFSWVPHQIFTFLDVLI
VIP_HUMAN	LIGMELVSVVLGFAVPFSIIAVFYFLLARAISA	EKHSSRKIIFSYVVVFLVCWLPYTHVAVLLDIF
S74767	LVGFVLYTFLMGFLLPVGAICLCYVLIIAKMRM	SERKITLMVMMVVMVFVICWMPFYVVQLVNVFA
S74769	YTGFIIYTFILGFLVPLTIICLCYLFIIIKVKS	SEKKVTRMVSIVVAVFIFCWLPFYIFNVSSVSM
MUSSRI2A	YTGFIIYAFILGFLVPLTIICLCYLFIIIKVKS	SEKKVTRMVSIVVAVFIFCWLPFYIFNVSSVSV
MUSSRI1A	LVGFVLYTFLMGFLLPVGAICLCYVLIIAKMRM	SERKITLMVMMVVMVFVICWMPFYVVQLVNVFA
S102315	YTGFIIYAFILGFLVPLTIICLCYLFIIIKVKS	SEKKVTRMVSIVVAVFIFCWLPFYIFNVSSVSV
BRB2_RAT	EVFTNMLLNLVGFLLPLSIITFCTVRIMQVLRN	TEKKATVLVLAVLGLFVLCWFPFQISTFLDTLL
TRFR_MOUSE	YSPIYLMDFGVFYVVPMILATVLYGFIARILFL	SRKQVTKMLAVVVILFALLWMPYRTLVVVNSFL

	Helix 7
ET1R_BOVIN	FLLLMDYIGINLATMNSCINPIALYFVSKKFKNC
ET1R_HUMAN	FLLLMDYIGINLATMNSCINPIALYFVSKKFKNC
ET1R_RAT	FLLLMDYIGINLATMNSCINPIALYFVSKKFKNC
ETBR_RAT	FLLVLDYIGINMASLNSCINPIALYLVSKRFKNC
ET1B_BOVIN	FLLVLDYIGINMASLNSCINPIALYLVSKRFKNC
ETBR_HUMAN	FLLVLDYIGINMASLNSCINPIALYLVSKRFKNC
GRPR_MOUSE	LHFVTSICARLLAFTNSCVNPFALYLLSKSFRKQ
A41007	LHFVTSICARLLAFTNSCVNPFALYLLSKSFRKQ
NMBR_RAT	GHMIVTLVARVLSFGNSCVNPFALYLLSESFRRH
B41007	GHMIVTLVARVLSFGNSCVNPFALYLLSESFRRH
NTR_RAT	FYHYFYMLTNALFYVSSAINPILYNLVSANFRQV
C5AR_HUMAN	LLNKLDSLCVSFAYINCCINPIIYVVAGQGFQGR
FMLR_HUMAN	EIGIAVDVTSALAFFNSCLNPMLYVFMGQDFRER
FML1_HUMAN	IILVLINPTSSLAFFNSCLNPILYVFMGRNFQER
A36309	EIGIAVDVTSALAFFNSCLNPMLYVFMGQDFRER
FML2_HUMAN	IIDILVNPTSSLAFFNSCLNPMLYVFVGQDFRER
IL8B_HUMAN	HIDRALDATEILGILHSCLNPLIYAFIGQKFRHG
IL8A_HUMAN	NIGRALDATEILGFLHSCLNPIIYAFIGQNFRHG
IL8A_RABIT	DIDRALDATEILGFLHSCLNPIIYAFIGQNFRNG
NK1R_HUMAN	FIQQVYLAIMWLAMSSTMYNPIIYCCLNDRFRLG
NK1R_RAT	FIQQVYLASMWLAMSSTMYNPIIYCCLNDRFRLG
NK2R_BOVIN	FIQQVYLALFWLAMSSTMYNPIIYCCLNHRFRSG
NK2R_RAT	FIQQVYLALFWLAMSSTMYNPIIYCCLNHRFRSG
NK2R_HUMAN	FIQQVYLALFWLAMSSTMYNPIIYCCLNHRFRSG
NK3R_RAT	YIQQVYLASFWLAMSSTMYNPIIYCCLNKRFRAG
S20435	YIQQVYLASFWLAMSSTMYNPIIYCCLNKRFRAG
NKR1_DROME	FIQELYLAIYWLAMSNSMYNPIIYCWMNSRFRYG
NKR2_DROME	YVQHMYLGFYWLAMSNAMVNPLIYYWMNKRFRMY
NYR_DROME	PLPYVWFAFHWLAMSHCCYNPIIYCYMNARFRSG
NY1R_HUMAN	NHNLLFLLCHLTAMISTCVNPIFYGFLNKNFQRD
NY1R_RAT	NHNLLFLLCHLTAMISTCVNPIFYGFLNKNFQRD
NY3R_BOVIN	TVHKWISITEALAFFHCCLNPILYAFLGAKFKTS
AG2R_RAT	IVDTAMPITICIAYFNNCLNPLFYGFLGKKFKKY
AG2R_BOVIN	IVDTAMPITICLAYFNNCLNPLFYGFLGKKFKKY
AG2R_HUMAN	IVDTAMPITICIAYFNNCLNPLFYGFLGKKFKKY
VIP_HUMAN	ALFTALHVTQCLSLVHCCVNPVLYSFINRNYRYE
S74767	DDATVSQLSVILGYANSCANPILYGFLSDNFKRS
S74769	ALKGMFDFVVVLTYANSCANPILYAFLSDNFKKS
MUSSRI2A	ALKGMFDFVVILTYANSCANPILYAFLSDNFKKS
MUSSRI1A	DDATVSQLSVILGYANSCANPILYGFLSDNFKRS
S102315	ALKGMFDFVVILTYANSCANPILYAFLSDNFKKS
BRB2_RAT	AVDIVTQISSYVAYSNSCLNPLVYVIVGKRFRKK
TRFR_MOUSE	QENWFLLFCRICIYLNSAINPVIYNLMSQKFRAA

Protein sequence is given in one-letter code for each of the seven putative transmembrane helices. Receptors are listed by codes that are indexed in Table 1.

One approach to solving these problems is the determination of either the availability of particular side-chains to modification by hydrophilic or hydrophobic probes that partition into the aqueous phase (Barclay and Findlay, 1984) or the fatty acid milieu of the bilayer (Davison and Findlay, 1986) respectively, or the environment of spin labels attached at particular residue positions (Altenbach *et al.*, 1990). Another approach is to make assumptions based on the more hydrophobic or variable face of the helices (exposed to lipid) and on the location of conserved or charged residues (internally orientated) as revealed by extensive sequence alignments. Analysis of the periodicity in such properties can confirm the presence of an α-helical region and define which face is internally orientated in the 3-D fold (e.g. Donnelly *et al.*, 1989, 1993). The relative direction of the hydrophobic and variable faces can also be used to determine the point at which a continuous helix enters or leaves the bilayer. These faces should be one and the same for regions within the bilayer, but will be opposite one another in the aqueous environment surrounding the membrane (Donnelly and Cogdell, 1993). Other common structural features include the presence of a single critical disulphide bond between the first and second extracellular loops, first suggested by Findlay and Pappin (1986) and subsequently confirmed by Karnik and Khorana (1990), phosphorylation sites in the *C*-terminus region (Thompson and Findlay, 1984) and esterified palmitic acid residues (Ovchinnikov *et al.*, 1988).

Probable ligand-binding regions in rhodopsin were indicated by the identification of the attachment sites on opsin for the chromophore 11–*cis* retinal (Wang *et al.*, 1980; Findlay *et al.*, 1981). Site-directed mutagenesis on the β-adrenergic receptor has provided additional ligand-binding data (Strader *et al.*, 1989, 1991). These studies suggested that the agonist-binding site is internally located between the transmembrane helices. Model building suggests that helices 3–7 may provide the principal sites for interactions. This may also be the case for the neuropeptide receptors, but since the ligands are much larger, it is likely that the extracellular loop regions will also play a part in ligand binding.

Phyletic relationships

Table 1 lists most of the neuropeptide receptors that have been cloned to date and some of the more closely related, non-neural peptide receptors. Sequence alignments of the seven transmembrane segments (see Fig. 1) can be used to generate distance scores from which a phyletic tree can be calculated. Fig. 2 shows a tree calculated using the program KITSCH (Felsenstein, 1985).

The tree is divided into two principal branches (Fig. 2). The upper branch includes the endothelin, bombesin and tachykinin receptors as well as the neuropeptide Y receptors from *Drosophila*, rat and man. The lower branch includes the bradykinin, angiotensin, vasoactive intestinal peptide and the bovine neuropeptide Y receptors, which are all closely related to the non-neural peptide receptors that have been included.

Various receptors have been omitted from Fig. 2 in order to prevent the tree from becoming too cluttered or ambiguous. The somatostatin receptors belong to the upper branch, while the neurotensin receptor belongs to the lower branch. A more precise placement for these receptors cannot be made on the sequence data alone. The closely related thyrotropin-releasing hormone receptor forms a separate outlier. The secretin, calcitonin and parathyroid hormone sub-family are not included since they are too distantly related from the other receptors to be aligned accurately.

Heterologous expression

Why use heterologous expression systems?

Heterologous expression systems facilitate the study of membrane receptors, especially those of human origin and those naturally present in low abundance. There are many different systems available, each with its special advantages and disadvantages. They can loosely be divided into two groups: those suitable solely for analysis, such as *Xenopus* oocytes and cultured mammalian cells, and those that may be appropriate for larger scale production of membrane proteins, such as yeast, bacteria and the insect larval cell system. Neuropeptide receptors have been expressed to high levels (up to 10^7 receptors per cell) in a variety of mammalian cells and there is one example in the literature of a receptor expressed in insect larval cells (the neurokinin-2 receptor; Aharony *et al.*, 1993), but as yet there are no examples where the production of milligram quantities of protein is practical. The overproduction of membrane proteins is central to structural investigations, but has proved more difficult than for soluble proteins. Perhaps this is not surprising, since membrane volume may be a major constraint and there can be the special problems of protein targeting and folding.

Expression levels can be quoted in several forms (e.g. receptor molecules per cell, moles per milligram membrane or total cell protein, or milligrams per litre culture) and it is important to note how the protein was detected. For example, a ligand-binding assay may assess functional receptor, whereas an SDS/PAGE cannot differentiate active from inactive protein. Signal transduction assays prove that expressed receptors are responsive to the binding of a signal peptide, but they are largely qualitative, since a minute number of activated receptors can evoke an intracellular cascade.

The following sections outline some of the more common expression systems used to study GPCRs, with particular reference to the tachykinin receptors.

Escherichia coli

The heterologous expression of integral membrane proteins in *Escherichia coli* has not been successful in many cases, and even then the protein has not always been obtained in a functional form. *E. coli* do not carry out most of the post-translational modifications which may be essential for receptor function. Similarly, they may not provide the conditions necessary for

Table 1. Cloned neuropeptide receptors and some of the more closely related non-neural peptide receptors

Code	Receptor	Species
ET1R_BOVIN	Endothelin type A	Cow
ET1R_HUMAN	Endothelin type A	Human
ETIR_RAT	Endothelin type A	Rat
ETBR_RAT	Endothelin type B	Rat
ET1B_BOVIN	Endothelin type B	Cow
ETBR_HUMAN	Endothelin type B	Human
GRPR_MOUSE	Bombesin/gastrin-releasing peptide	Mouse
A41007	Bombesin/gastrin-releasing peptide	Human
NMBR_RAT	Neuromedin-B	Rat
B41007	Neuromedin-B	Human
NTR_RAT	Neurotensin	Rat
C5AR_HUMAN	C5A anaphylatoxin chemotactic	Human
FMLR_HUMAN	fMet-Leu-Phe	Human
FML1_HUMAN	fMet-Leu-Phe — related type I	Human
A36309	*N*-formyl peptide	Human
FML2_HUMAN	fMet-Leu-Phe — related type II	Human
IL8B_HUMAN	High affinity interleukin-8 type B	Human
IL8A_HUMAN	High affinity interleukin-8 type A	Human
IL8A_RABIT	High affinity interleukin-8	Rabbit
NK1R_HUMAN	Substance P	Human
NK1R_RAT	Substance P	Rat
NK2R_BOVIN	Substance K	Cow
NK2R_RAT	Substance K	Rat
NK2R_HUMAN	Substance K	Human
NK3R_RAT	Neuromedin K	Rat
S20435	Neuromedin K	Human
NK1R_DROME	Tachykinin type 1	*Drosophila*
NKR2_DROME	Tachykinin type 2	*Drosophila*
NYR_DROME	Neuropeptide Y	*Drosophila*
NY1R_HUMAN	Neuropeptide Y type 1	Human
NY1R_RAT	Neuropeptide Y type 1	Rat
NY3R_BOVIN	Neuropeptide Y type 3	Cow
AG2R_RAT	Angiotensin II type 1	Rat
AG2R_BOVIN	Angiotensin II type 1	Cow
AG2R_HUMAN	Angiotensin II type 1	Human
VIP_HUMAN	Vasoactive intestinal peptide	Human
S74767	Somatostatin type 1	Human
S74769	Somatostatin type 2	Human
MUSSRI2A	Somatostatin type 2	Mouse
MUSSRI1A	Somatostatin type 1	Mouse
S102315	Somatostatin	Rat
BRB2_RAT	Bradykinin type 2	Rat
TRFR_MOUSE	Thyrotropin-releasing hormone	Mouse

correct folding, although it is sometimes possible to reconstitute inactive proteins with lipids to yield an active receptor.

Expression levels in *E. coli* are normally low, no higher than 300 receptors per cell for the human β-adrenergic receptor (Marullo *et al.*, 1989) and the production of a large biomass of cells still corresponds to only a very small quantity of protein. In addition, the subsequent necessity to purify the receptor from a concentrated protein extract presents significant technical problems.

The reasons for poor expression are subtle and require more complex solutions than simply using a stronger promoter or increasing the copy number of the expression vector. For example, the base composition of the first ten codons appears to be very important. Adding a bacterial signal sequence or fusing the desired receptor to a well-expressed, exported protein can improve expression levels of the neurotensin receptor from under 10 to over 400 copies per cell (Grisshammer *et al.*, 1993).

COS cells and *Xenopus* oocytes

COS cells are a derivative of the CV-1 monkey cell line. They contain a segment of SV40 DNA that expresses T antigen (an essential function for SV40 lytic growth) but which has been mutated to inactivate the SV40 origin of replication. Thus, plasmid vectors containing the SV40 origin replicate to very high copy numbers within these cells.

COS cells can be transfected in a variety of ways, including the use of calcium phosphate precipitation, DEAE-dextran, electroporation and liposome encapsulation. None of these methods is highly efficient (10–20% of cells typically express the desired protein), but transfected DNA replicates to high copy number in responsive cells and often leads to high level expression of functional protein. COS cells, however, can only be used as a transient expression system since massive vector replication ultimately leads to cell death. Another disadvantage of the COS system is that the cells grow as a monolayer which makes protein expression difficult to scale up.

Table 1. (contd.)

Most of the neurokinin-type receptors that have been cloned to date are shown. The non-neural peptide receptors are the C5A anaphylatoxin chemotactic receptor, the fMet-Leu-Phe receptor and related receptors, the N-formyl peptide receptor and the interleukin-8 receptors. Codes are for the OWL sequence database at Leeds (Akrigg et al., 1988; Bleasby and Wootton, 1990), except for those shown in italics, which are arbitary codes for the Drosophila tachykinin type 1 (Li et al., 1991), Drosophila tachykinin type 2 (Monnier et al., 1992) and human vasoactive intestinal peptide (Sreedharan et al., 1991) receptors.

Fig. 2. Phyletic tree based on percentage sequence identities from the alignment in Fig. 1

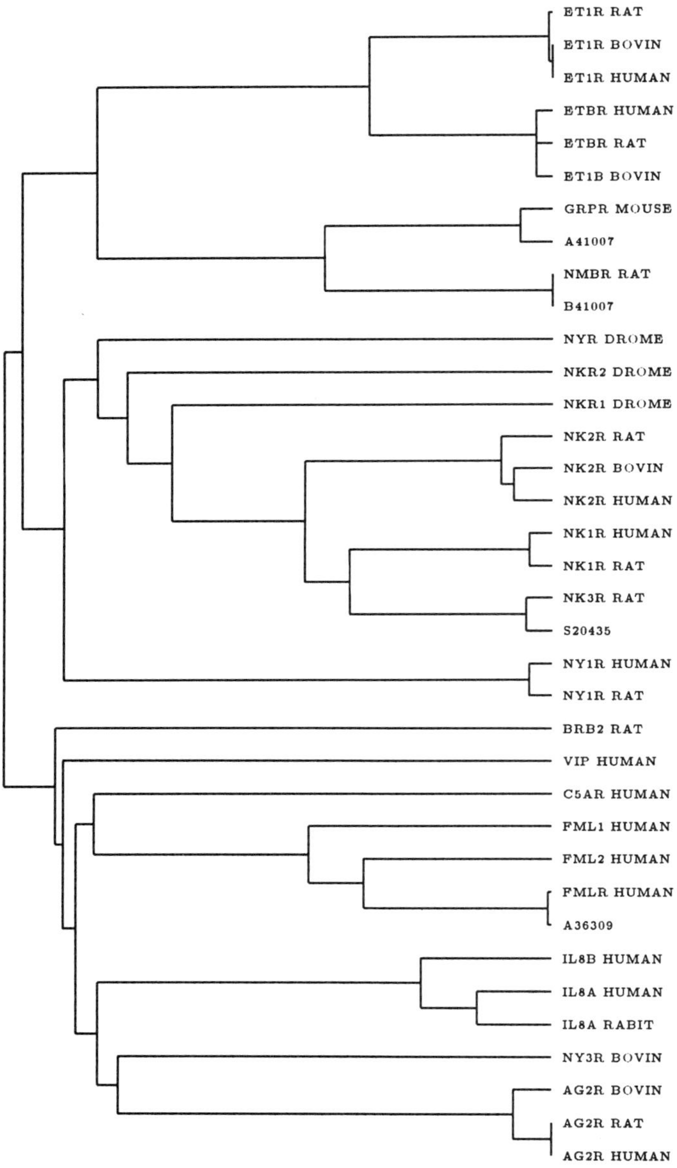

The secretin-type, somatostatin, neurotensin and thyrotropin-releasing hormone receptors have been omitted to avoid confusion (see text for details).

COS cells are used widely as an analytical expression system, particularly for the analysis of site-specific mutants. A variety of neuropeptide receptors have been examined in this way; receptors are generally assembled with high fidelity and expression levels can be high (greater than 10^5 receptors per cell). For example, the human substance P receptor has been expressed in COS-7 cells to 150 000 reeptors per cell (Takeda *et al.*, 1991). Transfected cells stimulated with 1 μM substance P gave a transient increase in cellular IP_3, suggesting coupling to an endogenous G-protein. In our laboratory, we express the human substance K receptor in COS-1 cells to approx. 10^6 receptors per cell ($K_d = 0.7$ nM) (N. Bhogal, D. Donnelly, C. R. Hurrell and J. B. C. Findlay, unpublished work). 72 h post-transfection using liposome encapsulation. Assuming a maximum transfection efficiency of 20%, expression levels may be greater than 5 million receptors per transfected cell. Upon stimulation with substance K, transfected cells are capable of mobilizing intracellular calcium, presumably due to activation of the IP_3 pathway through an endogenous G-protein.

As already indicated, *Xenopus* oocytes have been used successfully to express a variety of receptors. Since mRNA is injected into single cells, minute quantities of protein are manufactured and there is no way to substantially scale-up the process. Oocytes do, however, remain a powerful tool for expression cloning and for the measurement of intracellular calcium mobilization.

Stable mammalian cell lines

Foreign genes can be incorporated into mammalian cells such that the cells become permanently transformed. Stable cell lines of this sort can be difficult and time-consuming to establish, but do not require transfection each time receptor protein is required and give highly reproducible levels of expression. However, the suitability of stable cell lines for the large scale production of receptors has yet to be proven, since expression levels are generally much lower than with transient systems.

Permanently transformed cells are commonly set up by one of two methods; using bovine papillomavirus (BPV) or using Chinese hamster ovary (CHO) cells transfected with a vector containing a selectable marker [e.g. neomycin phosphotransferase-mediated resistance to the antibiotic G418 or dihydrofolate reductase (DHFR)]. BPV can oncogenically transform rodent cells. All the functions from BPV necessary for transformation can be combined with a mammalian transcription unit to produce a transforming vector that can exist as an episomal plasmid for long periods of time. Transformed cells can be selected for using a marker, or on the basis of changes associated with oncogenic transformation, such as the ability to grow in suspension.

CHO cell lines are commonly established from cells deficient in DHFR, an enzyme essential for growth in the absence of thymidine. Vectors carrying the DHFR gene can be transfected into CHO cells, where they integrate into the chromosome to form stable cell lines that are selectable on thymidine-free media. The rat tachykinin receptors have been expressed to 17 000–25 000 receptors per cell in CHO cells (Nakajima *et al.*, 1992). All three receptors were capable of stimulating both the IP_3 and cyclic AMP cascades. The human neurokinin-2 receptor has been expressed to greater levels (97 000 receptors per cell) in mouse 3T3 fibroblasts (Kris *et al.*, 1991). Even higher levels of expression are possible by exposing CHO cells expressing a plasmid-encoded DHFR gene to gradually increasing concentrations of DHFR. Drug resistance is effectively achieved by the production of tandem repeats (up to 1000 copies) of the DHFR gene with a concomitant replication of the foreign gene. This system has been used to express the human neurokinin-2 receptor to 7×10^5 receptors per cell (Turcatti *et al.*, 1993). Similarly, the β-adrenergic receptor has been expressed to 200 pmol per mg membrane protein using this approach (Lohse, 1992). Interestingly, it was necessary to stabilize the expressed receptor by including the receptor antagonist alprenolol in the culture medium.

Yeast

Yeast cells offer a eukaryotic expression system with many of the advantages of a bacterial system. The cells are easy to culture, grow rapidly to high cell densities, yet carry out eukaryotic post-translational modifications. However, yeast are intrinsically different from higher organisms. They possess a cell wall, which makes cell lysis difficult and they possess a vacuole rich in proteases, which may degrade expressed protein on cell lysis. Expressed protein may be inactive due to improper folding or unsuitable post-translational modifications. Similarly, yeast may not target a receptor correctly, or may have difficulty translating a foreign transcript.

Several vector systems have been developed for use with yeast, including integrative plasmids, centromeric plasmids and the transposable element Ty. Yeast episomal plasmids are *E. coli* shuttle vectors and are suitable for most applications. These replicate to approx. 100 copies per cell, utilize strong yeast promoters and are selectable on the basis of an auxotrophic marker. Many vectors utilize inducible promoters (e.g. the *Gal*1 or alcohol dehydrogenase promoter) which permit the expression of moderately toxic proteins. The promoter has a pronounced influence on gene expression, but the gene itself appears to be a critical expression factor. For example, the human m1 muscarinic acetylcholine receptor was expressed to 0.02 pmol per mg protein (Payette, 1990), with only 20 active receptors per cell, using the alcohol dehydrogenase promoter. In comparison, the human β_2-adrenergic receptor has been expressed from the *Gal*1 promoter to 115 pmol per mg protein (King *et al.*, 1990) as a fusion protein with 42 bp of coding sequence from the α-factor receptor (a putative GPCR). The use of protein fusions, signal sequences and similar tricks are all useful tools to enhance receptor expression. Thus, yeast offer a plausible system for the production of milligram quantities of receptor proteins.

Baculovirus system

The baculovirus *Autographa california* nuclear polyhedrosis virus has been successfully exploited to express a variety of membrane proteins in cultured insect larval cells. The system is attractive for several reasons; expressed protein can accumulate to high levels, eukaryotic protein modifications can occur and, with care, all of the cells in a culture can be expressing equivalent amounts of foreign protein. Furthermore, insect cells can be grown to high density in suspension culture, which offers the potential to produce large quantities of starting material for protein purification.

Each baculovirus possesses a large, circular, double-stranded DNA genome with a very strong promoter for the massively expressed coat protein polyhedrin (up to 1 mg per ml culture, containing 10^6 cells). The genome comprises about 128 kbp of DNA, which is too large to manipulate *in vitro*. The strategy employed to circumvent this problem lets the baculovirus do the genetic engineering. The gene of interest is cloned into a transfer vector between the polyhedrin promoter and essential downstream sequences. This construct is then co-transfected into insect cells with naked viral DNA. A rare homologous recombination event *in vivo* results in replacement of the viral polyhedrin transcription unit with that of the transfer vector. Traditionally, a time-consuming procedure was then necessary to isolate pure recombinant virus from a high background of wild-type virus. However, recent developments have simplified this procedure considerably. For example, it is now possible to purchase linearized baculovirus DNA that contains a lethal deletion, such that more than 99% of plaques formed contain pure recombinant virus.

A variety of membrane proteins have been expressed in the baculovirus system in a functional form, although often not all expressed protein is active and fully processed. Bovine opsin has been expressed to very high levels (3 mg per 10^6 cells $\simeq 4.5 \times 10^7$ receptors per cell) using the baculovirus system (Janssen *et al.*, 1988). The expressed protein had the same mobility on SDS/PAGE as opsin isolated from bovine retina, suggesting that the receptor had been correctly processed in the insect cells. The turkey β-adrenergic receptor and the m1 and m2 forms of human muscarinic acetylcholine receptor have been purified from recombinant baculovirus-infected insect cells (Parker *et al.*, 1991). Levels of expression varied from 5×10^5 receptors per cell for the β-adrenergic receptor to 2.5×10^6 for the m2 muscarinic receptor. All three receptors displayed typical binding characteristics for their respective agonists and antagonists, and the expressed β-adrenergic receptor stimulated endogenous adenylate cyclase suggesting coupling to an endogenous stimulatory G-protein. The human neurokinin-2 receptor has been expressed in Sf-21 cells to 10^5 receptors per cell (Aharony *et al.*, 1993). In our laboratory, we also use the baculovirus system to express the human neurokinin-2 receptor in Sf-9 cells. Binding studies using [^{125}I]substance K suggest we obtain up to 10^6 receptors per cell with an affinity constant (K_d) of 2.5 nM (A. Cox and J. B. C. Findlay, unpublished work). Thus, the baculovirus system has the potential to produce large quantities of functional neuropeptide receptors for purification and subsequent biochemical analysis. The limitations of the cell culture, however, suggest that only with exceptional efforts can milligram quantities of protein be obtained.

Analysis of receptors

Methods of detection

Sensitive detection methods, commonly employing radioactivity or enzyme-linked assays, are essential for the study of membrane proteins, since there is usually insufficient protein to visualize on a conventionally stained gel. If protein is scarce, specific antibodies can be raised against synthetic peptides corresponding to regions of a receptor. Peptide antibodies provide a powerful tool to study receptor protein, but they can be difficult to raise and may not recognize native receptor with high affinity. Expressed receptor can also be monitored by pulse-labelling, but a high background of endogenously labelled protein may mask the receptor of interest. These drawbacks have encouraged the development of detection methods based upon the innate ability of a receptor to specifically recognize a chemical ligand. Photoaffinity labelling and chemical cross-linking techniques have been used with limited success and are discussed below in more detail.

Chemical cross-linking and photoaffinity labelling

The side-chain carboxy- and amino-groups of peptide ligands lend themselves to being covalently cross-linked to corresponding groups on receptors. Ligands are typically labelled with ^{125}I and, in a few cases, have allowed the identification of particular components by autoradiography of SDS/PAGE gels. This approach has been used with samples taken from both heterologous expression systems and native tissues. Commonly used, bi-functional, cross-linking agents have been disuccinimidyl reagents (e.g. disuccinimidyl suberate), carbodiimides (e.g. 1-ethyl, 1-3-(3-dimethyl-aminopropyl) carbodiimide:HCl) and 1,5-difluoro-2,4-dinitrobenzene; however, the most suitable reagent for a particular receptor–ligand complex can only be determined empirically. For the endothelin receptors, it has proved possible to visualize receptor protein without cross-linking (Takasuka *et al.*, 1991). Radiolabelled ligand–receptor complexes were visible on autoradiographs of SDS/PAGE gels run at low temperature, presumably because the receptor retains or regains its native conformation and its affinity for endothelin is extremely high (K_d approx. 25 pM).

The mobility of a particular radioligand-tagged receptor on SDS/PAGE has been used to estimate the extent of *N*-linked glycosylation. Principally, this is done by comparing the mobility of a receptor before and after treatment with *N*-glycanase. For example, the bombesin/gastrin-releasing peptide receptor expressed in NIH-3T3 cells is extensively glycosylated (Kris *et al.*, 1987), with estimated M_rs of 65 000 and 45 000 before and after *N*-glycanase treatment, respectively. In comparison, the human substance K receptor expressed in the same cell-line appeared to have little or no *N*-linked glycosylation (Kris *et al.*, 1991), despite the presence of two potential *N*-glycosylation sites in the predicted *N*-terminal domain. However, these patterns of glycosylation may not be an accurate model for what is happening in nature. Protein modifications are cell-type specific and, in contrast to expression in NIH-3T3 cells, the human substance K receptor appears to be highly glycosylated in CHO cells; bands of M_r 64 000 and 39000, respectively, have been observed before and after treatment with *N*-glycosidases (Turcatti *et al.*, 1993). Expression of the human neurokinin-2 receptor in our laboratory suggests that all the expressed protein in COS-1 cells is highly glycosylated. Covalent attachment of the ligand, by chemical cross-linking and immunoblotting with antibody, gives a molecular species on SDS/PAGE with an apparent M_r in the region of 66 000 (N. Bhogal and J. B. C. Findlay, unpublished work).

A more reliable way, than mutagenesis, to identify residues involved in ligand binding is to use radiolabelled analogues of neuropeptides that contain photoreactive cross-linking moieties. These ligands can be covalently bound to expressed receptor by excitation with u.v. light. The position and nature of the photoreactive group is critical, since it must not disrupt high-affinity binding, and must be in the correct environment to cross-link to the receptor. The attachment point for each specific ligand can be determined by cleaving purified receptor protein into fragments that can be resolved by SDS/PAGE, followed by protein sequencing of the radiolabelled band. Using a range of ligands containing photoreactive groups in different positions, it may be possible to build up a picture of a particular receptor–ligand complex. A radio-iodinated substance P derivative containing the photoreactive species *p*-benzoylphenylalanine has been used to label the substance P receptor (Boyd *et al.*, 1991). Similarly, a *p*-benzoylbenzoyl moiety on cholecystokinin-8 has been used to label the cholecystokinin type-B receptor (Thiele and Fahrenholz, 1993). In neither case has the attachment point been identified. Therefore, there is no empirical model of a neuropeptide receptor–ligand complex based upon this approach.

Mutagenesis

Site-directed mutagenesis is a popular method of investigating GPCR structure and function (reviewed in Savarese and Fraser, 1992). However, although mutagenesis can provide interpretable data, results can be misleading and great care must be taken to consider the effect of point mutations on the protein as a whole.

The mammalian tachykinin receptors possess a number of interesting features. First, the three receptor subtypes show exquisite specificity (about 1000-fold) between the three natural agonists, whose sequences vary substantially only in the *N*-terminal half of the peptide. We can, therefore, speculate that this region primarily determines receptor specificity. In contrast, the *C*-terminal consensus sequence is likely to be mainly responsible for receptor activation. In a hypothetical model based on this idea, the *N*-terminal would interact with residues nearer the surface of the receptor while the *C*-terminal associated with the classical agonist-binding site would be buried in the transmembrane segments where it could induce a conformation change capable of stimulating G-protein activation.

Secondly, the ligands are large (10–11 amino acid residues) and there are a myriad of ways in which they can bind to the receptor. It is probable that multiple epitopes scattered throughout the receptor structure coalesce in 3-D space to form a ligand-binding pocket. Data from chimeric neurokinin-1/neurokinin-3 receptors support this hypothesis; substance P shows a gradual increase in affinity for the neurokinin-3 receptor as helices from the *N*-terminal are replaced with those of the neurokinin-1 receptor (Gether *et al.*, 1993*a*). Mutagenesis of residues in the putative binding site of the human neurokinin-2 receptor reveals residues distributed throughout the receptor, Ser-274, Leu-292 and Met-297 (N. Bhogal, C. R. Hurrell and J. B. C. Findlay, unpublished work), that have small, but significant effects on agonist binding. Residues much closer to the membrane surface, Gly-190 and Leu-194 (N. Bhogal and J. B. C. Findlay, unpublished work) and in the *N*-terminal segment (Yokota *et al.*, 1992) are clearly also important. The situation appears quite different for non-peptide antagonists, where single amino acid substitutions can dramatically reduce receptor–antagonist affinity. For example, Val-349 is critical in binding non-peptide antagonists of the cholecystokinin-B receptor (Beinborn *et al.*, 1993*a*; 1993*b*), and Val-116 and Ile-290 are critical residues in the human neurokinin-1 receptor for binding the rat selective antagonist RP67580 (Fong *et al.*, 1992*c*). Furthermore, in a number of cases, non-peptide antagonists appear not to mimic the subtle interactions between a peptide ligand and its cognate receptor. His-197 in the fifth transmembrane segment of the neurokinin-1 receptor is a critical residue for binding the non-peptide antagonist CP96345, but it is not important in binding substance P (Fong *et al.*, 1993; Gether *et al.*, 1993*b*). Similarly, Fong *et al.* (1992*b*) have used the antagonist L703606 to show that mutant receptors that do not bind peptide ligands are still expressed and folded correctly in the bilayer.

Mutagenesis has also been used to investigate regions of the neurokinin-1 and -2 receptors important in G-protein activation (Blount and Krause, 1993). As expected for a GPCR, the third cytoplasmic loop of the neurokinin-1 receptor was found to be important for IP_3 turnover and cyclic AMP production. Surprisingly, this was not the case for the neurokinin-2 receptor. Conversely, the *C*-terminal tail was important in the neurokinin-2 receptor, but not in the neurokinin-1 receptor. This correlates to some extent with existence of a naturally occurring *C*-terminally truncated neurokinin-1 receptor (Fong *et al.*, 1992*a*). Similar constructs have not been found for the neurokinin-2 receptor. Furthermore, the domains found to be important in G-protein activation could not be substituted by the analogous domain of the other receptor.

Conclusion

There are, at present, over 20 neuropeptide receptors that are recognizable as GPCRs. They are all found in low abundance in nature and there is no example to date where sufficient protein can be produced to attempt crystallization trials. It appears likely, therefore, that high-resolution structural information will not be available for a number of years. Undoubtedly, the key to obtaining sufficient material for crystallization relies heavily upon efficient over-expression systems and novel techniques to produce pure recombinant protein. The development of new production systems and an increased understanding of the rules that govern protein expression should lead to the isolation of sufficient protein for detailed biochemical analysis. Meanwhile, the results of mutagenesis experiments and photoaffinity labelling studies can be used to enhance our knowledge of receptor structure and function.

References

Aharony, D., Little, J., Powell, S., Hopkins, B., Bundell, K. R., McPheat, W. L., Gordon, R. D., Hassall, G., Hockney, R., Griffin, R., *et al.* (1993) Pharmacological characterization of cloned human NK-2 (neurokinin A) receptor expressed in a baculovirus/Sf-21 insect-cell system. Mol. Pharmacol. **44**, 356–363

Akrigg, D., Bleasby, A. J., Dix, N. I. M., Findlay, J. B. C., North, A. C. T., Parry-Smith, D., Wootton, J. C., Blundell, T. L., Gardner, S. P., Hayes, F., *et al.* (1988) A protein/structure database. Nature (London) **335**, 745–746

Altenbach, C., Marti, T., Khorana, H. G. and Hubell, W. L. (1990) Transmembrane protein-structure–spin-labelling of bacteriorhodopsin mutants. Science **248**, 1088–1092

Arai, H., Hori, S., Aramori, I., Ohkubo, H. and Nakanishi, S. (1990) Cloning and expression of a cDNA encoding an endothelin receptor. Nature (London) **348**, 730–732

Attwood, T. K., Eliopoulos, E. E. and Findlay, J. B. C. (1991) Multiple sequence alignment of protein families showing low sequence homology — a methodological approach using database pattern-matching discriminators for G-protein-linked receptors. Gene **98**, 153–159

Attwood, T. K. and Findlay, J. B. C. (1993*a*) Design of a discriminating fingerprint for G-protein-coupled receptors. Protein Eng. **6**, 167–176

Attwood, T. K. and Findlay, J. B. C. (1993*b*) Fingerprinting G-protein-coupled receptors. Protein Eng. **7**, 195–203

Barclay, P. L. and Findlay, J. B. C. (1984) Labelling of the cytoplasmic domains of ovine rhodopsin with hydrophilic chemical probes. Biochem. J. **220**, 75–84

Battey, J. F., Way, J. M., Corjay, M. H., Shapira, H., Kusano, K., Harkins, R., Wu, J. M., Slattery, T., Mann, E. and Feldman, R. I. (1991) Molecular-cloning of the bombesin gastrin-releasing peptide receptor from Swiss 3T3 cells. Proc. Natl. Acad. Sci. U.S.A. **88**, 395–399

Beinborn, M., Lee, Y.-M., McBride, E. W., Quinn, S. M. and Kopin, A. S. (1993*a*) A single amino acid of the cholecystokinin-B/gastrin receptor determines specificity for non-peptide antagonists. Nature (London) **362**, 348–350

Beinborn, M., Lee, Y.-M., McBride, E. W., Quinn, S. M. and Kopin, A. S. (1993*b*) A single amino acid of the cholecystokinin-B/gastrin receptor determines specificity for non-peptide antagonists. Nature (London) **364**, 362

Bleasby, A. J. and Wootton, J. C. (1990) Construction of validated, nonredundant composite protein-sequence databases. Protein Eng. **3**, 153–159

Blount, P. and Krause, J. E. (1993) Functional nonequivalence of structurally homologous domains of neurokinin-1 and neurokinin-2 type tachykinin receptors. J. Biol. Chem. **268**, 16388–16395

Boyd, N. D., White, C. F., Cerpa, R., Kaiser, E. T. and Leeman, S. E. (1991) Photoaffinity labeling the substance P receptor using a derivative of substance P containing *p*-benzoylphenylalanine. Biochemistry **30**, 336–342

Buck, S. H., Fanger, B. O. and Van Giersbergen, P. L. M. (1991) Pharmacological and biochemical-evidence for multiple types of tachykinin NK_2 receptors. Ann. N.Y. Acad. Sci. **632**, 112–115

Buell, G., Schulz, M. F., Arkinstall, S. J., Maury, K., Missotten, M., Adami, N., Talabot, F. and Kawashima, E. (1992) Molecular characterisation, expression and localisation of human neurokinin-3 receptor. FEBS Lett. **299**, 90–95

Burbach, J. P. H. and Meijer, O. C. (1992) The structure of neuropeptide receptors. Eur. J. Pharmacol. **227**, 1–18

Carstairs, J. R., Maggi, C. A., Andersson, Per O. and Nimmo, A. J. (1993) A novel tachykinin NK2 receptor associated with capsaicin-sensitive afferent nerves in rat bladder. Neuropeptides **24**, 200

Chang, M. M., Leeman, S. E. and Niall, H. D. (1971) Amino-acid sequence of substance P. Nature (London) New Biol. **232**, 86–87

Christians, S., Neubauer, K. H. and Ulrich, H. (1993) Purification and characterization of the head-activator receptor from a multiheaded mutant of *Chlorohydra viridissima*. FEBS Lett. **316**, 141–146

Corjay, M. H., Dobrzanski, D. J., Way, J. M., Viallet, J., Shapira, H., Worland, P., Sausville, E. A. and Battey, J. F. (1991) Two distinct bombesin receptor subtypes are expressed and functional in human lung-carcinoma cells. J. Biol. Chem. **266**, 18771–18779

Corless, J. M., McCaslin, D. R. and Scott, B. L. (1982) Two-dimensional rhodopsin crystals from disk membranes of frog retinal rod outer segments. Proc. Natl. Acad. Sci. U.S.A. **79**, 1116–1120

Davison, M. D. and Findlay, J. B. C. (1986) Identification of the sites in opsin modified by photoactivated azido-[^{125}I]-iodobenzene. Biochem. J. **236**, 389–395

Donnelly, D. and Cogdell, R. J. (1993) Predicting the point at which transmembrane helices protrude from the bilayer. Prot. Eng. **6**, 629–635

Donnelly, D., Johnson, M. S., Blundell, T. L. and Saunders, J. (1989) An analysis of the periodicity of conserved residues in sequence alignments of G-protein coupled receptors — implications for the 3-dimensional structure. FEBS Lett. **251**, 109–116

Donnelly, D., Overington, J. P., Ruffle, S. V., Nugent, J. H. A. and Blundell, T. L. (1993) Modeling alpha-helical transmembrane domains — the calculation and use of substitution tables for lipid-facing residues. Protein Sci. **2**, 55–70

Felsenstein, J. (1985) Confidence-limits on phylogenies — an approach using the bootstrap. Evolution **39**, 783–791

Findlay, J. B. C., Brett, M. and Pappin, D. J. C. (1981) Primary structure of C-terminal functional sites in ovine rhodopsin. Nature (London) **293**, 314–316

Findlay, J. B. C. and Pappin, D. J. C. (1986) The opsin family of proteins. Biochem. J. **238**, 625–642

Fong, T. M., Anderson, S. A., Yu, H., Huang, R. R. C. and Strader, C. D. (1992*a*) Differential activation of intracellular effector by two isoforms of human neurokinin-1 receptor. Mol. Pharmacol. **41**, 24–30

Fong, T. M., Cascieri, M. A., Yu, H., Bansal, A., Swain, C. and Strader, C. D. (1993) Amino-aromatic interaction between histidine 197 of the neurokinin-1 receptor and CP96345. Nature (London) **362**, 350–353

Fong, T. M., Huang, R. R. C. and Strader, C. D. (1992*b*) Localisation of agonist and antagonist binding domains of the human neurokinin-1 receptor. J. Biol. Chem. **267**, 25664–25667

Fong, T. M., Yu, H. and Strader, C. D. (1992*c*) Molecular basis for the species selectivity of the neurokinin-1 receptor antagonists CP96345 and RP67580. J. Biol. Chem. **267**, 25668–25671

Gerard, N. P., Eddy, R. L., Jr, Shows, T. B. and Gerard, C. (1990) The human neurokinin A (substance K) receptor. J. Biol. Chem. **265**, 20455–20462

Gerard, N. P., Garraway, L. A., Eddy, R. L., Jr, Shows, T. B., Iijima, H., Paquet, J.-L. and Gerard, C. (1991) Human substance P receptor (NK-1): organization of the gene, chromosome localization, and functional expression of cDNA clones. Biochemistry **30**, 10640–10646

Gether, U., Johansen, T. E. and Schwartz, T. W. (1993*a*) Chimeric NK_1 (substance P)/NK_3 (neurokinin B) receptors. J. Biol. Chem. **268**, 7893–7898

Gether, U., Johansen, T. E., Snider, R. M., Lowe, J. A., Nakanishi, S. and Schwartz, T. W. (1993*b*) Different binding epitopes on the NK_1 receptor for substance P and a non-peptide antagonist. Nature (London) **362**, 345–348

Grisshammer, R., Duckworth, R. and Henderson, R. (1993) Expression of a rat neurotensin receptor in *Escherichia coli.* Biochem. J. **295**, 571–576

Henderson, R., Baldwin, J. M., Ceska, T. A., Zemlin, F., Beckman, E. and Downing, K. H. (1990) Model for the structure of bacteriorhodopsin based on high-resolution electron cryomicroscopy. J. Mol. Biol. **213**, 899–929

Henderson, R. and Unwin, P. T. N. (1975) Three-dimensional model of purple membrane obtained by electron microscopy. Nature (London) **257**, 28–32

Jackson, T. R., Blair, L. A. C., Marshall, J., Goedert, M. and Hanley, M. R. (1988) The *mas* oncogene encodes an angiotensin receptor. Nature (London) **335**, 437–440

Janssen, J. J. M., Van de Ven, W. J. M., Van Groningen-Luyben, W. A. H. M., Roosien, J., Vlak, J. M. and de Grip, W. J. (1988) Synthesis of functional bovine opsin in insect cells under control of the baculovirus polyhedrin promoter. Mol. Biol. Rep. **13**, 65–71

Karnik, S. S. and Khorana, H. G. (1990) Assembly of functional rhodopsin requires a disulfide bond between cysteine residue-110 and residue-187. J. Biol. Chem. **265**, 17520–17524

King, K., Dohlman, H. G., Thorner, J., Caron, M G. and Lefkowitz, R. J. (1990) Control of yeast mating signal transduction by a mammalian b_2-adrenergic receptor and G_s alpha-subunit. Science **250**, 121–123

Kris, R. M., Hazan, R., Villines, J., Moody, T. W. and Schlessinger, J. (1987) Identification of the bombesin receptor on murine and human cells by cross-linking experiments. J. Biol. Chem. **262**, 11215–11220

Kris, R. M., South, V., Saltzman, A., Felder, S., Ricca, G. A., Jaye, M., Huebner, K., Kagan, J., Croce, C. M. and Schlessinger, J. (1991) Cloning and expression of the human substance K receptor and analysis of its role in mitogenesis. Cell Growth Differ. **2**, 15–22

Li, X. J., Wolfgand, W., Wu, Y. N., North, R. A. and Forte, M. (1991) Cloning, heterologous expression and developmental regulation of a *Drosophila* receptor for tachykinin-like peptides. EMBO J. **10**, 3221–3229

Libert, F., Parmentier, M., Lefort, A., Dinsart, C., Vansande, J., Maenhaut, C., Simons, M. J., Dumont, J. E. and Vassart, G. (1989) Selective amplification and cloning of four new members of the G-protein coupled receptor family. Science **244**, 569–572

Lin, H. Y., Harris, T. L., Flannery, M. S., Aruffo, A., Kaji, E. H., Gorn, A., Kolakowski, L. F., Lodish, H. F. and Goldring, S. R. (1991*a*) Expression cloning of an adenylate-cyclase coupled calcitonin receptor. Science **254**, 1022–1024

Lin, H. Y., Kaji, E. H., Winkel, G. K., Ives, H. E. and Lodish, H. F. (1991*b*) Cloning and functional expression of a vascular smooth-muscle endothelin-1 receptor. Proc. Natl. Acad. Sci. U.S.A. **88**, 3185–3189

Lohse, M. J. (1992) Stable overexpression of human β_2-adrenergic receptors in mammalian-cells. Naunyn-Schmiedebergs Arch. Pharmakol. **345**, 444–451

Marullo, S., Delavierklutchko, C., Guillet, J. G., Charbit, A., Strosberg, A. D. and Emorine, L. J. (1989) Expression of human β_1 and β_2 adrenergic-receptors in *Escherichia coli* as a new tool for ligand screening. Bio-technol. **7**, 923–927

Masu, Y., Nakayama, K., Tamaki, H., Harada, Y., Kuno, M. and Nakanishi, S. (1987) cDNa cloning of bovine substance-k receptor through oocyte expression system. Nature (London) **329**, 836–838

McEachern, A. E., Shelton, E. R., Bhakta, S., Obernolte, R., Bach, C., Zuppan, P., Fujisaki, J., Aldrich, R. W. and Jarnagin, K. (1991) Expression cloning of a rat B_2-bradykinin receptor. Proc. Natl. Acad. Sci. U.S.A. **88**, 7724–7728

McIntosh, R. P. and Catt, K. J. (1987) Coupling of inositol phospholipid hydrolysis to peptide-hormone receptors expressed from adrenal and pituitary mRNA in *Xenopus laevis* oocytes. Proc. Natl. Acad. Sci. U.S.A. **84**, 9045–9048

Monnier, D., Colas, J.-F., Rosay, P., Hen, R., Borrelli, E. and Maroteaux, L. (1992) NKD, a developmentally regulated tachykinin receptor in *Drosophila.* J. Biol. Chem. **267**, 1298–1302

Murphy, T. J., Alexandra, R. W., Griendling, K. K., Runge, M. S. and Bernstein, K. E. (1991) Isolation of a cDNA encoding the vascular type-1 angiotensin-II receptor. Nature (London) **351**, 233–236

Nakajima, Y., Tsuchida, K., Negishi, M., Ito, S. and Nakanishi, S. (1992) Direct linkage of three tachykinin receptors to stimulation of both phosphatidylinositol hydrolysis and cyclic AMP cascades in transfected Chinese hamster ovary cells. J. Biol. Chem. **267**, 2437–2442

Nakamuta, M,. Takayanagi, R., Sakai, Y., Sakamoto, S., Hagiwara, H., Mizuno, T., Saito, Y., Hirose,

S., Yamamoto, M. and Nawata, H. (1991) Cloning and sequence-analysis of a cDNA encoding human nonselective type of endothelin receptor. Biochem. Biophys. Res. Commun. **177**, 34–39

Nathans, J. and Hogness, D. S. (1983) Isolation, sequence-analysis, and intron exon arrangement of the gene encoding bovine rhodopsin. Cell **34**, 807–814

Nawa, H., Hirose, T., Takashima, H., Inayama, S. and Nakanishi, S. (1983) Nucleotide sequences of cloned cDNAs for two types of bovine brain substance P precursor. Nature (London) **306**, 32–36

Ovchinnikov, Y. A., Abdulaev, N. G. and Bogachuk, A. S. (1988) Two adjacent cysteine residues in the C-terminal cytoplasmic fragment of bovine rhodopsin are palmitylated. FEBS Lett. **230**, 1–5

Parker, E. M., Kameyama, K., Higashijima, T. and Ross, E. M. (1991) Reconstitutively active G-protein-coupled receptors purified from baculovirus-infected insect cells. J. Biol. Chem. **266**, 519–527

Payette, P., Gossard, F., Whiteway, M. and Dennis, M. (1990) Expression and pharmacological characterization of the human M1 muscarinic receptor in *Saccharomyces cerevisiae*. FEBS Lett. **266**, 21–25

Saito, Y., Mizuno, T., Itakura, M., Suzuki, Y., Ito, T., Hagiwara, H. and Hirose, S. (1991) Primary structure of bovine endothelin ET_B receptor and identification of signal peptidase and metal proteinase cleavage sites. J. Biol. Chem. **266**, 23433–23437

Sakurai, T., Yanagisawa, M., Takuwa, Y., Miyazaki, H., Kimura, S., Goto, K. and Masaki, T. (1990) Cloning of a cDNA encoding a non-isopeptide-selective subtype of the endothelin receptor. Nature (London) **348**, 732–735

Sandberg, K., Markwick, A. J., Trinh, D. P. and Catt, K. J. (1988) Calcium mobilization by angiotensin-II and neurotransmitter receptors expressed in *Xenopus laevis* oocytes. FEBS Lett. **241**, 177–180

Sasaki, K, Yamano, Y., Bardhan, S., Iwai, N., Murray, J. J., Hasegawa, M., Matsuda, Y. and Inagami, T. (1991) Cloning and expression of a complementary-DNA encoding a bovine adrenal angiotensin-II type-1 receptor. Nature (London) **351**, 230–233

Savarese, T. M. and Fraser, C. M. (1992) *In vitro* mutagenesis and the search for structure-function-relationships among G-protein-coupled receptors. Biochem. J. **283**, 1–19

Schertler, G. F. X., Villa, C. and Henderson, R. (1993) Projection structure of rhodopsin. Nature (London) **362**, 770–772

Shigemoto, R., Yokota, Y., Tsuchida, K. and Nakanishi, S. (1990) Cloning and expression of a rat neuromedin-K receptor cDNA. J. Biol. Chem. **265**, 623–628

Spindel, E. R., Giladi, E., Brehm, P., Goodman, R. H. and Segerson, T. P. (1990) Cloning and functional-characterization of a complementary-DNA encoding the murine fibroblast bombesin gastrin-releasing peptide receptor. Mol. Endocrinol. **4**, 1956–1963

Sreedharan, S. P., Robichon, A., Peterson, K. E. and Goetzl, E. J. (1991) Cloning and expression of the human vasoactive-intestinal-peptide receptor. Proc. Natl. Acad. Sci. U.S.A. **88**, 4986–4990

Strader, C. D., Candelore, M. R., Hill, W. S., Sigal, I. S. and Dixon, R. A. F. (1989) Identification of two serine residues involved in agonist activation of the β-adrenergic-receptor. J. Biol. Chem. **264**, 13572–13578

Strader, C. D., Gaffney, T., Sugg, E. E., Candelore, M. R., Keys, R., Patchett, A. A. and Dixon, R. A. F. (1991) Allele-specific activation of genetically engineered receptors. J. Biol. Chem. **266**, 5–8

Straub, R. E., Frech, G. C., Joho, R. H. and Gershengorn, M. C. (1990) Expression cloning of a cDNA encoding the mouse pituitary thyrotropin-releasing hormone receptor. Proc. Natl. Acad. Sci. U.S.A. **87**, 9514–9518

Takasuka, T., Satoh, N., Horii, I., Furuichi, Y. and Watanabe, T. (1991) Heterogeneity of endothelin receptor. J. Cardiovasc. Pharmacol. **17**, S109–S112

Takeda, Y., Chou, K. B., Takeda, J., Sachais, B. S. and Krause, J. E. (1991) Molecular cloning, structural characterization and functional expression of the human substance P receptor. Biochem. Biophys. Res. Commun. **179**, 1232–1240

Tanaka, K., Masu, M. and Nakanishi, S. (1990) Structure and functional expression of a cloned rat neurotensin receptor. Neuron **4**, 847–854

Thiele, C. and Fahrenholz, F. (1993) Photoaffinity labelling of central cholecystokinin receptors with high efficiency. Biochemistry **32**, 2741–2746

Thompson, P. and Findlay, J. B. C. (1984) Phosphorylation of ovine rhodopsin — identification of the phosphorylated sites. Biochem. J. **220**, 773–780

Turcatti, G., Ceskowski, K. and Chollet, A. (1993) Biochemical characterization and solubilization of human NK_2 receptor expressed in Chinese hamster ovary cells. J. Recept. Res. **13**, 639–652

Van Giersbergen, P. L. M., Shatzer, S. A., Harbeson, S. L., Roussi, N., Nantel, F. and Buck, S. H.

(1991) Multiple NK_2 receptor subtypes are suggested by physiological and biochemical studies with neurokinin-A (NKA) analogs and antagonists. Ann. N.Y. Acad. Sci. **632**, 483–484

von Euler, U. S. and Gaddum, J. H. (1931) An unidentified depressor substance in certain tissue extracts. J. Physiol. (London) **72**, 74–87

Wang, H. Y., Lipfert, L., Malbon, C. C. and Bahouth, S. (1989) Site-directed anti-peptide antibodies define the topography of the β-adrenergic-receptor. J. Biol. Chem. **264**, 14424–14431

Wang, J. K., McDowell, J. H. M. and Hargrave, P. A. (1980) Site of attachment of 11-*cis*-retinal in bovine rhodopsin. Biochemistry **19**, 5111–5117

Wank, S. A., Harkins, R., Jensen, R. T., Shapira, H., Deweerth, A. and Slattery, T. (1992*a*) Purification, molecular-cloning, and functional expression of the cholecystokinin receptor from rat pancreas. Proc. Natl. Acad. Sci. U.S.A. **89**, 3125–3129

Wank, S. A., Pisegna, J. R. and Deweerth, A. (1992*b*) Brain and gastrointestinal cholecystokinin receptor family — structure and functional expression. Proc. Natl. Acad. Sci. U.S.A. **89**, 8691–8695

Williams, J. A., McChesney, D. J., Calayag, M. C., Lingappa, V. R. and Logsdon, C. D. (1988) Expression of receptors for cholecystokinin and other Ca^{2+}-mobilizing hormones in *Xenopus* oocytes. Proc. Natl. Acad. Sci. U.S.A. **85**, 4939–4943

Yokota, Y., Akazawa, C., Ohkubo, H. and Nakanishi, S. (1992) Delineation of structural domains involved in the subtype specificity of tachykinin receptors through chimeric formation of substance-P substance-K receptors. EMBO J. **11**, 3585–3591

Yokota, Y., Sasai, Y., Tanaka, K., Fujiwara, T., Tsuchida, K., Shigemoto, R., Kakizuka, A., Ohubo, H. and Nakanishi, S. (1989) Molecular characterization of a functional cDNA for rat substance-P receptor. J. Biol. Chem. **264**, 17649–17652

Subject index